D1237468

Sinte Gleska University

3 2958 00030 1433

TECHNIQUES AND TOPICS IN BIOINORGANIC CHEMISTRY

ASPECTS OF INORGANIC CHEMISTRY

General Editor: C. A. McAuliffe

TRANSITION METAL COMPLEXES OF PHOSPHORUS, ARSENIC
AND ANTIMONY LIGANDS

Edited by C. A. McAuliffe

Techniques and Topics in Bioinorganic Chemistry

edited by

C. A. McAULIFFE

University of Manchester Institute of Science and Technology

A HALSTED PRESS BOOK

SINTE GLESKA
COLLEGE LIBRARY

JOHN WILEY & SONS
New York - - Toronto

13517

Selection and editorial matter © C. A. McAuliffe 1975

Part 1 © Marvin W. Makinen 1975
Part 2 © J. M. Pratt 1975
Part 3 © F. L. Bowden 1975
Part 4 © John Webb 1975
Part 5 © S. J. Ferguson 1975

All rights reserved. No part of this publication may be reproduced or transmitted, in any form or by any means, without permission.

First published in the United Kingdom 1975 by
THE MACMILLAN PRESS LTD

Published in the U.S.A. and Canada by
Halsted Press, a division of
John Wiley & Sons, Inc.,
New York

Library of Congress Cataloging in Publication Data

McAuliffe, Charles Andrew.
Techniques and topics in bioinorganic chemistry.

"A Halsted Press book."
Includes bibliographical references.
1. Biological chemistry. 2. Chemistry, Inorganic.
I. Title.
QP531.M3   574.1'914   74-5074
ISBN 0-470-581190

Printed in Great Britain

*This Volume is dedicated to
Professor James V. Quagliano
an early worker in the field*

PREFACE

The maintenance of life depends on two types of chemical process: (1) the photochemical conversion of solar energy into the electrochemical energy necessary to convert carbon dioxide and water into reduced organic material and oxygen; (2) the reverse of this process, that is the oxidation of organic matter to produce carbon dioxide, water and energy. Metal ions are involved in both of these processes. Radiant solar energy enters the biosphere via magnesium porphyrin and chlorophyll. Electron transfer subsequently may occur through a series of carriers such as the cytochromes (Fe^{2+}/Fe^{3+}), ferredoxin (Fe^{2+}/Fe^{3+}) and plastocyanin (Cu^{+}/Cu^{2+}); the production of molecular oxygen involves a manganese complex. Process (2) involves enzymes, which control synthesis and degradation in biochemistry. As biological systems are thermodynamically unstable, it is the controlled release of energy, frequently involving metalloenzymes, which enables life to exist.

About twenty elements have been shown to be essential to life, although living matter frequently contains traces of all the elements in its surroundings. The major constituents of living systems (2–60 atoms per cent) are hydrogen, carbon, nitrogen and oxygen. All of the trace elements (0.02–0.1 atoms per cent) phosphorus, sulphur, chlorine, sodium, potassium, magnesium and calcium have been shown to be necessary constituents of living systems, as indeed have some of the ultra-trace elements (<0.001 atoms per cent) manganese, iron and copper. It is quite probable that vanadium, cobalt, molybdenum, boron and silicon are also of universal biological importance, but it is an extremely arduous and involved task to demonstrate that an ultra-trace element is biologically required. Occasionally the almost complete absence of an element from soils has indicated their biological necessity to plants and animals. Thus, the absence of copper from areas in Australia affected the nervous systems of sheep and also caused anaemia and wool deterioration. Boron deficiency in soil has been claimed to be the cause of 'heart-rot' in beets, 'cracked stems' in celery and 'internal cork' in apples. Addition of boron to the soil appears to cure these abnormalities. However, too much boron kills the plants. The relationship between detrimental and beneficial effects of ultra-trace elements is probably a constantly changing one, and this phenomenon of an essential element becoming toxic at higher than normal concentrations is not rare. Selenium is an element essential to mammals yet selenium-rich plants such as 'locoweed' are poisonous to livestock.

The cations Na^{+}, K^{+}, Mg^{2+} and Ca^{2+} are associated with a number of biological control and trigger mechanisms. Protein and nucleic acids are polyvalent anions and thus require cations for electrical neutrality; there are specific relationships between certain cations and macromolecules. Thus the state of aggregation of the ribosomes is known to be affected by Mg^{2+} concentration. The phospholipid constituent of cell-membrane systems contains Ca^{2+}, and this cation can affect the threshold of nerve excitation.

The ultra-trace elements are the necessary 'cofactors' of a large number of enzymes. There are about 900 known enzymes and of these about 30 per cent either contain complexed metal ions, require the addition of metal ions for their activity, or may have their activity enhanced by the addition of metal ions.

The interest that inorganic chemists and biochemists have in each others' field has increased dramatically during the last decade. This relationship has been cemented somewhat by X-ray crystallographic analysis of metalloproteins, but the inorganic chemists' interest in the concepts of complex stabilisation by ligand-field effects, hard and soft acid–base theory, catalysis by metal complexes, the effects of complexation on e.m.f., and general thermodynamic and kinetic effects in complexation naturally seem to lead to a prime role in explaining the behaviour (and the eventual mimicing in 'model', or even industrial, processes?) of metallo-enzymes.

One excellent two-volume work has already appeared on inorganic biochemistry, and no doubt others may be in preparation; there is certainly a need for such comprehensive texts. In keeping with the stated aims of the series 'Aspects of Inorganic Chemistry' the present volume was solicited in order to offer to interested workers: (a) a volume based largely on the biochemistry of iron and molybdenum; (b) to allow experts to relate how some physical techniques are actually used in bioinorganic investigations and (c) to expound on the principles of catalysis involving iron- and molybdenum-containing enzymes. This book is not intended to be a comprehensive treatise on either the biochemistry of iron and molybdenum or on physical investigative techniques. While there is a central theme, the authors were allowed a fairly free hand in what they wrote. This inevitably allows for a certain amount of overlap, but I judged that each separate contribution made more interesting reading as a whole if it was not edited too severely.

Part 1, on structural and electronic aspects of metal ions in proteins, initially deals with the resolving power of protein X-ray crystallography and places before both inorganic chemists and biochemists the problems associated with crystal-structure determinations of macromolecules. In all examples dealt with an attempt has been made to relate crystallographic data to that obtained from such techniques as magnetic susceptibility, paramagnetic resonance and polarised single-crystal absorption spectra. Haem proteins, zinc-containing metalloenzymes (and the cobalt-, copper-, cadmium-, mercury-, nickel- and manganese-substituted carboxy-peptidases) and calcium-binding proteins are discussed at some length in this part.

Part 2, on principles of catalysis by metalloenzymes, attempts to pinpoint some of the ways in which the presence of protein can affect the reactivity of transition-metal complexes. Three examples are detailed in order to show how the protein can alter (a) the thermodynamics of a particular step, for example the equilibrium constant for the co-ordination of dioxygen by iron (II) in haemoglobin and myoglobin; (b) the kinetics of individual steps, as in the reactions of the iron in peroxidase and catalase with hydrogen peroxide and other substrates; and (c) the thermodynamics of the overall reaction, as in driving the endothermic reduction of dinitrogen to hydrazine by means of a second, thermodynamically favourable, reaction. The discussion is not restricted to iron and molybdenum; a postscript is added on the isomerase reactions of vitamin B_{12}. Earlier in this part it is pointed out that the cofactor that co-ordinates dioxygen may be an iron porphyrin, nonhaem iron, copper and even vanadium—nature has come up with several solutions to the co-ordination-chemistry problem of how to bind dioxygen.

Part 3 concerns the more general aspects of molybdenum biochemistry, which illustrates in particular the large volume of work being done on so-called model compounds. Thus, the use of spectroscopic and magnetic techniques to yield information about the structure and bonding in xanthine oxidase, aldehyde oxidase, nitrate reductase and nitrogenase is in turn used to design model compounds for biological behaviour.

The polynuclear iron (III) proteins phosvitin, gastroferrin, ferritin, haemosiderin and iron-dextran are not usually discussed to the same extent as the iron–sulphur proteins and hemerythrin. The next contribution, Part 4, attempts to remedy this imbalance. It concentrates more on structure than function, and thus the structural information obtained from magnetic, electron paramagnetic, electronic absorption, Mössbauer and X-ray diffraction and infrared spectroscopic techniques is outlined.

Metal ions as n.m.r. probes in biochemistry is outlined in a short article on one of the most powerful techniques recently to emerge for inorganic biochemical investigations. In Part 5 the type of information that can be obtained and an outline of the different approaches available are discussed. For instance, in cases where the natural metal ion does not possess useful n.m.r. properties it is often possible to substitute a metal with more attractive n.m.r. properties. The principles of the method are described and, because of the total emphasis of this volume, specific examples are drawn from haem protein and iron–sulphur protein investigations.

I hope that this volume will prove to be a useful reference source to those already working in the field, and to others so they may enjoy seeing how an area drawn from almost opposite chemical disciplines is developing because of investigations by workers employing many different techniques to tackle very complicated systems.

I am grateful for the contributions of the authors, which represent so much work. I acknowledge with thanks the help given by Kathryn Cotterrell, Ian Nuttall, William Levason and Margaret McAuliffe during the various stages of editing and production.

September 1973   CHARLES A. McAULIFFE
                                                              Manchester

CONTENTS

Contents

PART 2   PRINCIPLES OF CATALYSIS BY METALLO-ENZYMES

J. M. Pratt

PART 4   POLYNUCLEAR IRON(III) PROTEINS

PART 5   METAL IONS AS N.M.R. PROBES IN BIOCHEMISTRY

S. J. Ferguson

ACKNOWLEDGMENTS

PART 1

I am indebted to numerous colleagues who have provided valuable discussions and advice during the preparation of this article. I thank Drs T. L. Blundell, H. A. O. Hill, and R. J. P. Williams for critical evaluation of the manuscript during its preparation. I thank Dr S. Lindskog and Professor J. T. Edsall for helpful discussions about the chemistry of carbonic anhydrase. I am particularly indebted to Drs W. A. Eaton and H. Kon for providing helpful suggestions and advice, for the use of unpublished data, and for earlier collaborative studies that served to deepen my interests in and understanding of the spectroscopy and chemistry of haem proteins. I thank Professor D. C. Phillips for advice and facilities provided during the preparation of the manuscript.

MARVIN W. MAKINEN

PART 4

The studies on polynuclear iron(III) proteins in which the author was actively involved were carried out in the laboratories of Dr Harry B. Gray at the California Institute of Technology and Dr Paul Saltman at the University of California, San Diego, while a Fellow of the Commonwealth Scientific and Industrial Research Organization, Australia. This award of an overseas postgraduate fellowship is gratefully acknowledged. It is a pleasure to also acknowledge many stimulating discussions with colleagues of these two laboratories, and in particular, Drs Jagir Multani, Dana Powers, George Rossman and Harvey Schugar. The helpful criticisms of this review by Drs Harry Gray, Paul Saltman, Harvey Schugar, Tom Spiro and Gerry Vigee are deeply appreciated.

JOHN WEBB

PART 5

I am most grateful to Drs R. A. Dwek and G. K. Radda for introducing me to the applications of n.m.r. in biochemistry and for help in writing this article. I should also like to thank the Medical Research Council and St John's College Oxford for the award of scholarships.

S. J. FERGUSON

PART 1
Structural and Electronic Aspects of Metal Ions in Proteins

MARVIN W. MAKINEN[†]

Laboratory of Molecular Biophysics, Department of Zoology, University of Oxford

† Supported by a National Institutes of Health Special Fellowship (1–F03–GM50802–01) from The National Institute of General Medical Sciences, National Institutes of Health, Bethesda, Maryland, U.S.A. Present address, Department of Biophysics and Theoretical Biology, University of Chicago, Cummings Life Science Center, 920 E 58th Street, Chicago, Illinois 60637, U.S.A.

1 INTRODUCTION

1.1 GENERAL REMARKS

Metal ions in proteins and enzymes exhibit a variety of roles of either catalytic or structural importance. The catalytic importance of metal ions in enzymes is readily attested by the fact that approximately one third of the enzymes known in biochemistry require metal ions for activity[1]. The other category of metal ions in biological systems—those associated with numerous types of proteins, without direct catalytic action but none the less essential for biological function—further emphasises the structural role of metal ions. In a general sense, therefore, metal ions in proteins and enzymes may be divided roughly into two classes—'chemical' metals and 'structural' metals. Chemical metals in enzymes and proteins are those that enter directly into biological reactions in a chemical manner, for instance in the oxidation–reduction reactions of peroxidases and ferredoxins or the binding of oxygen by iron(II) in haemoglobin. Structural metals, on the other hand, stabilise the protein conformation necessary for biological function—for example, calcium(II) in thermolysin—or indirectly promote catalysis by inducing required orientation of substrates or catalytic groups of the protein—for example, magnesium(II) in phosphoglucomutase.

Williams[2] has pointed out that metal ions in compounds of biological interest, especially proteins and enzymes, appear to have the ability to act in two ways: as an indispensable part of the protein, removable only by extreme chemical attack and with high metal-ion specificity in function, or loosely bound to the enzyme or substrate, readily dialysable and having lower metal-ion specificity for catalytic function. In an incisive review Malmström and Rosenberg[3] stated that the claim for a true metal enzyme rests on the demonstration that the presence of the metal changes the reaction mechanism. Since the activity of certain purified enzymes can be enhanced on addition of numerous types of metal ions, and many biochemical reactions proceed with measurable rates in the absence of any catalyst, this criterion is important in defining and demonstrating true metal–enzyme systems. On this basis true metal enzymes[3] can be separated into two categories, quantitatively based on the affinity constants governing the binding relationships: metalloenzymes (enzymes with built-in metal ions) and metal-activated enzymes (enzymes to which metal ions must be added for activity).

The chemistry and concomitant stereochemistry of metal-ion co-ordination complexes are greatly diversified, a circumstance which no doubt permits the wide variety of biological functions that characterise metal ions in proteins.

An examination of the various metal ions utilised by metalloenzymes and metalloproteins alone, as illustrated in table 1, indicates that a relatively small number of the metallic elements is found naturally in association with proteins. Inclusion of

TABLE 1 COMPARISON OF METAL IONS IN METALLOENZYMES AND METALLOPROTEINS†

Metal	Protein	Source	Function
Mo	nitrogenase	*Cl. pasteurianum*	electron transfer
	nitrate reductase	*N. crassa*	electron transfer
Mo, Fe	xanthine oxidase	milk	Mo, electron transfer; Fe, ?
Co	methionine synthetase	micro-organisms	methionine biosynthesis, transfer of $-CH_3$ group
	methane synthetase	*Methonosarcina barkerii*	methane production from H_2 and CO_2
Fe	haemerythrin	blood cells of marine worms	oxygen storage and/or transport
	haemoglobin	mammalian erythrocytes	
	myoglobin	mammalian muscle	
	cytochrome c	animal and plant tissue	electron transfer in mitrochondrial respiration
	cytochrome b		
	catalase	liver, erythrocytes	oxidation of H_2O_2 to O_2
	peroxidase	plant roots	reduction of H_2O_2 to H_2O
	cytochrome P-450	liver, adrenal glands	electron transfer in drug and steroid hydroxylation
	ferritin	spleen, liver	iron storage
	ferredoxin	chloroplasts	electron transfer in photosynthesis
	NADH-dehydrogenase	yeast and animal tissue	electron transfer in mitrochondrial respiration
Fe, Cu	cytochrome oxidase	yeast and animal tissue	electron transfer in reduction of oxygen
Fe, Zn	cytochrome b_2 (lactic dehydrogenase)	yeast	Fe, electron transfer in lactate oxidation; Zn?
Cu	haemocyanin	blood of molluscs	oxygen transport
	ceruloplasmin	serum	unknown
	tyrosinase	mushroom	mixed-function oxidase
	dopamine-β-hydroxylase	adrenal medulla	mixed-function oxidase
	monoamine oxidase	serum	oxidase
Cu, Zn	erythrocuprein	erythrocytes	superoxide dismutation
Zn	carboxypeptidase	pancreas	protein hydrolysis
	carbonic anhydrase	erythrocytes	CO_2 hydration–dehydration; acid-base control
	glutamic dehydrogenase	liver	metabolism and oxidation of glutamic acid
	alcohol dehydrogenase	liver	metabolism and oxidation of ethanol
	aspartate transcarbamylase	*E. coli*	pyrimidine biosynthesis
	alkaline phosphatase	serum, placenta, *E. coli*	organic phosphate-ester hydrolysis
	aldolase	yeast	cleavage of fructose disphosphate
	insulin	pancreatic β-cells	Zn, ?; glucose metabolism
Zn, Cd	metallothionein	mammalian kidney	unknown
Zn, Ca	thermolysin	*B. thermoproteolyticus*	Zn, protein hydrolysis; Ca, thermal stability
Mn	pyruvate carboxylase	liver	formation of pyruvic acid from acetylCoA
Mn, Ca	conconavalin A	jack bean	cell mitogen
Ca	α-amylase	*B. subtilis*, saliva	hydrolysis of carbohydrates

metal-activated enzymes would add only sodium, potassium and magnesium to this list of elements. Since it is known that the biological role of a metal ion in a protein is highly specific and yet a variety of functions are performed by the same metal ion depending on the type of protein, the diversity of functions promoted is clearly the result of constraints imposed by the protein environment. On this basis the biological specificity of metal function can be considered correctly to be of stereo-chemical origin. Being the principal themes of this review, the importance of structural and stereochemical considerations in elucidating the behaviour of metal ions in proteins and enzymes, and the importance of defining in detail the structure of their metal–ligand co-ordination centres, thus require little further justification.

A discussion of stereochemical and structural aspects of metal–enzyme and metal–protein function in biology is limited by the proteins that have received detailed structural descriptions on the basis of X-ray diffraction studies. Table 2 gives a summary of the stage of structural determination of various metal-ion-requiring enzymes and proteins. These proteins can be separated into three general categories according to the control of metal-ion reactivity by metal–protein interaction: (i) proteins in which the reactivity of the metal ion incorporated into a prosthetic group is controlled primarily by interaction of the protein chain with the skeleton of the prosthetic group, such as the haem proteins; (ii) proteins in which the reactivity of the metal ion is controlled directly through the donor ligands of amino-acid side chains to which the metal ion is tightly bound, and with catalytic function aided by binding of the substrate to nearby protein residues, such as carboxypeptidase; and (iii) proteins in which binding of the metal ion occurs with less affinity but nonetheless, with specific amino-acid side chains, and is required for stabilisation of structure or is essential for catalytic function, such as in thermolysin and staphylococcal nuclease.

Absolute demarcation lines obviously cannot be drawn between these categories. Also, the affinity of metal-ion binding is often further modified on interaction of the enzyme with the substrate molecule. However, these three categories provide a reasonable basis for selection of representative, structurally defined metal-ion-requiring protein and enzyme systems in our discussion of the structural origins of metal-ion reactivity and function. The proteins selected on this basis for the purpose of this review are respectively (i) myoglobin and haemoglobin, (ii) carboxypeptidase and carbonic anhydrase, and (iii) carp albumin and staphylococcal nuclease. We shall consequently limit our discussion to the co-ordination chemistry of these proteins for which high-resolution crystallographic studies have made detailed descriptions of the structure and stereochemistry of the metal–ligand centres

---

† Extensive reviews on the biochemistry of metalloproteins is given by Vallee and Wacker[4]; on Cu-containing enzymes by Malkin and Malmström[5]; on Mo-containing enzymes by Bray and Swann[6] and on Co-containing enzymes by Wood and Brown[7]. Certain enzymes also require *added* metal ions for activity and are known as metal-activated enzymes. Examples include sodium or potassium-activated ATPases, magnesium-activated phosphoglutomutase, as well as the large number of manganese(II) or magnesium-activated enzymes. Reviews of metal-activated enzymes are given by Mildvan[8] and Mildvan and Cohn[9].

TABLE 2 X-RAY CRYSTALLOGRAPHIC STUDIES OF METAL-ION-REQUIRING ENZYMES AND PROTEINS

Class	Metal	Molecule	Source	Molecular weight	No. of subunits	No. of subunits per asymmetric unit	Space group	Stage (pm)	Ref.
oxygen-binding proteins	Fe	metmyoglobin	sperm-whale muscle	17 500	1	1	$P2_1$	200	10, 11
			seal muscle	17 800	1	1	$C2$	500	12
			yellow-fin tuna muscle	17 800	1	1	$P2_12_12_1$	500	13
		erythrocruorin	fly, Chironomus	17 000	1	1	$P3_1$	250	14
		methaemoglobin cyanide	sea lamprey	16 600	1	1	$P2_12_12_1$	200	15
		methaemoglobin	horse	64 500	4	2	$C2$	280	16
		deoxyhaemoglobin	horse	64 500	4	2	$C222_1$	280	17
		deoxyhaemoglobin	human	64 500	4	2	$P2_1$	350	18
electron-transport proteins	Fe	ferricytochrome c	horse heart	12 400	1	1	$P4_3$	280	19
		ferrocytochrome c	bonito tuna	12 400	1	1	$P2_12_12_1$	245	20
		ferricytochrome b_5	calf liver	10 280	1	1	$P2_12_12_1$	200	21
		ferricytochrome c_2	R. rubrum		1	1	$P2_12_12_1$	500	22
		ferricytochrome c peroxidase	baker's yeast	40 000	1	1	$P2_12_12_1$	5000	23, 24‡
		catalase	bovine liver	250 000	4	1	$P3_12_1$	250	25
		ferredoxin (oxidised)	M. aerogenes	6 000	1	1	$P2_12_12_1$	225	26
		high-potential iron protein† (oxidised)	Chromatium	10 000	1	1	$P2_12_12_1$		27
		rubredoxin† (oxidised)	C. pasteurianum	6 000	1	1	$R3$	150	28

Class	Metal	Protein	Source	M_r			Space group	Resolution	References
proteases	Zn	carboxypeptidase A	bovine pancreas	34 600	1	1	$P2_1$	200	29
	Zn, Ca	thermolysin	*B. thermolyticum*	37 500	1	1	$P6_122$	230	30, 31
	Ca	neutral protease	*Rhizopus chinensis*	35 000	1	1	$P2_12_12_1$		32‡
nucleases and phosphate-ester hydrolases	Ca	nuclease	*S. aureus*	16 800	1	1	$P4_1$	200	33
	Zn	alkaline phosphatase	*E. coli*	80 000	2	1	$P3_12_1$		34‡
oxidoreductases	Zn	alcohol dehydrogenase (apoenzyme)	horse liver	84 000	2	1	$C222_1$	550	35
	Zn	alcohol dehydrogenase (+ 2 NADH)	horse liver		2	1	$P2_1$	550	35
other proteins	Zn	aspartate transcarbamylase	*E. coli*	310 000	12	2	$R32$	600	36
		carbonic anhydrase C	human	34 000	1	1	$P2_1$	200	37
		carbonic anhydrase B	human	34 000	1	1	$P2_12_12_1$		38‡
		insulin (2-Zn)	porcine pancreas	5 780	1	2	$R3$	280	39
		insulin (4-Zn)	porcine pancreas	5 780	2	2	$R3$	190	40
	Mg	glutamine synthetase	*E. coli*	582 000	12	6	$P4_2$	400	41
		phosphoglycerate kinase	horse muscle	48 000	1	1	$P2_1$	700	42
	Mn, Ca	concanavalin A	jack bean	210 000	8	1	$I222$	425, 300	43, 44, 45
	Ca	carp albumin	carp fish muscle	11 000	1	1	$C2$	200	288

† probable electron-transport proteins.
‡ survey for heavy-atom derivatives.

feasible. Other detailed aspects of protein structure and conformation are discussed in the references cited. This article represents a survey of the literature up to the beginning of 1973.

1.2   RESOLVING POWER OF PROTEIN X-RAY CRYSTALLOGRAPHY

1.2.1 Problems in protein-structure determination

Quantitative descriptions of the catalytic metal–ligand centre of an enzyme both before and after substrate binding are obviously essential for formulating the critical steps of a plausible mechanism of catalytic function. To this end stereo-chemical characterisation of the co-ordination environment of a metal ion and its structural relationships to nearby amino-acid residues involved in substrate binding is essential. Detailed elucidation of the chemical basis of metal-ion reactivity in enzymes requires, furthermore, correlation of molecular structure, stereochemistry and electronic structure with biological function: formulations of the critical steps of enzyme action deduced on the basis of structural information must be consistent with the results of kinetic studies indicating substrate affinities, the probable nature of reaction intermediates and rate-determining processes; moreover, these results must be in agreement with spectroscopic studies defining the electronic and atomic rearrangements involving the enzyme and substrate molecules. As with simple co-ordination complexes, detailed molecular structural data permit assessment of the electronic structure and bonding of metal ions and ligands in proteins. Further descriptions of changes in the stereochemistry of metal–ligand centres during catalysis then allows an understanding of the alterations in electronic structure responsible for catalytic action. On this basis, the detailed relationships of molecular structure, stereochemistry and electronic structure of metal co-ordination centres in proteins acquire importance for understanding the control of their biological reactivity and function. The experimental means through which this understanding becomes possible rests on the precise, detailed characterisation of protein molecular structure and the understanding that detailed structural data impart to the inter-pretation of spectroscopic phenomena.

Precise quantitative specification of the stereochemistry of metal–ligand co-ordination centres of enzymes and proteins requires the chemical identification and accurate positioning of every atom of the molecule. It might be supposed that this need would be met by applying the powerful methods of three-dimensional X-ray crystallography. It remains doubtful whether any such determination can achieve a sufficiently accurate definition of protein molecules that would be required for detailed theoretical interpretations of structure and electronic bonding. Diffraction methods are, however, the only method of defining the spatial relation-ships of all, or nearly all, of the atoms in a protein molecule. Although the methods for refinement of protein structures are being constantly improved, the best attainable descriptions of protein stereochemistry at present are only roughly quantitative in contrast to the high level of precision attainable with X-ray crystallography of small molecules.

Since there is relatively poor precision for X-ray diffraction methods applied to protein molecules, it is necessary to provide some background in the resolving power of the crystallographic techniques. In this review there will be no attempt to give an exhaustive background on the theory and application of diffraction methods; texts such as those by Woolfson[46] and by Lipson and Cochran[47] should be referred to. Thorough reviews of the X-ray crystallography of proteins have been written by Phillips[48] and North and Phillips[49]. A recent general survey of results of protein structural studies has been given by Blundell and Johnson[50]. Reviews such as those of Blake[51], Eisenberg[52] and Davies and Segal[53] should be consulted for methods and techniques in crystallisation, preparation and crystallographic examination of protein derivatives.

(1) Phase determination

The end product in the crystallographic study of molecules is a plot of the electron density of the crystal $\rho(x,y,z)$ as a function of the three co-ordinate axes. This can be computed by use of the well-known Fourier inversion

$$\rho(x,y,z) = \frac{1}{V} \sum_h \sum_k \sum_l F(hkl) \exp i\alpha(hkl) \exp[-2\pi i(hx + ky + lz)] \quad (1.1)$$

where $F(hkl)$ is the observable structure amplitude of the crystal reflection hkl, and $\alpha(hkl)$ is its phase, which cannot be observed directly. V is the volume of the unit cell of the crystal structure, and the summations are over the observable X-ray reflections. In studies of small molecules this calculation is often made to yield a structure with no prior knowledge of the arrangement or composition of the atoms in the molecule. This is not possible with protein molecules at present.

In the analysis of small-molecule structures, the crystallographer obtains an initial electron-density map by the application of direct methods of structure determination or, for instance, by locating the positions of a few of the heavier atoms in the molecule from Patterson syntheses. This initial Fourier map is generally ambiguous and may indicate electron-density peaks for only a fraction of the atoms in the molecule. The positions of these atoms are interpreted from the initial map and are used to calculate a set of phases. This set is then utilised to calculate an electron-density map with the aid of the observed structure amplitudes, resulting in further clarification of the positions of other atoms. The contribution of these atoms can then be used to calculate an improved set of phases, leading to an improved electron-density map. This iterative procedure is carried out through numerous consecutive cycles, until all atomic parameters are defined. The use of alternate cycles of phase refinement and electron-density map calculations can be carried out because of the large number of experimentally observed parameters, that is the structure amplitudes, in comparison to the number of variable parameters; in fact, the structures of small molecules, are generally highly overdetermined. Protein molecules, however, have a significantly greater number of atoms. Since each must be specified by three positional parameters with generally six others to indicate thermal vibration, the number of independently observable X-ray reflections does not usually outnumber the variables by a sufficiently large factor. In addition a much higher number of observable structure amplitudes is necessary for a crystallographic least-squares adjustment of parameters to fit the data. Most importantly, however, Patterson syntheses of native crystalline proteins have remained largely uninterpretable because of the large number of atoms in the unit cell, and direct methods of structure determination have not been successfully developed for large molecules. These methods consequently cannot be employed to determine phases for crystalline protein molecules.

Therefore, for proteins other information is necessary in order to calculate and interpret the electron-density map. It is in this connection that only approximate specification of atomic positions can be achieved by interpretation of the resultant electron-density map. The fundamental problem in the whole analysis is concerned with the determination of the phases $\alpha(hkl)$. This problem was first successfully solved by the demonstration[54] that the method of isomorphous replacement with use of heavy-metal salts can be employed to determine the structures of crystalline proteins. On this basis the phase angles for corresponding structure factors can be determined from the changes in structure amplitude due to the replacement of one atom in a structure by another of different scattering power, provided the structure remains unchanged with substitution. As is well established by the systematic analysis of errors by Blow and Crick[55], the accuracy of determining the phase angles becomes the single most important factor in assessing the resulting structure analysis. The relative importance of phase determination as compared to the accuracy of amplitudes is indicated by the study of Srinivasan[56] in which the phases from one structure were combined with the amplitudes from a second very different structure: the resulting Fourier synthesis represented the first structure closely. In contrast to small molecules, therefore, the crystallographer arrives immediately at the best attainable map with the determined set of phases. An image of the structure is calculated on this basis, limited in detail only by the resolution of the diffraction data used. This limitation is the fundamental restriction on the precision with which atomic positions of protein electron-density maps can be interpreted.

(2) Resolution of electron-density maps

As X-ray diffraction data are collected corresponding to sets of parallel reflecting planes within the crystal, each set having a mutual separation d_i, the theoretical resolution of the Fourier map improves as the value of d_i decreases. When X-ray data are collected for a complete set of X-ray reflections down to a corresponding interplanar spacing of d_{min}, then the limit of resolution in the three-dimensional Fourier map is given by the relation[57]

$$d_{lim} = 0.71\, d_{min} \tag{1.2}$$

Thus, in a Fourier synthesis of a crystalline protein for which X-ray data are collected to 250 pm resolution, structural detail can be theoretically resolved and interpreted in the three-dimensional electron-density map between atoms no closer than 180 pm apart. This is somewhat less than most metal–ligand bonds encountered in biological systems but is still greater than C–C, C–N and C–O bond lengths in organic substances[58].

There are numerous limitations on the theoretically attainable resolution. For instance, although the theoretical resolution of the Fourier map improves as more and more reflections are included in the Fourier series, the effect of most sources of error, for example phase indeterminancy and isomorphism, also increases[59] as X-ray data are collected to increasingly higher resolution. In addition, the computational problems involved in handling the larger quantities of data required for high-resolution studies may become limiting. For instance, for a crystalline protein, which may require the measurement of 4000 independent reflections at 300 pm resolution, the number of independent reflections to be measured at 150 pm resolution will have increased to approximately 4000 x 2^3 or 32 000. With the usual methods of protein structure determination, all of these reflections must be measured, in addition, for each isomorphous derivative, a procedure which also becomes increasingly time consuming. Computational problems as well as man-hours of experimental effort can thus enormously complicate the solution of a protein structure when high-resolution data are desired. In addition, for most proteins studied by X-ray methods thus far, the optimal value of d_{min} is between 280 and 200 pm and is dictated by the general properties of the crystal itself, that is, the limiting resolution to which diffraction data can be accurately measured. For this reason the diffracting properties of the crystal become markedly important. With most protein crystals the accuracy falls rapidly for phases determined as the 200 pm limit is reached. Also the X-ray intensities are weaker and more difficult to measure accurately. These considerations provide some indication of why the theoretical limit of resolution d_{lim} is never in practice achieved in X-ray diffraction studies of crystalline protein molecules. Since electron-density maps at different levels of resolution are expected to reveal different levels of organisation of protein structure and because the effect of errors becomes more serious as d_{min} decreases, it is general practice simply to report the limiting value of d_{min} for a protein structure analysis without implying a more detailed assessment of the limit of resolution,

Once the electron-density map of the molecule has been calculated, the protein structure is interpreted by construction of a skeleton model of the protein molecule, usually with the aid of an optical comparator[60]. This procedure entails fitting the wire skeleton images of amino-acid residues to the electron-density map of the protein. Refinement methods must be then applied in order to determine the best fit of the co-ordinate positions measured on the skeleton model with the electron-density map. The refinement procedure is carried out on the basis of the best fit of the experimentally determined electron-density map to the images of the amino-acid residues either by variation of certain dihedral and bond angles[61] or on the basis of minimisation of the potential-energy function of the molecule[62]. The methods thus allow comparatively little variation of bond angles and bond lengths from those determined as the average values for simple compounds in completely different crystal environments. For most residues of a given type in a protein, comparable geometrical relationships of the nuclear positions probably exist. The possibility of structural perturbations resulting in straining of bonding relationships must, however, be constantly considered. Especially in the region of active metal–ligand co-ordination centres, such changes may be of catalytic importance in view of the distorted metal-ion co-ordination geometries observed in crystallographic studies of numerous metalloenzymes.

However, at the present time it is quite likely in most cases that even if greater flexibility of bond angles and bond lengths were permitted, the electron-density data may not be sufficiently well defined to distinguish between standard and non-standard conformations. Furthermore, since it is difficult even at high resolution to determine atomic co-ordinates of protein molecules with an accuracy better than 25 pm, standard deviations of bond lengths of 10–20 pm for the amino-acid residues of the polypeptide chain are generally obtained by refinement methods[61]. The assessment of the accuracy and precision of refined co-ordinates, when stereochemical information is imposed on the interpretation of the electron-density data, is clearly difficult to judge in an objective manner. For these reasons the results of protein X-ray crystallographic studies yield considerably less quantitative stereochemical specification for atomic positions than in comparable small-molecule studies, in which bond angles and bond lengths are objectively assessed individually on the basis of the calculated standard deviation.

It is not certain that diffraction studies of crystalline proteins in the future will in all cases successfully achieve significant improvement of structural resolution. Crystalline proteins differ from crystalline small molecules through a high solvent content ($\geqslant 40$ per cent) of the unit cell[63,64]. Protein molecules, therefore, are not subjected by intermolecular contacts in the crystalline state to lattice forces comparable to those that hold small molecules fixed in a considerably more rigid state. For this reason the results of X-ray studies of proteins often are associated with regions of the electron-density map which do not visualise well. A classic example of this phenomenon is the motional freedom of the terminal regions of the α and β subunits of horse methaemoglobin[16]. Under the high salt conditions used for crystallisation, these regions are prevented from entering into stabilising polar contacts with each other. In addition, the role of polypeptide flexibility characteristic of protein chains, which, in fact, may be of enzymatic importance, cannot be discounted. Similar factors relevent to the description of ordered solvent structure and interactions of solvent molecules with protein residues prevent collection of data necessary to describe molecular structure precisely. It is readily apparent that regions of the polypeptide chain of protein molecules that cannot be seen clearly in the Fourier map cannot be subjected to refinement methods.

1.2.2 Problems of special importance to metal-ion-requiring proteins

There are, in addition, numerous other complicating factors which are generally associated with the analysis of protein structures and which can assume greater importance in the case of metal-ion-requiring proteins. It is well known in protein

X-ray crystallography that errors associated with various stages of the structure analysis, for example, phase determination, refinement of heavy-atom positions, etc., can result in distorted regions of negative electron density in the calculated Fourier map at the sites of the heavy-metal atoms. This phenomenon was already evident from the early work on the structure of sperm-whale metmyoglobin[65]. Such perturbations on the electron-density features can significantly hinder structural interpretations in those regions. In fact, Banaszak *et al.*[66] have pointed out that the negative electron-density regions associated with the Fourier map of sperm-whale metmyoglobin have probably contributed to the difficulties in interpreting structural features of the ligands co-ordinating the $Zn(II)$ and $Cu(II)$ ions that can be bound under certain conditions. The sites of heavy-atom derivatives in this case are close to both metal-ion binding sites. Comparable situations can arise in the case of metal-activated enzymes, which bind metal ions of catalytic importance.

Errors from the use of isomorphous derivatives, prepared by exchange of the metal ion found in the native protein for heavier elements, may arise from the general absence of *identical* metal–ligand centres on metal substitution. Crystallographic studies of carboxypeptidase A[29,67], carbonic anhydrase C[68,69] and staphylococcal nuclease[33] show that the heavier metal ion is displaced from the site of the bound metal in the native protein, a condition expected on the basis of the markedly different co-ordination tendencies and ionic radii of the metal ions concerned. Since such changes must be expected to result in distortion of the configurations of amino-acid side chains acting as ligands, the resultant calculated phases are associated with some error by the loss of exact isomorphism. Furthermore, changes in solvation and co-ordination geometry within the metal-ion binding site must be expected under conditions of metal exchange; for example, $Hg(II)$ exchanged for tetrahedral $Zn(II)$. These changes may further add to the loss of true isomorphism. In general this problem can be resolved in part by utilising phases from several isomorphous derivatives in the Fourier calculation. When a heavy metal has been substituted for the cation bound by the native protein, the phases from that derivative need not be applied in calculating the electron density within the region of the active metal-binding site in preference to the phases determined with other derivatives. In this manner, errors due to loss of exact isomorphism can be minimised in determining the stereochemistry of the metal–ligand centre.

While the exchange of the native-bound metal ion for a heavier atom of greater scattering power remains an attractive method of preparation of heavy-metal derivatives (in view of the general need to resort otherwise to time-consuming trial-and-error methods[51]), it may, nonetheless, be preferable in the case of metal-ion-requiring proteins to avoid this method whenever possible. Since precise stereochemical description of the metal–ligand co-ordination centre essential to the biological functioning of the protein is a particularly important objective of such structure analyses, the use of heavy metals attached to other parts of the protein should be preferred whenever possible, either during preparation of suitable derivatives or during the resultant computation of the corresponding part of the Fourier map.

1.2.3 Improved stereochemical descriptions of proteins

The preceding parts have provided a brief account of the origins of the considerably more approximate stereochemical detail obtainable from X-ray diffraction studies of proteins as compared to that of small, low-molecular-weight compounds. However, structural data of higher precision will be undoubtedly required for further advances in the study of structure–activity relationships. As we seek to elucidate in ever-increasing detail the behaviour of the active site or functional regions of these biological macromolecules, the resolving power of diffraction studies must accordingly increase. Especially in the study of metal-ion-requiring enzymes and proteins, small configurational changes of amino-acid residues in the metal binding-site region and stereochemical alterations of metal–ligand geometries during catalytic action must be precisely characterised. As pointed out earlier (section 1.2.1), characterisation of these structural alterations allows assessment of the stereochemical origin of the electronic rearrangements that occur in interacting substrate. and enzyme molecules and are responsible for catalytic action.

(1) The difference Fourier technique

To some extent under favourable conditions stereochemical detail associated with small configurational changes of protein residues can be improved. By use of the difference Fourier method, changes in the electron-density distribution between two isomorphous crystal structures can be observed and measured beyond the limit of resolution (d_{lim}). In the difference Fourier technique, a Fourier synthesis is carried out such that the difference electron-density distribution between two structurally similar molecules is computed. This computation utilises, in contrast to equation 1.1, the coefficients

$$m(|F_H| - |F|)\exp(i\alpha) \qquad\qquad (1.3)$$

summed over the observable reflections, where F_H and F are the measured structure amplitudes of two similar noncentrosymmetric projections, α is the phase of F (belonging to the native reference-crystal structure) and m is a weighting factor[65] used to weight down terms for which α is not well determined or for which the expression above is not valid.

Successful application of the difference Fourier method in the study of large protein molecules was first demonstrated in the study of Stryer et al.[70] in which the stereochemical relationships of the azide anion to the iron–porphyrin group and nearby amino-acid residues in metmyoglobin were demonstrated (see section 2.1.6 (1). A recent analysis[71] of the origin of errors in the use of the difference Fourier technique indicates that a difference Fourier map will normally have a much lower error level than the corresponding Fourier map of the parent structure and that the difference Fourier map is able to detect much smaller features of electron density than those revealed by the normal Fourier map with the same phases. This undoubtedly accounts for its widespread, successful application in the study of crystalline proteins. Under favourable conditions in applying the difference Fourier method the limit of resolution of structural changes can approach 10 pm.

The difference Fourier method, however, is applicable only when closely similar crystal structures are to be compared. In those cases in which the protein of interest crystallises in different space groups on chemical modification, such as sea-lamprey haemoglobin[72], or where the structural alterations become too large on binding of small molecules to allow direct application of the method, as for triose phosphate isomerase[73], a new structure analysis must be undertaken for the modified derivative. Comparison of two independently solved protein structures under such conditions remains associated with less precise stereochemical quantitation. Such comparative results cannot be expected to approach the stereochemical detail that in principle can be obtained by the difference Fourier method.

The power of the difference Fourier method, therefore, lies in its use of phases of only one structure to investigate a large number of similar structures for which only the amplitudes of the structure factors need be measured. For metal-ion-requiring proteins the method clearly acquires importance not only because of elimination of the need to determine independently a set of phases for the structure of a chemically altered protein, but also because of the improved precision in defining the new metal–ligand stereochemistry and co-ordination geometry associated with the altered protein. The difference Fourier method has found particularly successful application to problems in protein–substrate interactions for defining stereochemical relationships of substrates with the catalytically important amino-acid residues in the active-site regions of enzymes (see for instance reference 49). It has also been valuable for defining changes in the co-ordination geometry of the catalytically functional metal ions of numerous enzymes[29,33,37] on binding substrates and inhibitors.

(2) Least-squares refinement of protein structures

While the precision of stereochemical definition of the altered protein derivative is increased with respect to that of the parent structure by application of the difference Fourier technique, the specification of the three-dimensional structure of the parent derivative remains associated with less precise quantitation. Recent advances, however, suggest that the level of refinement of the three-dimensional structures of protein molecules may approach that associated with the least-squares refinement method of the structures of small molecules in the true crystallographic sense. Such refinement methods, as first suggested by the early refinement studies on sperm-whale myoglobin (see reference 74), coupled with the improved precision afforded by the difference Fourier method, can be expected to allow significant improvement in the precision of stereochemical detail of functional metal enzyme and protein co-ordination centres.

These refinement methods were developed during the recent determination of the structure of oxidised clostridial rubredoxin[28,75]. This protein of approximate molecular weight 6000 contains an iron atom co-ordinated by four sulphur atoms of cysteine residues in an approximate tetrahedral configuration. In this study, refinement of the structure was initiated with a set of atomic co-ordinates and phases determined in the conventional manner at 200 pm resolution. Structure amplitudes measured only for the native structure at a resolution of d_{min} = 150 pm were then utilised in the resulting cycles of refinement calculations with use of the initial set of phases determined at d_{min} = 200 pm. While a detailed discussion of the parameter adjustments made at each stage of the refinement is beyond the scope of this review and must await further application of the method to other proteins for thorough evaluation, the results of this study are important in that they demonstrate the utility of protein structure refinement. The precision of structural features such as bond lengths and angles, the planarity of peptide residues and the precision of especially the Fe–S bond lengths and angles of the metal co-ordination site could be assessed with certainty. These results are to be contrasted with those at 250 pm resolution[76], at which stage of structure analysis the asymmetry of the tetrahedral co-ordination of the Fe(III) cation could not be unambiguously determined.

The improved precision of stereochemical data obtained in this study demonstrates that there are compelling reasons for comparable precision of stereochemical detail in the structure analyses of all proteins and that such refinement methods should be applied. Successful application of refinement of protein structures coupled with the use of difference Fourier techniques will provide in the future a more powerful method of defining the precise stereochemical detail of the co-ordination centres of metal proteins and enzymes. The results of such studies and the increased understanding they will impart to the study of structure and bonding of metal-ligand co-ordination centres and structure–activity relationships are eagerly anticipated. It is consequently of some value to evaluate the results of X-ray diffraction studies of proteins that have been studied at high resolution by more standard methods and to inquire to what extent the results help to demonstrate the basis of their biological function. A thorough evaluation of X-ray studies and the results of newer refinement methods after about another decade undoubtedly will indicate more accurately the value of high-precision stereochemical data in the study of biological macromolecules. Only then can the value of such data be fully assessed in relation to the results of other related chemical and physical studies.

1.3 STEREOCHEMISTRY AND GEOMETRIES OF METAL–AMINO-ACID COMPLEXES

A detailed examination of the co-ordination geometries and concomitant electronic structures of metal ions in proteins invariably requires reference to the corresponding properties of simple metal-ion co-ordination complexes. While not all co-ordination complexes of metal–amino acid and metal–peptide systems will find direct applicability, there are among these metal complexes certain geometrical relationships which are remarkably constant[77]. It is therefore important to determine whether these relationships, for example relative metal-ligand bond lengths, geometries of metal–imidazole interaction, etc., exist in the more complex biological systems. Major deviations from these constant geometrical relationships would then indicate a significant perturbation on the structural and electronic properties of the co-ordination centre.

Table 3 gives a list of the ionic radii of metal ions commonly found in proteins. Comparison is also made to the radii of metal ions that generally can be specifically substituted for them. Of more relevance perhaps, since these ionic radii pertain to simple inorganic salts, are the corresponding metal–ligand bond distances observed with donor-ligand atoms of amino acids, which commonly

TABLE 3 COMPARISON OF CRYSTAL IONIC RADII[†] OF METALS COMMONLY FOUND OR SUBSTITUTED IN PROTEINS

Ion	Radius (pm)	Ion	Radius (pm)
Mg(II)	65	Mn(II)	80
Ca(II)	99	Fe(II)	76[‡]
Sr(II)	113	Co(II)	74
Ba(II)	135	Ni(II)	72
		Cu(II)	72
La(III)	101.6	Zn(II)	74
		Cd(II)	97
Nd(III)	99.5		
		Hg(II)	110
Y(III)	89.3		
Eu(III)	95.0	Fe(III)	64[§]
		Co(III)	63

† ionic radii are from Pauling[58] except for Cu(II)[78] and for lanthanides[79].
‡ corresponds to a high-spin Fe(II) cation.
§ corresponds to a high-spin Fe(III) cation.

serve to co-ordinate metal ions in proteins. These are tabulated in table 4 and are largely summarised from the extensive review of Freeman[77]. This comparison serves to illustrate the variations in metal–ligand bond lengths dependent on metal-ion radius and the varying geometries of interaction of metal ions with imidazole rings. Additional pertinent information has been added for comparative purposes to indicate other small geometrical differences which metal–amino acid complexes exhibit. Co-ordination complexes of amino acids with Mn(II)[77b] or lanthanide(III) cations are almost unknown.

Metal ions in proteins and enzymes can be specifically substituted for one another according to strict requirements determined by ionic radii and stereochemical demands on co-ordination. Unfortunately the structures of metal-substituted protein derivatives have remained on the whole less well defined on the basis of X-ray diffraction studies in comparison to the corresponding native protein structures. The structural basis through which co-ordination centres of metal-ion-requiring proteins can accommodate other cations, often with markedly different ionic radii and co-ordination tendencies, without resultant widespread structural distortions of the polypeptide chain, often remains an unexplained circumstance. To this end, the stereochemical data of metal–amino-acid interaction supplied in table 4 can be of value in assessing the effects of metal-ion substitution.

In view of the prominence of imidazole side chains of histidine residues in co-ordinating metal ions, especially for the first transition series, and their roles in co-ordinating metal ions within the active-site regions of metalloproteins and metalloenzymes, it is instructive to point out some implications of the stereochemical data of table 4. Freeman[77] has pointed out that the metal–nitrogen bond lengths formed with nitrogen donor ligands of primary amines and from the N_1-position of histidine residues are equivalent. This structural relationship implies little dπ–pπ interaction in the binding of metal ions by imidazole derivatives. In short, the plane of the co-ordinating imidazole ring appears to exhibit no definite orientational preference with respect to the d-orbital system of the central metal ion[77]. Moreover, in no model metal–imidazole complex does the metal ion occupy a position precisely coplanar with the imidazole ring. The data in table 4 serve to underline the variation in the out-of-plane displacement of various metal ions co-ordinated by imidazole derivatives.

Although these considerations need not be expected to apply rigorously in all metal–protein systems, they help to explain the structural basis through which the active sites of certain enzymes such as carboxypeptidase and carbonic anhydrase can accommodate metal ions with widely different co-ordination tendencies. For instance, the tetrahedrally co-ordinated Zn(II) ion in carboxypeptidase is readily substituted by numerous transition-metal ions[84,85] (see section 3.2.2.), and two of the four ligands are the N_1-positions of histidine residues[29,67]. The substitution of Co(II) ions for Zn(II) may be expected on the basis of numerous well-defined, isostructural tetrahedral complexes of Co(II) and Zn(II). However, the accommodation of a Cd(II) ion remains somewhat unexpected on the basis of its greater ionic radius and generally preferred six-co-ordinate geometry[86] since strict stereochemical relationships of amino-acid residues in proteins are necessary in maintaining structural integrity. Table 4, however, indicates a significant difference between tetrahedral Zn(II) and octahedral Cd(II) complexes in the angle that the metal–nitrogen bond makes with the plane of the imidazole ring, although the corresponding bis(L-histidinato)–metal complexes have similar crystal structures. A simple trigonometric calculation in this case indicates that the displacement of the metal ions from the co-ordinating N_1-position projected onto the plane of the imidazole ring is nearly identical for both. In a comparable manner, a histidine residue in the active site of an enzyme could accommodate a metal ion of increased radius without widespread structural distortions on nearby residues because of an otherwise necessary translational displacement. On the other hand, in the case of carboxypeptidase the greater perpendicular displacement of the Cd(II) ion from the plane of the imidazole ring and its greater ionic radius may alter structural relationships with

TABLE 4 COMPARISON OF AVERAGE METAL–LIGAND BOND LENGTHS OF METAL–AMINO-ACID COMPLEXES AND GEOMETRIES OF METAL–IMIDAZOLE INTERACTION

Bond type (M–L)	Metal–ligand bond length (pm)								
	Mn(II)[†]	Fe(II)	Co(II)[‡]	Ni(II)[‡]	Cu(II)[§]	Cu(II)[‖]	Zn(II)[¶]	Zn(II)[††]	Cd(II)
M–N (amino)	223	–	215	211	200	–	209	205	231
M–N$_1$ (imidazole)	–	199[‡‡]	218	209	200	197	220	204	225
M–O (carboxylate) in chelate ring	217	–	212	213	198	–	215	–	231
M–OH$_2$	–	–	209	212	197	–	211	–	–
range of deviation from plane of imidazole ring (pm)	–	–	18–38	12	17–53	–	–	15	–
range of angle between M–N bond and imidazole ring (degrees)	–	4–11	4.7–10	3.6	5.0–15	–	–	0.3	29

† octahedral Mn(II) chelate of pyridoxylidenevaline[80]. Nitrogen is a secondary amine.

‡ octahedral co-ordination, no tetrahedral amino-acid complexes known.

§ octahedral, square pyramidal, or square planar.

‖ flattened tetrahedral.

¶ octahedral Zn(II).

†† tetrahedral Zn(II). A Co–N bond distance of 195 pm in tetrahedral bis(p-toluidine)CoCl$_2$[81] is given as comparison.

‡‡ see reference 82 for bis(dimethylglyoximato)diimidazoleiron(II)-dimethanol. This structure contains an octahedrally co-ordinated low-spin ferrous cation. This is to be compared with the value of 205 pm estimated for the analogous bis(cyclohexane-1,2-dioximato)diimidazoleiron(II) dihydrate[83].

the co-ordinating carboxylate group of glutamate-72 or may change the site occupied by the co-ordinating water molecule[29] with respect to other nearby residues.

Furthermore, these considerations place no requirements on the retention of comparable co-ordination geometries and co-ordination numbers for metal-ion substitution. Indeed, since the co-ordinating ligands supplied by the protein generally remain identical, and thus the contributions of ligand polarisabilities in determining co-ordination number are unchanged, identical co-ordination geometries need not be expected to be retained on metal-ion substitution. Since the stereochemical relationships of metal–imidazole interaction vary significantly (table 4), small rotational displacements of histidine side chains may be sufficient for the co-ordinating geometrical requirements characteristic of the substituted cation while accommodating a change in co-ordination number. Additional co-ordination sites can be readily occupied either by other nearby amino-acid chains or by solvent molecules. Unfortunately there is little detailed structural data of metal–ligand co-ordination centres for metalloenzymes and metalloproteins with substituted cations, and stereochemical assessment of the effects of metal-ion substitution cannot be meaningfully pursued further. The comparative aspects of the data in table 4, however, supply, some structural basis for assessing relative changes expected on metal-ion substitution.

Since the chemical and structural properties of metal-ion-requiring proteins have been studied to a large extent on the basis of comparative aspects of metal-ion substitution, evaluation of these results within the framework provided by the three-dimensional structure of the native protein can supply important chemical correlations of structure–function relationships. Changes in biological reactivity on metal-ion substitution can result on the basis of numerous factors. Alterations in co-ordination geometry as well as the chemical nature of the metal ion and ligands themselves must be evaluated in determining the origin of the change. Since alterations in enzymatic activity can occur for numerous reasons, it is equally important to explain the basis for increased activity as well as loss of activity on metal-ion substitution.

2 HAEM PROTEINS

2.1 HAEMOGLOBIN AND MYOGLOBIN

Haem proteins are widely distributed in cellular systems, catalysing a variety of reactions, primarily oxidation–reduction processes essential for respiration and production of metabolic energy. A variety of haem proteins with widely different biological reactivities and specificities are known (see reference 87). However, three-dimensional structures have been determined by diffraction methods only for myoglobins[10-13,88], haemoglobins of both mammalian and lower-order species[14-18,89], calf-liver cytochrome b_5[21,90], and oxidised horse heart[19] and reduced bonito-tuna[20] cytochrome c. Structural investigations have been initiated on yeast cytochrome c peroxidase[23,24]. Also, a variety of difference Fourier studies have been carried out on sperm-whale myoglobin[66,70,91-3] and haemoglobins[94-7].

Since these haem proteins differ considerably from each other in function despite the common haem prosthetic group, iron–protoporphyrin IX, from which they are synthesised, it is consequently not feasible to discuss in adequate detail aspects of molecular structure and haem electronic structure pertinent to the functional properties of each. For this reason the relationships of iron–porphyrin interaction important in haemoglobin and myoglobin function will be examined in detail, since for these two proteins not only high-resolution structural information but also a large body of spectroscopic studies pertinent to the electronic structure of the iron–porphyrin group can be utilised to point out relationships of stereochemistry, electronic effects and functional properties. The interaction of the prosthetic group with its surrounding protein environment will be examined in detail on this basis. Relationships of iron–porphyrin stereochemistry pertinent also to the functional behaviour of other types of haem proteins will become evident through this discussion. Because of closely similar structural properties and related biological functions, the two proteins, haemoglobin and myoglobin, will be discussed together.

2.1.1 Haem crevice structure, iron electron configuration and functional behaviour

The structures of haemoglobin[16] and myoglobin[10] demonstrate that both proteins have a common characteristic property of retaining the haem prosthetic group within a fold or crevice formed by the protein chain. The haem crevice structure in the β subunit of haemoglobin is illustrated in figure 1. This structural property appears to prevent complete exposure of the haem group to solvent molecules except through channels formed by the protein chain and has led to the 'oil drop' concept of haem proteins, pictured as maintaining the iron–porphyrin prosthetic group within a highly hydrophobic environment. Figure 2 illustrates the general quaternary structural relationships that the subunits of tetrameric haemoglobins

SINTE GLESKA
COLLEGE LIBRARY

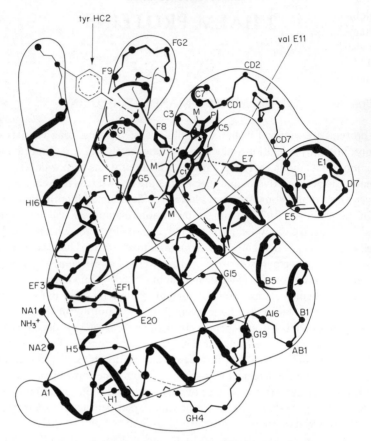

Figure 1 Diagrammatic illustration of the course of the polypeptide chain in the β-subunit of haemoglobin. In going from the
amino to the carboxyl end, helical regions are denoted A to H,
non-helical residues at the amino end are NA and at the
carboxyl end HC. Non-helical regions between helices are
denoted as AB, BC, etc. The haem prosthetic group lies in a
pocket between helices E and F and is covalently attached via
the iron to histidine-92 of the F helical region. The conformation of the α-subunit is closely similar except for the addition
and deletion of certain residues. From Perutz and TenEyck[89],
with permission.

have to each other, that is, the packing relationships of the subunits to each other.
It is a result of the protein environment that the chemical and electronic properties
of the iron–porphyrin group are modified in the control of its physiological function.
Functional properties of tetrameric or polymeric haemoglobins in contrast to those
of myoglobins are largely determined according to whether the subunits act as individual, single-protein chains or whether the subunit chains in contact with each
other mutually modify the chemical reactivity of the iron-porphyrin group. Since

myoglobins[10-12], single-chain haemoglobins[14,15,97] and the individual subunits of tetrameric haemoglobins[16,99] have grossly similar tertiary structures but still differ widely according to oxygen affinity and kinetics and pH dependence of ligand binding, the constraints placed on the reactivity of the iron–porphyrin group must be sought on a detailed structural level.

Figure 2   Model of horse methaemoglobin molecule derived from the three-dimensional Fourier synthesis at 550 pm resolution[146]. View is perpendicular to the a axis showing the general structural relationships of the α(white) and β(black) subunits to each other and the projections of the haem groups onto the ab face of the crystal. The helical regions are labelled as in figure 1. The label HS indicates the site of the β-93 cysteine sulphydryl residue in the α1–β2 subunit interface region. From Perutz[99], with permission.

The iron–protoporphyrin IX prosthetic group of haem proteins may be regarded as a tetraco-ordinated complex of near square-planar configuration with two ligands approaching the iron perpendicular to the porphyrin plane. In haem proteins for which structures are now known, the fifth position of the resulting octahedral complex is occupied by the N_3-position of the so-called *proximal* histidine residue attached to the surrounding polypeptide chain. The sixth position in the ferric form of haemoglobin and myoglobin is occupied by a water molecule (figure 3). It is with respect to the sixth co-ordination position, where rapid exchange or binding of appropriate ligands occurs, that the important physiological and chemical properties of these haem proteins are manifested[†]. Electronic and stereochemical constraints on ligand binding at the sixth co-ordination position thus become of prime importance in their chemical and physiological behaviour.

In the physiological function of haemoglobin and myoglobin, the ferrous iron–porphyrin moiety is directly concerned in reversible oxygenation. The oxygen-binding behaviour of haemoglobin is characterised by a sigmoidal relationship between the fraction of haems co-ordinated and the logarithm of the concentration of dissolved oxygen[98,100]. In contrast, the binding behaviour of myoglobin-type molecules is characterised by a hyperbolic relationship. These differences are respectively

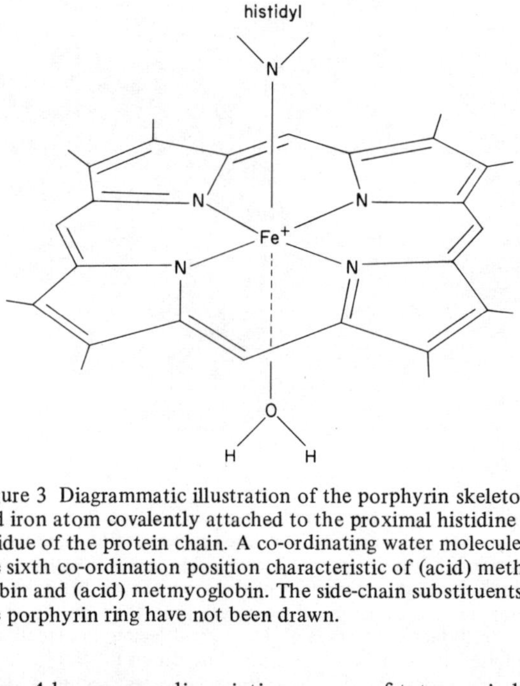

Figure 3 Diagrammatic illustration of the porphyrin skeleton and iron atom covalently attached to the proximal histidine residue of the protein chain. A co-ordinating water molecule occupies the sixth co-ordination position characteristic of (acid) methaemoglobin and (acid) metmyoglobin. The side-chain substituents of the porphyrin ring have not been drawn.

illustrated in figure 4 by oxygen-dissociation curves of tetrameric human haemoglobin having two pairs of α and β subunits, and haemoglobin H, a genetic mutant of human haemoglobin comprised of tetramers of β subunits. The binding and release of oxygen by the haemoglobin molecule is associated with a change in the quaternary structure of the protein[101,102]. This structural alteration underlies the well-known sigmoidal binding behaviour, known as *co-operative ligand binding* (or co-operative interaction of subunits) in which the binding (or release) of a ligand by one haem facilitates the binding (or release) of a second ligand by a neighbouring haem group. The co-operative interaction of subunits has been postulated[103] to result from an intricate mechanism of formation and breakage of amino-acid

† An excellent review of the chemical behaviour of haemoglobin and myoglobin with respect to ligand binding has been written by Antonini and Brunori[98].

contacts between the α and β subunits (see figure 2) of the tetrameric haemoglobin molecule. These changes are postulated[103,104] to result from the changes in iron electron configuration that accompany reversible oxygen binding.

Changes in the electron configuration of the iron d orbitals are intimately associated with ligand binding in both ferrous and ferric derivatives of haemoglobin and myoglobin. Pauling and Coryell[105] were the first to demonstrate on the basis of magnetic susceptibility studies that deoxyhaemoglobin contained a high-spin Fe(II)

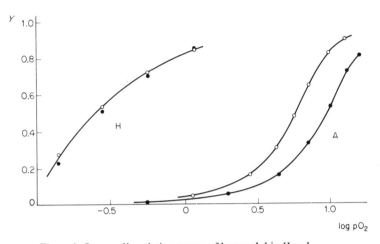

Figure 4 Oxygen dissociation curves of haemoglobin H and normal human haemoglobin A. HbH consists of a tetramer of β-subunits and displays no co-operative interaction of subunits, as characterised by the hyperbolic titration curve. HbA is representative of tetrameric mammalian haemoglobins which display co-operative subunit interaction characterised by the sigmoidal titration curve. The Bohr effect, a shift in the position of the titration curve along the horizontal axis as a function of pH, indicating ionisation of amino-acid residues linked to oxygenation, is indicated by two titration curves for HbA. Data obtained in 0.1 M phosphate buffer at pH 7.30 (○) and pH 6.88 (●). *Y* indicates the fraction of haems oxygenated. From Benesch *et al.*[319], with permission.

cation with four unpaired electrons, and oxyhaemoglobin was characteristic of a low-spin Fe(II) cation with no unpaired electrons. This information in conjunction with the crystallographic studies of the quaternary structure of haemoglobin for corresponding co-ordinated[16,106] and unco-ordinated[101,102] conformations indicates that the structural alteration is associated with a change in iron spin state, that is, the spectroscopic ground state of the haem iron. Furthermore, the exchange of ligands in ferric forms of haemoglobin and myoglobin also has been demonstrated to be accompanied by changes in haem-iron spin state depending on the ligand and temperature[107-12]. Ligand exchange in the ferric form of haemoglobin, however, is not associated with the gross alterations in quaternary protein

conformation characteristic of oxygen binding. Exchange of ligands at the sixth
co-ordination position of the Fe(III)ion under these conditions involves protein
derivatives of identical crystal structure[113].

Since the spin state, that is the magnetic susceptibility, of the haem-iron cation
has become of central interest in the study of haem proteins in characterising
protein derivatives of different iron oxidation states, it is necessary to make clear
the relationships of high- and low-spin electron configurations of the iron d orbital

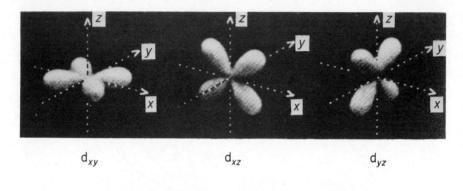

$$d_{xy} \qquad d_{xz} \qquad d_{yz}$$

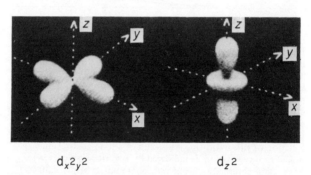

$$d_{x^2y^2} \qquad d_{z^2}$$

Figure 5 Diagrammatic illustration of the electron-density
distribution in d-orbitals. From Malkin and Malmström[5],
with permission.

system before proceeding to a discussion of iron–porphyrin stereochemistry. The
electron-density distribution of the valence shell of a transition-metal cation is
directly related to the angular dependence of the corresponding wave functions
that describe the d orbital configuration. In a transition-metal cation such as
those of Fe(II) and Fe(III) there are five d orbitals designated $d_{x^2-y^2}$, d_{z^2}, d_{xy}, d_{xz}
and d_{yz}, which have the electron-density distributions indicated in figure 5. In the
absence of surrounding ligands the orbitals exhibit fivefold degeneracy. Because
of the different shapes of the orbitals, the degeneracy is removed by interaction
with surrounding ligands (see Ballhausen[114] for a detailed discussion of theoretical
aspects of metal-ion co-ordination).

In an octahedral ligand field, as approximated in iron–porphyrin complexes, the $d_{x^2-y^2}$ and d_{z^2} orbitals are concentrated close to the ligands. These orbitals are consequently raised in energy relative to the other three as a result of the electrostatic interaction with each ligand electron cloud. The orbital energy levels are split according to the strength of the ligand field, as illustrated in figure 6,

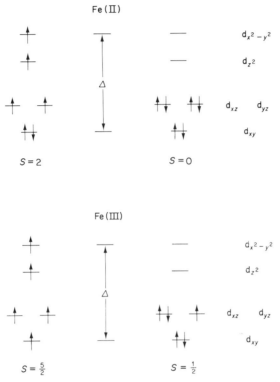

Figure 6 Diagrammatic illustration of relative energy levels of d orbitals of the high-spin and low-spin electron configurations of Fe(II) and Fe(III) cations for approximate square-planar D_{4h} symmetry. S is the net spin for the electron configuration; Δ represents the energy splitting between the e_g and t_{2g} orbital levels; Π represents the electron-pairing energy. The requirement for equilibrium between high-spin and low-spin configurations for a given oxidation state is that Δ (high spin) $\leqslant \Pi \leqslant \Delta$ (low spin).

resulting in the energy separation Δ between the e_g and t_{2g} levels. In the case of iron–porphyrin systems, high- and low-spin states refer to electron configurations with maximum numbers of unpaired and paired electrons respectively. For the Fe(II) and Fe(III) ions with six and five d electrons respectively, these configurations are illustrated in figure 6. Accordingly, the various commonly encountered ferrous and ferric derivatives of haemoglobin and myoglobin corresponding to high- and low-spin configurations of the haem iron are tabulated in table 5.

The change in spin state of the haem iron has become of central interest in the study of haem proteins not only because the magnetic states characterise the various haem complexes but also because haem proteins exhibit an unusually delicate equilibrium between states of maximum and minimum spin multiplicity. The initiation of a change in iron spin state is thought to be stereochemically derived. The change in spin state on transfer of electrons between the e_g and t_{2g} orbital sets is associated with a change in the ionic radius of the iron cation with a resultant change in metal–ligand bond length. As demonstrated on the basis of simple inorganic complexes[58] the ionic radius of the Fe(II) or Fe(III) ion increases by approximately 20 per cent on change from low-spin to high-spin

TABLE 5 COMPARISON OF OXIDATION, LIGAND AND SPIN STATES OF
HAEMOGLOBIN AND MYOGLOBIN DERIVATIVES

Iron oxidation state	Complex	Ligand at sixth co-ordination position	Net spin[†] (S)
Fe(II)	deoxyhaemoglobin	none	2
	oxyhaemoglobin	O_2	0
	carboxyhaemoglobin	CO	0
	nitrosylhaemoglobin[‡]	NO	1/2
Fe(III)	(acid) methaemoglobin	H_2O	5/2
	methaemoglobin fluoride	F^-	5/2
	methaemoglobin cyanide[§]	CN^-	1/2
	methaemoglobin azide[§]	N_3^-	1/2
	(alkaline) methaemoglobin[§]	OH^-	5/2 (50%)
			1/2 (50%)
	methaemoglobin imidazole[§]	imidazole	1/2

† from Scheler *et al.*[110]. Haemoglobin complexes correspond to identical oxidation and ligand states of myoglobin complexes.

‡ on basis of electron paramagnetic resonance absorption.

§ a higher proportion of high-spin character is associated with the corresponding metmyoglobin complex[112].

states (see table 3). The stereochemical importance of a given spin state for an iron–porphyrin complex, therefore, is that the relationship of the iron cation to the plane of the co-ordinating porphyrin pyrrole nitrogen atoms is dependent on the iron–porphyrin bond lengths as modified by the ionic radius of the metal ion and metal–ligand bonding interactions. Furthermore, since the binding of oxygen is associated with a change in spin state[105] and the position of the iron atom relative to the porphyrin plane must be immediately related in time and space to the binding of the oxygen molecule, alterations in iron–porphyrin stereochemistry have been invoked[103,104] to initiate the conformational events responsible for the co-operative binding of oxygen. Herein lies the biological importance of the electron configuration of the iron atom in the physiological function of haemoglobin.

2.1.2 Iron–porphyrin stereochemistry in model complexes

The important stereochemical relationships relevant to iron–porphyrin interaction in haem proteins have been determined on the basis of high-resolution X-ray diffraction studies of numerous low molecular weight, model metalloporphyrin complexes. These stereochemical considerations were formulated on the basis of detailed crystallographic investigations of metalloporphyrins with metal ions of different ionic radii and of iron–porphyrin complexes of different spin states. Detailed reviews and discussions of these points have been well presented by Hoard[104,115-17]. It is of importance to this discussion to recapitulate only those details pertinent to the discussion of iron–porphyrin stereochemistry in haem proteins. To this end figure 7 illustrates a diagram of the carbon–nitrogen skeleton of the porphinato core with the extreme values determined crystallographically for bond lengths and bond angles of various metalloporphyrins[116,117]. The radius of the central hole defined by the pyrrole nitrogen atoms is constrained by the electronic structure of the porphinato core to a narrow range of variation. The value of the Ct . . . N radius that corresponds most closely to *minimisation of radial strain*, determined by examination[116,117] of a variety of metalloporphyrin structures, is approximately 201 pm. This value is characteristic of iron–porphyrin stereochemistry, as is evident by comparison of pertinent bond lengths of iron-porphyrin complexes in table 6. The stereochemical data for the porphinato core with minimal radial strain are compared in figure 7 to those for metalloporphyrin derivatives with maximally strained in-plane metal–nitrogen bonding relationships.

The detailed geometry of model high-spin ferric complexes has been determined primarily from an analysis of the high-spin five-co-ordinate ferric porphyrin derivatives methoxy-iron(III)-*meso*-porphyrin(IX) dimethylester[120] and chlorohaemin[121]. A diagram of their characteristic square-pyramidal co-ordination geometry is illustrated in figure 8. From figure 7 it is seen that a position of an iron atom coplanar with the pyrrole nitrogen (N_p) atoms is stereochemically incompatible for $Fe-N_p$ bond distances > 201 pm. In these high-spin ferric complexes the $Fe-N_p$ bond distances are ≈ 207 pm, resulting in a displacement of the metal atom from the plane of pyrrole nitrogen atoms. Thus, the iron atom lies 49 pm from the mean porphyrin plane in the five-co-ordinate Fe(III) complexes. Furthermore, in a model square-pyramidal complex the porphyrin skeleton is slightly domed, a configuration which may allow a more favourable orientation for bonding to the Fe(III) ion.

Since the difference in the ionic radii of a high-spin and low-spin iron cation (table 3) approximates to 12 pm for octahedral co-ordination[58], a coplanar configuration of the iron atom and pyrrole nitrogen plane would be expected for low-spin complexes. A coplanar configuration is indeed observed in the low-spin bis(imidazole)-$\alpha,\beta,\gamma,\delta$-tetraphenylporphinatoiron(III) chloride (Imid$_2$TPPFe(III)) complex[118]. Furthermore, since Fe–N bond lengths are predicted to differ by ≈ 2 pm for otherwise chemically analogous Fe(III) and Fe(II) complexes[58], a coplanar iron–porphyrin stereochemistry is similarly expected for a low-spin ferrous compound. The predicted similar configurations for low-spin ferrous and

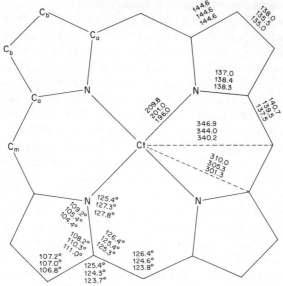

Figure 7 Diagram of the carbon–nitrogen skeleton in the
porphinato core of a metalloporphyrin drawn for the real or
effective retention of D_{4h} symmetry. The metal atom is
centred at Ct. Values of the principal radii (pm), bond lengths
(pm) and bond angles (degrees) of three metalloporphyrins
are shown on the diagram and refer from top to bottom to:
Ni(II)-2,4-diacetyldeuteroporphyrin-IX dimethyl ester; the
reference metalloporphyrin of *least strain* corresponding
closely to iron-substituted porphyrin derivatives as in
$Pip_2TPPFe(II)$[119], and to α,β,γ,δ-tetraphenylporphinato-
dichlorotin(IV). The Ni(II) and Sn(IV) derivatives correspond
to the metalloporphyrins identified with the smallest and
largest Ct . . . N radii observed thus far. Since these derivatives
have metal ions coplanar with the pyrrole nitrogen atoms, the
stereochemical data indicate the resultant effects in the
porphinato core exerted by maximum strain through the
(in-plane) shortest and longest M . . . N bond lengths respectively. C_a, C_b and C_m are the symbols used for three
chemically and structurally distinct types of carbon atoms
The porphinato core is observed to exhibit a narrow range of
variation in bond lengths and bond angles. From Hoard[117],
with permission.

ferric systems has been recently confirmed[119] from the determination of the
structure of bis(piperidine)-α,β,γ,δ-tetraphenylporphinatoiron(II) ($Pip_2TPPFe(II)$).
These derivatives both have $Fe-N_p$ bond distances < 201 pm with coplanar iron
and pyrrole nitrogen stereochemistries. In view of the Ct . . . N distances for these
complexes in table 6, the movement of an iron cation with respect to the pyrrole
nitrogen plane on alteration in spin state by addition of axial ligands can occur
with negligible effects in radial strain of the porphinato core. The critical Ct . . . N
distance of 201 pm characteristic of minimal radial strain in the porphyrin ring is
not appreciably altered on change between high and low spin states.

TABLE 6 PARAMETERS OF IRON–PORPHYRIN DERIVATIVES CORRESPONDING TO THE FOUR COMMON VALENCE STATES OF THE HAEM IRON IN HAEMOGLOBIN AND MYOGLOBIN

Type of distance	Length (pm)			
	Low-spin ferric[†]	High-spin ferric[‡]	Low-spin ferrous[§]	High-spin ferrous[‖]
Ct . . . M	0.9	48	0.0	(70)
Ct . . . N	198.9	201.9	200.4	(202)
M–N pyrrole	198.9	207.4	200.4	(214)
M–A	195.7[¶]	184.2[††]	212.7[‡‡]	(190)[§§]
	199.1			

† from Collins *et al.*[118] for bis(imidazole)-α,β,γ,δ-tetraphenylporphinatoiron(III).

‡ from Hoard[116]. Values correspond to those determined for the high-spin methoxy-iron(III) derivative of mesoporphyrin-IX dimethyl ester, chlorohaemin and μ-oxo-bis-(tetraphenyl-porphinato)iron(III).

§ from Radonovich *et al.*[119] for bis(piperidine)-α,β,γ,δ-tetraphenylporphinatoiron(II).

‖ calculated on the basis of stereochemical arguments by Hoard[116].

¶ axial ligands are imidazole groups and iron cation is displaced towards the more distant group.

†† axial ligand is a methoxy group.

‡‡ axial ligand is piperidine and bond length is considerably increased due to steric interaction with porphyrin the ring[119].

§§ axial ligand is considered as an imidazole or pyridine group.

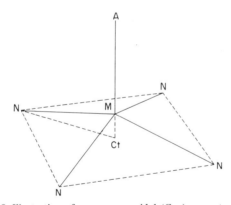

Figure 8 Illustration of square-pyramidal (C_{4v}) geometry characteristic of the co-ordination of Zn(II), Mg(II) and high-spin Fe(III) porphyrins. From Hoard[116], by permission.

Since no high-spin ferrous derivative has been thus far crystallisable, an estimate of the out-of-plane displacement of a high-spin, five-coordinate ferrous iron–porphyrin derivative must be made from stereochemical considerations. These have been formulated by Hoard[115,116,120]. A five-co-ordinate, high-spin Fe(II) ion has an ionic radius approximately 12 pm larger than that of a high-spin Fe(III) cation for octahedral co-ordination. The maximum corresponding $Fe-N_p$ bond lengths for a high-spin ferrous iron–porphyrin complex, therefore, are expected to be $\leqslant 219$ pm. Under the assumption that a Ct . . . N distance of ≈ 201 pm is maintained, an out-of-plane displacement of $\leqslant 87$ pm of the high-spin Fe(II) ion is expected as an approximate upper limit. While the stereochemistry of the high-spin ferrous complex awaits direct experimental confirmation, the predicted stereochemistry is based on reasonable values of metal–ligand bond distances observed in model co-ordination complexes. As will be pointed out later (section 2.1.3(1)) the stereochemistry of a high-spin Fe(II) complex in proteins is most probably in close agreement with prediction[115,116] despite the assumptions inherent in the formulation of the stereochemical arguments.

The structural principles deduced on the basis of the stereochemistry of model metalloporphyrins and especially of the iron–porphyrin complexes illustrated in table 6 have been of invaluable aid in demonstrating the expected iron–porphyrin geometries for corresponding spin states of haem-protein complexes. Their detailed stereochemical and geometrical relationships cannot, however, be applied directly to haem configurations in proteins without some modification. In particular, as pointed out earlier by Hoard and coworkers[120], the model high-spin ferric complexes are five-co-ordinate while the high-spin ferric complexes of haem-proteins are six-co-ordinate. Some differences with respect to metal–ligand bond lengths and metal out-of-plane displacement, are, therefore, to be expected (section 2.1.3). Furthermore, the Ct . . . N distance of 201 pm in a high-spin five-co-ordinate Fe(II) complex may not obtain precisely. The Ct . . . M distance for this protein derivative may be an overestimate of the actual value. We wish, therefore, to emphasise that the values of bond lengths in table 6 define only the general structural principles in iron–porphyrin stereochemistry which are to be expected with spin-state transitions in haem proteins. None the less, as will be pointed out in the following section (section 2.1.3), the deductions of Hoard[115–17] and the bond lengths in table 6 provide an accurate guide of the stereochemical geometry of the central iron–porphyrin group of haem proteins for each of the commonly found oxidation states of the haem iron in haemoglobin.

2.1.3 Iron–porphyrin stereochemistry in proteins

(1) Haemoglobin studies

Of the crystalline haem proteins studied by X-ray diffraction methods, the stereochemistry of the iron–porphyrin group has been most precisely determined by refinement methods in sea-lamprey methaemoglobin cyanide[15]. As indicated in table 5, this derivative is a ferric low-spin haem complex and corresponds to the $\text{Imid}_2\text{TPPFe(III)}$ derivative described in table 6. Structural resolution to 20 pm has been achieved, and the polypeptide structure has been refined by the Diamond method[61]. An illustration of a portion of the model atomic structure corresponding to the haem environment and the iron–porphyrin group is given in figure 9. The haem is covalently bonded to the globin moiety by a bond between the iron atom and the N_3 atom of histidine-104. A cyanide ion occupies the sixth co-ordination site.

The position of the iron atom in sea-lamprey methaemoglobin cyanide was quantitatively estimated by the separation of the electron density due to the iron atom from the density of the surrounding porphyrin ring. A weighted centroid position for the iron atom

$$\bar{x}_{Fe} = \sum_i (\bar{x}_i)\, \rho(\bar{x}_i) / \sum_i \rho(\bar{x}_i) \qquad (2.1)$$

was computed from the electron densities $\rho(\bar{x}_i)$ at grid points $(\bar{x}_i)$. A least-squares plane was similarly calculated for the iron-devoid haem. The centroid position was

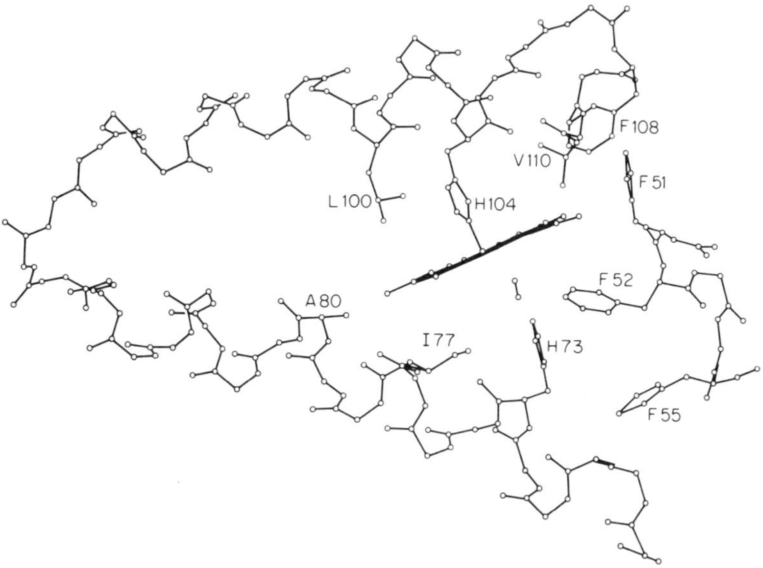

Figure 9 Illustration of the haem environment of sea-lamprey methaemoglobin cyanide: A portion of the backbone of the model atomic structure corresponding to the electron density map. The haem and parts of the CD region, residues 50(C6) to 55(CD4) and the E, EF, F, and FG regions, residues 67(E1) to 110(FG5), are shown. The stick drawing includes the side chains phenylalanine-51(C7), phenylalanine-52(CD1), phenylalanine-55(CD4), histidine-73(E7), isoleucine-77(E11), alanine-80(E14), leucine-100(F4), histidine-104(F8), phenylalanine-108(FG5) as well as atoms of the polypeptide backbone. The figure is viewed parallel to the haem plane. From Hendrickson and Love[15] with permission.

calculated to lie no more than 1 pm from the mean porphyrin plane, being displaced towards the proximal histidine-104 residue[15]. This result fully confirms the predictions of Hoard[115,116] in close analogy to the iron–porphyrin stereochemistry[118] observed in the low-spin complex of $Imid_2 TPP Fe(III)$. Difference electron-density maps of sea-lamprey methaemoglobin cyanide and carboxyhaemoglobin in projection suggested, furthermore, no evidence of haem movement or change in iron position. These studies establish the position of the iron atom to be essentially coplanar with the pyrrole nitrogen atoms in low-spin haem protein complexes. Sufficient resolution of X-ray diffraction data is not present, however, to determine the extent of ruffling of the porphyrin skeleton. At present the porphyrin skeleton can only be described as planar in sea-lamprey methaemoglobin cyanide[15].

Comparable precision in the determination of the position of the iron atom with respect to the porphyrin plane has not been determined for the high-spin ferric and ferrous complexes of horse haemoglobin and of the insect haemoglobin erythrocruorin. There is little doubt, however, that the haem stereochemistry in these proteins remains consistent with the structural principles established earlier by Hoard[115,116]. Perutz[103] has pointed out in the case of horse methaemoglobin that on construction of the skeleton model of the molecule to a scale of 100 pm = 2 cm, measurements of the distance between the iron and the average porphyrin plane with the aid of superposed electron-density maps[60] indicate that the distance is 30 pm for the haem groups of both α and β subunits. An identical displacement was observed similarly in meterythrocruorin[14]. As pointed out originally by Hoard and coworkers[120], the out-of-plane displacement of 49 pm observed in model *five-co-ordinate* high-spin ferric complexes need not apply strictly to the *six-co-ordinate* high-spin ferric complexes of haem proteins. Since a water molecule occupies the sixth co-ordination position in high-spin ferric haem-protein complexes, the ferric ion is expected to lie slightly closer to the pyrrole nitrogen plane. Because of the steric repulsions by the pyrrole nitrogen atoms, the bonding between the Fe(III) ion and the water molecule is expected to allow only weak complexing. The Fe—OH distance of $\leqslant 205$ pm estimated for a hydrated iron-porphyrin complex[122] serves as an approximate estimate of the corresponding Fe—O bond length in the high-spin ferric haem proteins.

In addition, measurements[103] with the help of the electron-density maps of horse deoxyhaemoglobin, also determined at 280 pm resolution, showed that the perpendicular distance of the five-co-ordinate, unco-ordinated Fe(II) cation above the average porphyrin plane was 75 pm. These results thus establish the essential correctness of Hoard's predictions about the changes in the position of the iron atom relative to the porphyrin plane. Precise quantitation of iron—porphyrin relationships in these derivatives still awaits further refinement studies, however, as carried out for methaemoglobin cyanide[15]. The position of the iron relative to the porphyrin plane in haem proteins comparable to those for model iron-porphyrin complexes, strongly suggests that the geometry of the porphyrin skeleton in proteins must correspond closely to structural data for corresponding spin states indicated in table 6. These results furthermore, are, generally confirmed by three-dimensional difference Fourier syntheses of high-spin versus low-spin haem-protein complexes of erythrocruorin[96] and of *glycera* haemoglobin[95,97]. These studies demonstrate that in haemoglobin-type molecules the metal ion occupies a position essentially coplanar with the porphyrin ring in low-spin derivatives and is displaced relative to the porphyrin ring towards the proximal co-ordinating histidine residue in high-spin derivatives.

(2) Myoglobin studies

Although initial investigations of sperm-whale metmyoglobin[10] utilised the stereochemistry of the planar Ni(II)etioporphyrin complex[123] to fit a haem skeleton model to the electron-density map, subsequent reconsideration[74] indicated that the iron atom must lie at least 25 pm from the porphyrin plane, analogous to the high-spin ferric complex of chlorohaemin[121,124]. The most recent description of the atomic parameters of sperm-whale metmyoglobin[11] yields a calculated iron out-of-plane displacement of 30 pm. Although the method of parameter refinement of the iron–porphyrin group is not stated, this calculated distance most likely reflects the essential correctness of the position of the high-spin Fe(III) ion relative to the porphyrin plane in view of the stereochemical considerations of Hoard[115,116]. The recent preliminary assessment of the displacement of the Fe(III) ion as 60 pm from the pyrrole-nitrogen plane determined by neutron-diffraction methods[88] must await refinement studies. Under the assumption that little change occurs in the Ct . . . N distance of ≈ 201 pm, the iron out-of-plane displacement remains unduly large for a six-co-ordinate complex in comparison to the Ct . . . M distance of high-spin five-co-ordinate complexes (table 6).

Alterations in iron–porphyrin stereochemistry accompanying changes in the haem-iron spin state were not reported in the early difference Fourier studies of sperm-whale myoglobin derivatives. With respect to the structure of high-spin (aquo) metmyoglobin, these studies have included three-dimensional difference Fourier syntheses of metmyoglobin azide at 200 pm resolution[70] and deoxymyoglobin at 280 pm resolution[91], and a series of high- and low-spin metmyoglobin derivatives at 280 pm resolution calculated for centrosymmetric projections only[92]. Because of the alterations of iron–porphyrin stereochemistry observed in haemoglobin-type molecules, it is difficult to simply rationalise the myoglobin results as reflecting insufficient precision since the difference Fourier method has had remarkably successful application with a variety of crystalline proteins.

It is possible, however, that an explanation of these apparent discrepancies of myoglobin studies can be found from the more recent work of Bretscher[93]. He found that in the three-dimensional difference Fourier synthesis of low-spin metmyoglobin cyanide versus high-spin metmyoglobin, small difference electron-density peaks could be attributed to movement of a pyrrole nitrogen atom and the associated C_a and C_b atoms of its pyrrole ring. While the extent of movement could be estimated only as ≈ 10 pm, it is significant that re-examination[93] of the difference electron-density map of metmyoglobin azide indicated similar peaks. These peaks were not reported for the metmyoglobin azide derivative since they were not greater than six times the root-mean-square value of the electron density chosen as the significance level by Stryer and co-workers[70]. Furthermore, there was a strong similar correlation of features for the low-spin cyanide and the mixed-spin metmyoglobin hydroxide derivative described by Watson and Chance[92]. Bretscher also observed that the peaks were absent in the high-spin metmyoglobin fluoride versus high-spin (aquo) metmyoglobin difference map[92]. The significance level chosen by Bretscher was three times the root-mean-square electron density.

While the small difference electron-density peaks observed by Bretscher[93] cannot
be readily interpreted to yield precise estimates of shifts of porphyrin skeletal
atoms, the correlation found for low-spin derivatives—but not in a re-examination[93]
of the difference maps of high-spin metmyoglobin fluoride versus high-spin
metmyoglobin[92]—suggests that the peaks have not arisen from systematic errors.
It must therefore be concluded that small and as yet imprecisely defined, positional
shifts of the porphyrin skeleton and pyrrole nitrogens also occur on changes in
haem-iron spin state in sperm-whale myoglobin. Presumably these changes consist
essentially in a decrease of the displacement of the metal ion from the average
pyrrole nitrogen plane on the change from a high-spin to a low-spin derivative.

Furthermore, no change in the position of the iron atom relative to the porphyrin
plane was observed on reduction of sperm-whale metmyoglobin to deoxymyoglobin[91]. No change in iron out-of-plane displacement was similarly observed on
reduction of the ferric ion in meterythrocruorin[96]. Both reduced proteins were
characterised by diffraction studies to contain a five-co-ordinate haem complex.
These results are not necessarily incompatible with the stereochemical considerations
of Hoard[115,116]. The iron out-of-plane displacement of ≈ 70 pm predicted[116] for
a high-spin ferrous complex represents an estimated upper limit and presumes no
change in the co-ordinate positions of the pyrrole nitrogen atoms with respect to
the rest of the porphyrin skeleton. A small increase in the Ct . . . N distance on
reduction of the Fe(III) ion may well occur to accommodate the high-spin Fe(II)
metal ion on an approximate 12 pm increase in radius. Such slight positional shifts
of the nuclei, including an estimated minimum detectable shift of 10 pm for the
iron atom as the heaviest nucleus in the complex[91,125], may have remained
unobserved because of the problems inherent in application of difference Fourier
methods to proteins[115].

It must be emphasised, however, that the changes in iron–porphyrin stereochemistry in myoglobin are considerably smaller than those observed in haemoglobin molecules. While it is not possible to assess quantitatively the contributions
of various errors, such as inaccuracy of phases to defining haem stereochemistry
in the sperm-whale myoglobin studies as compared to comparable effects in the
numerous investigations of haemoglobin molecules, the X-ray studies clearly suggest
that haem configurational and stereochemical changes are more pronounced in
haemoglobin. Padlan[95] and Padlan and Love[97] have observed changes in iron–
porphyrin stereochemistry with spin-state alterations of the monomeric *glycera*
haemoglobin from difference Fourier studies calculated only for centrosymmetric
projections at 300 pm resolution. Virtually no change was observed in difference
Fourier projection studies of sperm-whale myoglobin at 280 pm resolution[92].
Furthermore, although there appears to be no change in the displacement of the
iron atom from the porphyrin plane in meterythrocruorin and deoxyerythrocruorin,
readily observed alterations in iron–porphyrin stereochemistry and porphyrin configuration have been described[96] by three-dimensional difference Fourier studies
at 250 pm resolution for binding of carbon monoxide by the deoxy derivative.
Moreover, horse methaemoglobin, modified covalently with bis(N-maleimidomethyl)-

ether (BME) to prevent the change in quaternary protein structure[126], shows peaks due to a tilt of the porphyrin rings at only 350 pm resolution after chemical reduction of the Fe(III) cation[89]. It is possible that the more marked tendency of the haem group in myoglobin to remain less altered stereochemically on change of iron oxidation and ligand state may reflect a fundamental structural characteristic differentiating myoglobins from haemoglobin-type molecules. The more pronounced tendency of the porphyrin plane in sperm-whale myoglobin to remain unchanged has been recently confirmed by spectroscopic investigations. The results of these studies will be discussed in more detail later (section 2.1.4(1)).

(3) Iron–imidazole interaction

The iron-porphyrin group is tightly bonded to the protein in both haemoglobin and myoglobin via a covalent link between the metal cation and the N_3-position of the proximal histidine residue[11,16]. On the basis of the changes in Fe–N bond lengths observed with spin-state alterations, comparable effects might be anticipated in the axially directed Fe–N bond formed with the proximal co-ordinating imidazole ring (Fe–N_{Imid}). The nature of the axial Fe–N_{Imid} bond in haem complexes on change in spin state, however, remains an unsettled question. While changes in iron–porphyrin stereochemistry resulting from spin-state induced Fe–N_p bond-length changes have been observed in haem proteins as well as model iron–porphyrin complexes, the postulated[127,128] corresponding alteration of the Fe–N_{Imid} bond length has not been convincingly demonstrated on an experimental basis. According to Williams[127,128], the transfer of two electrons between the iron t_{2g} and e_g orbitals, for instance, on deoxygenation of haemoglobin or exchange of a cyanide ligand for the water molecule in methaemoglobin would be expected to result in a change in the axial metal–ligand bond length as observed for the Fe–N_p bonds.

Hoard[115], however, has pointed out that in the five-co-ordinate, high-spin ferric haem complexes, the axially directed Fe–ligand bond length is abnormally short, and is in fact shorter (see table 6) than what would be expected for an octahedrally co-ordinated low-spin Fe(III) cation. On the basis of the stereochemistry of model five-co-ordinate complexes, Hoard[115,116] has concluded that combination of the d_{z^2} orbital of the high-spin Fe(III) ion with the appropriate σ-type orbital of the ligand yields a strongly bonding orbital. This orbital accepts the electron pair, and a weakly antibonding orbital accommodates the unpaired electron. Substitution of a high-spin Fe(II) ion for a high-spin Fe(III) ion is considered to result in an enhancement of these effects. On this basis Hoard[115,116] anticipates that the axial Fe–N_{Imid} bond length for a five-co-ordinate ferrous iron–porphyrin complex should be $\leqslant 200$ pm, closely comparable to the corresponding Fe–N_{Imid} bond length[82] for a low-spin six-co-ordinate complex (see tables 4 and 6). On this basis Hoard[115,116] has concluded that the axial Fe–N_{Imid} bond length in the high-spin ferrous protein is already at or near its low-spin value.

There are, however, two possible factors which could contribute to shortening the Fe–O and Fe–Cl bond lengths of the model five-co-ordinate complexes on the basis of which these deductions were formulated[115,116]. Both iron–porphyrin complexes contain an Fe(III) ion resulting in a net positive charge associated with the porphyrin group. The axial ligand in both cases is anionic, and consequently electrostatic contributions[†] may be of importance in shortening the bond length. In ferrous haem-protein complexes both the haem group and imidazole ring of the proximal histidine are neutral species. However, the general tendency of co-ordination complexes towards metal–ligand bond shortening on a decrease in co-ordination number, in this case from six to five, may be involved also, and this factor may be of importance for the protein. Indeed, the Fe–N bond lengths of model five-co-ordinate Fe(III) haem complexes (table 6) are shorter than the 216–233 pm observed for Fe–N bond distances in the six- and seven-co-ordinate high-spin Fe(III) amine–polycarboxylate complexes[129–32]. This factor could provide a restraining effect against the tendency for an increase in the axial Fe–N bond length on change in spin state and co-ordination number and furthermore, would be enhanced by Fe–imidazole d_π–p_π interactions. The hypothesis[127,128] of an increased axially directed Fe–N bond length for a high-spin iron–porphyrin complex thus awaits a rigorous test. Only then can the relative importance of the change in the axial Fe–N_{Imid} bond length relative to the Fe–N_p bonds be quantitatively assessed with respect to the functional properties of haemoglobin and myoglobin (see section 2.1.4).

2.1.4 Structural and functional importance of iron–porphyrin stereochemical alterations in haem proteins

(1) Transmission of the spin-state induced change in iron radius
The preceding discussion leads to the expectation that the reversible oxygenation of haemoglobin (and myoglobin), a process marked by a reversible high-spin to low-spin transition[105], should be associated with a non-trivial alteration in the detailed configuration and dimensions of the iron–porphyrin prosthetic group. It is then immediately of interest to determine whether the stereochemical alterations associated with the haem centre of one subunit are propagated—through co-operative movements of protein groups in the case of haemoglobin—such that they are sensed in one of the neighbouring chains. Experimental evidence that haem reactivity is, indeed, modified by such co-operative interactions has been thoroughly discussed[98]. The changes have been interpreted as arising from alterations in the contacts of amino-acid side chains in the subunit interface regions[103]. The stereochemical alterations of the haem centre on this basis are postulated to serve as an initiating event[104] for the structural mechanism that must account for the co-operative nature of reversible oxygenation in haemoglobin[98,100] and for the large accompanying movements of the subunits of haemoglobin relative to each other[101,102].

The change in iron–porphyrin stereochemistry[103,104] and the postulated change in Fe–N_{Imid} bond length[127,128] have both been implicated as initiating the chain

† The possible importance of electrostatic contributions has been particularly stressed by Prof. R. J. P. Williams pers. com.

of conformational events responsible for co-operative interaction of haemoglobin subunits and the alteration in quaternary protein structure. Structural data clearly indicate that iron-porphyrin stereochemical changes are associated with spin-state transitions of the centrally co-ordinated metal ion in haem proteins. No quantitative assessment of the relative contributions of changes in iron–porphyrin structure as opposed to that of the postulated change in $Fe-N_{Imid}$ bond length can be made, however, in view of the absence of sufficiently precise stereochemical data for the latter. Such a quantitative assessment has no doubt been hindered by the fact that the deoxy and co-ordinated quaternary structures of tetrameric mammalian haemoglobins crystallise in different space groups[113], preventing a direct comparison of the pertinent structures by, for instance, difference Fourier methods. Furthermore, both mechanisms have been postulated[103,128] as a triggering event, initiating the change in quaternary structure on release or binding of an oxygen molecule, for the induction of a structural alteration in the F helix of the protein subunits containing the co-ordinating histidine residue. This structural alteration is then expected to be propagated to other regions of the protein. The nature of the alteration in the F helix has remained undefined, however.

To be able to relate changes in Fe–ligand bond lengths to co-operative structural events defined in sequence of their occurrence, an assessment of the stereochemical alterations of the co-ordination centre and its environment with respect to a set of fixed co-ordinate axes as reference points is clearly required. It is not possible at present to precisely define kinetically the series of co-operative structural events, beginning with the reduction in ionic radius of the Fe(II) cation on oxygen binding and culminating in an alteration[99,103] of $\alpha_1-\beta_2$ subunit-interface contacts of amino-acid side chains; but it is possible to demonstrate the probable manner in which the change in iron radius is propagated to a more distant region of the protein.

Hoard[104,115] clearly pointed out that since the iron atom is attached rigidly to the globin moiety, the shifting of atomic positions is best considered relative to axes fixed in the molecule or in the crystal during the high-spin to low-spin transition. This structural change must contain a translational component altering the position of the porphyrin ring with respect to the iron nucleus by ≈ 80 pm. Since the postulated change in $Fe-N_{Imid}$ bond length[127,128] should result in no more than a 20 pm displacement of the iron with respect to the co-ordinating histidine N_3-position, the major contribution to the initiation of co-operative structural events would appear to have its origin in iron–porphyrin stereochemical alterations. Indeed, it appears that the transmission of structural changes results from the nonbonded protein–porphyrin interactions between the porphyrin skeleton and nearby amino-acid side chains upon the change in iron–porphyrin stereochemistry.

Recently on the basis of spectroscopic studies the importance of iron–porphyrin stereochemical configuration in the transmission of structural changes to more distant regions of the protein has been quantitatively assessed by a detailed analysis[133-6] of the polarised Soret absorption properties of single crystals of haemoglobin and myoglobin. Haem proteins and iron–porphyrin derivatives exhibit an intense absorption band near 400 nm known as the Soret transition. The high intensity of this transition compared to that of other haem transitions is illustrated in figure 10 in which the crystal, solution and polarisation-ratio spectra of horse oxyhaemoglobin are shown. The Soret transition has been demonstrated to be polarised in the plane of the porphyrin ring[137,138]. The line labelled *x,y*-polarised gives the polarisation ratio for a transition polarised in the plane of the porphyrin ring and is obtained from the ratio of the integrated intensities of

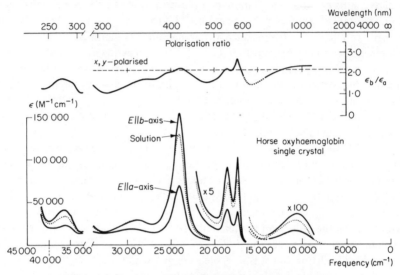

Figure 10 Polarised single-crystal absorption spectrum of horse oxyhaemoglobin. Of particular reference to the discussion in the text is that the absorption intensity of the Soret band (ϵ_{max} at 420 nm) is greater in the *b* polarised state than in the *a* polarised state. The $\epsilon_b : \epsilon_a$ Soret polarisation ratio of the integrated absorption intensities for oxyhaemoglobin is 2.27, significantly different from that of methaemoglobin (2.45) and carboxyhaemoglobin (2.11) or cyanomethaemoglobin (2.18). From Makinen and Eaton[134], with permission.

the *b*- and *a*-axis polarised Soret bands. This ratio serves as a monitor of porphyrin orientation, since it defines the projections onto the crystallographic axis system of the transition-moment vectors of a doubly degenerate transition, which are polarised in the plane of the porphyrin ring. The simplest explanation of a change in the polarisation ratio observed on a change in iron oxidation state and ligand state is that the orientation of the porphyrin planes relative to the crystallographic axes has changed[134-6]. More detailed experimental and theoretical justification of this interpretation has been put forth by Makinen and Eaton[135,136].

A detailed examination and analysis of the Soret polarisation properties of oriented single crystals of haemoglobin derivatives showed[133-6] that significant changes in

Soret polarisation ratio result on exchange or loss of ligands bound at the sixth co-ordination position of the iron. The minimum change in porphyrin orientation compatible with the alteration in Soret polarisation ratio was then calculated for changes between the high-spin and low-spin ligand and oxidation states. A comparison of the calculated minimum changes in porphyrin orientation compatible with polarisation data for myoglobin derivatives and the average minimum change in porphyrin orientation of an α and β subunit pair in haemoglobin[†] is illustrated in figure 11. In this figure the change in porphyrin orientation has been correlated with the iron out-of-plane displacement deduced on the basis of X-ray crystallographic studies for corresponding spin states[15,103]. The single-crystal spectroscopic studies (compare figure 11) show that in addition to the change in position of the iron relative to the porphyrin ring, a change in porphyrin orientation also occurs on a change in iron oxidation state and ligand state. The change in iron–porphyrin stereochemistry and porphyrin orientation within the haem crevice can be considered to occur by two different types of structural events: a purely translational component bringing the porphyrin ring and iron atom in or out of near coplanarity, and a rotational component resulting in a different orientation of the porphyrin plane with respect to axes fixed in the crystal (and, therefore, in the protein itself).

Of considerable importance in supplying direct structural confirmation of the spectroscopic interpretations of Makinen and Eaton[133-6] is that an increase in Soret polarisation ratio is observed on the change from methaemoglobin to the deoxy derivative of haemoglobin modified with BME (figure 11). This change in polarisation ratio corresponds[135] to an average minimum change in orientation of the porphyrin rings of the α and β subunits of approximately 4°. Recent difference Fourier studies of BME-modified deoxyhaemoglobin versus methaemoglobin at 350 pm resolution[89,103] show electron-density peaks corresponding to a tilt of the haem groups of both α and β subunits. Although no estimate has been made of the extent of the change in orientation of the porphyrin rings, the tilt in both subunits results in an increase in the parallelness of the haems to the *b* crystallographic axis and a decrease in the projection of the haems onto the *a* axis[103]. This change in heme tilt is consistent with the observed change in Soret polarisation ratio. The results of the single-crystal spectroscopic studies (figure 11) indicate, in addition, that, since small Soret polarisation changes are observed between ferric high-spin and ferric or ferrous low-spin states, the porphyrin groups in haemoglobin do not have identical orientations for all co-ordinated states.

The structural basis for transmission of changes in iron–porphyrin stereochemistry to more distant regions of the protein by a change in porphyrin

† The optical asymmetric unit in the studies of horse haemoglobin derivatives of space group C2 consists of an α and β subunit pair. Only the average change in orientation for both haem groups can therefore be estimated. Furthermore, the deoxyhaemoglobin derivative in figure 11 is that of deoxyhaemoglobin modified with the bifunctional reagent bis(N-maleimidomethyl)-ether. Covalent attachment of this reagent to the β subunits prevents the change in quaternary structure to that of unmodified deoxyhaemoglobin[126]. Only in this manner can the Soret polarisation properties co-ordinated haemoglobin derivatives be directly compared to those of an unco-ordinated protein in the crystal state. See ref. 135 for future details.

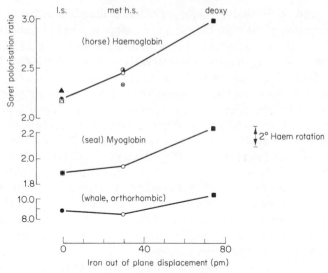

Figure 11 Graphical illustration of change of porphyrin
orientation in haemoglobin and myoglobin as a function of
iron out-of-plane displacement. The ordinate scale gives the
Soret polarisation ratios of various derivatives plotted so that for
each protein the changes correspond to a constant scale of
rotation in degrees. The data for iron out-of-plane displacement
are taken from X-ray diffraction studies of haemoglobin-type
molecules as explained in the text. Symbols used for each
oxidation and ligand state are

(■) deoxy;	(○) met-aquo	(◐) met-fluoride;
(⊗) met-formate;	(▲) oxy;	(●) met-cyanide;
(□) met-azide;	(△) carboxy-haem	

derivatives (see table 5 for corresponding spin and oxidation
states). The line connecting points is drawn for corresponding
changes between low-spin met-cyanide, high-spin aquomet- and
high-spin deoxy states. The sperm-whale myoglobin data are for
the crystal of space group $P2_1 2_1 2_1$, which serves as an unusually
sensitive system for detecting changes in porphyrin orientation
because of the tenfold greater projection of the porphyrin plane
onto the *a* axis than onto the *b* axis. The data for deoxyhaemoglobin are given for the BME-modified derivative[126] of deoxyhaemoglobin, which crystallises in space group C2. From Makinen
and Eaton[135].

orientation can be readily appreciated in the light of the environment of the haem
groups determined by X-ray diffraction studies[16,99]. In the region of the haem
group there are approximately sixty atoms of nearby amino-acid residues which
make van der Waals contact with carbon atoms of the porphyrin skeleton. A
rotational change in porphyrin orientation of 4° corresponds to a positional shift
of ≈ 30 pm for a carbon nucleus on the periphery of the porphyrin ring. Such a
structural change would readily result in tertiary conformational changes in the
immediate protein environment. These data, therefore, suggest that a change in

porphyrin orientation of the magnitude observed in spectroscopic studies[133-6] must be accompanied by tertiary structural changes of the protein. Such structural changes, occurring, for instance, on release or binding of oxygen, could then be transmitted to more distant regions of the protein subunits such as the α_1-β_2 contact regions (see figure 2) by nonbonded interactions between the porphyrin skeleton and amino-acid side chains in the immediate haem environment.

The changes in porphyrin orientation observed[133-6] on reduction of the Fe(III) ion in BME-modified methaemoglobin, would correspond to release of an oxygen molecule in the co-ordinated conformation of the protein. That comparable structural alterations induced by changes in porphyrin orientation can occur on binding of oxygen by deoxyhaemoglobin is supported by the recent observations of Anderson[139]. He has observed at 350 pm resolution that in deoxyhaemoglobin crystals, in which change in lattice structure is prevented by treatment with polyacrylamide gels, oxidation of the ferrous haems is accompanied by protein and porphyrin structural alterations in the reverse direction to those observed for the change between methaemoglobin and BME-modified deoxyhaemoglobin. Thus, the reversible nature of the structural changes associated with changes in porphyrin orientation on ligand binding and removal is indicated, on the basis of difference Fourier studies of the haemoglobin molecule confined to both the co-ordinated[89,103] and unco-ordinated[139] quaternary conformations. The spectroscopic studies[133-6] demonstrate that changes in porphyrin orientation accompany a change in iron oxidation state and ligand state. In view of their magnitude and association with changes in spin state of the haem iron in haemoglobin and in view of the less prominent changes in sperm-whale myoglobin (see figure 11), they probably reflect an important component in transmitting the conformational changes underlying co-operative ligand binding. Interpretation of the difference Fourier maps at 350 pm resolution by Anderson[139] indicates that residues of the F helix adjacent to the co-ordinating proximal histidine are predominantly affected by the change in protein–porphyrin van der Waals contacts.

(2) Comparison of co-operative interactions in haemoglobin

Figure 11 indicates only small differences in porphyrin orientation between high- and low-spin ferric derivatives of haemoglobin and between methaemoglobin and the ferrous carboxy- and oxy-co-ordinated species. These small differences in porphyrin orientation and consequently in the associated tertiary structure of the immediate haem environment cannot be disregarded. Comparison of the pH dependence of co-operativity of haemoglobin for the oxygenation and oxidation-reduction reactions respectively indicates the functional importance of such small structural differences.

X-ray diffraction studies of mammalian haemoglobin in the co-ordinated quaternary conformation have been carried out on methaemoglobin[16,103,140], since suitable crystals with a sufficiently high content of oxyhaemoglobin cannot be experimentally achieved. The origins of the lack of structural data on the oxygen-co-ordinated molecule of haemoglobin derive from the oxidation of oxyhaemoglobin to methaemoglobin during the period of 1–2 months necessary to grow crystals of suitable size[†] for X-ray diffraction studies[113] and the catalysis of the oxidation of oxyhaemoglobin by incident X-ray beams analogous to the oxidation of oxymyoglobin[142] (see section 2.1.6(3)). While the oxygenation of haemoglobin has long been known to proceed in a co-operative manner[100] essentially independent of pH, the oxidation–reduction reaction of the half-cell couple

$$\text{methaemoglobin} + e^- \rightleftharpoons \text{deoxyhaemoglobin} \qquad E_{\frac{1}{2}} = 0.15 \text{ V} \qquad (2.2)$$

exhibits co-operativity markedly dependent on pH[98,143]. This aspect of haemoglobin chemistry has remained unexplained, since the oxidation–reduction

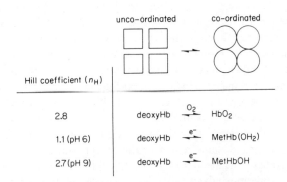

Figure 12 Diagrammatic illustration of reversible reactions of tetrameric haemoglobins, which exhibit co-operative inter-action of protein subunits. The quaternary conformational states of haemoglobin are the unco-ordinated or deoxyhaemoglobin molecule characterised by a high-spin, five-co-ordinate, ferrous haem complex, and the co-ordinated conformation characterised by a six-co-ordinate ferrous or ferric haem complex. Co-ordinated derivatives do not necessarily exhibit the same precise tertiary structural characteristics. The oxidation–reduction reactions involve the co-ordinated derivatives of (acid) methaemoglobin (MetHb(OH₂)) at pH 6 and (alkaline) methaemoglobin (MetHbOH) at alkaline conditions. The diagrammatic illustration of squares and circles for unco-ordinated and co-ordinated conformational states indicates that different inter-subunit amino-acid contracts are present in each state[103].

reaction clearly involves the protein in both unco-ordinated and co-ordinated quaternary conformational states in a manner analogous to the oxygenation reaction. The reactions involving the unco-ordinated and co-ordinated quaternary conformational states of the tetrameric protein molecules are schematically compared in figure 12.

† Spectrophotometric measurement of the oxidation of oxyhaemoglobin under the crystal-lisation conditions described by Perutz[113] indicates that at least 50 per cent of the oxy derivative has been already oxidised after two weeks[141].

In the oxidation–reduction reaction the pH dependence of the Hill coefficient, a measure of co-operativity, which changes from $n_H = 1.1$ at pH 6 to $n_H = 2.7$ at pH 9 [143] is, however, readily explained on the basis of the spin states of the oxidised co-ordinated species involved [111] and the conclusion [134-6] that changes in protein–porphyrin nonbonded interactions accompany spin-state transitions of the iron. At pH 6 methaemoglobin is characterised by a six-co-ordinate haem complex in which a water molecule occupies the sixth co-ordination position [16]. This derivative is characterised on the basis of magnetic susceptibility [110] and e.p.r. absorption studies [144] by only one spectroscopic nondegenerate ground state of the haem iron corresponding to a high-spin Fe(III)ion. On the other hand, under high pH conditions the bound water molecule ionises with a $pK_a \approx 8.1$ forming the well-known methaemoglobin hydroxide (alkaline methaemoglobin) derivative [145]. This derivative exhibits a high-spin–low-spin temperature-dependent equilibrium such that at room temperature equal concentrations of high- and low-spin methaemoglobin hydroxide derivatives are formed [111].

This spin-state equilibrium suggests that the increase in co-operative interaction of haemoglobin subunits in the oxidation–reduction reaction under alkaline conditions results from reduction of the low-spin methaemoglobin hydroxide derivative to deoxyhaemoglobin. On the basis of the changes in porphyrin orientation dependent on iron spin state illustrated in figure 11, it is evident that $\approx 2°$ rotation of the porphyrin ring occurs on change from a ferric high-spin to a ferric low-spin state and is accompanied by the iron atom assuming a position coplanar with the pyrrole nitrogen atoms. This would result in a translational displacement of the porphyrin ring towards the iron atom of ≈ 30 pm. This small change in the orientation of the porphyrin ring occurring as a result of its translational displacement and rotation emphasises the importance of the small tertiary conformational changes of the protein accompanying the change in iron–porphyrin stereochemistry.

The absence of co-operative subunit interaction in the oxidation–reduction reaction at pH 6, however, cannot be considered to result only from the difference in porphyrin orientation with respect to the immediate protein environment of the haem groups. Co-operative interaction is the result of transmission of structural changes beginning in a change in radius of the iron and culminating in transmission of structural changes across subunit interface regions. As pointed out earlier, the interaction of the porphyrin skeleton with its protein environment via nonbonded interactions appears to be predominantly responsible for transmitting the effect of a change in ionic radius. An increase in alkalinity not only results in the increased formation of a low-spin oxidised species with a (presumably) coplanar iron–porphyrin configuration but can also result in ionisation of amino-acid side chains, bringing them into positions favourable for nonbonded interactions with the porphyrin ring of a low-spin species. These small changes are none the less necessary for the full expression of co-operative interaction of haemoglobin subunits. They cannot be considered to be structurally represented in the comparison of the three-dimensional structures of the high-spin methaemoglobin and deoxy-haemoglobin derivatives. This simply emphasises the fact that although the co-ordinated haemoglobin derivatives have a similar crystal structure characterised by the space group C2 for horse haemoglobin[113], they need not be expected to retain identical molecular structures on a detailed tertiary structural level.

It is unfortunate that these structural differences are small and beyond the present level of resolution of X-ray diffraction studies of mammalian haemoglobin. As also pointed out by Williams[128] the comparison of oxygenation and oxidation-reduction reactions on the basis of the presently known three-dimensional structures of high-spin methaemoglobin and deoxyhaemoglobin[99,103] does not permit a complete view of structural changes responsible for co-operative interaction. The low-spin co-ordinated and high-spin unco-ordinated haemoglobin species are the only ones associated with co-operative subunit interaction. In view of the unfortunate condition that horse methaemoglobin crystals exhibit a doubling of the b axis under solvent conditions of pH > 7[146], preventing X-ray diffraction studies of the methaemoglobin hydroxide derivative, and the experimental difficulties inherent in growing suitable crystals of oxyhaemoglobin, the structural properties of the low-spin co-ordinated protein involved in co-operative phenomena should be approximated by use of a stable derivative such as methaemoglobin cyanide. Without doubt studies of low-spin co-ordinated derivatives at higher resolution than the present 280 pm achieved in mammalian haemoglobin studies[99,103] will uncover structural differences in the haem environment when compared to the high-spin met- and deoxyhaemoglobin derivatives. These results will then define somewhat more precisely the nature of the interaction of the porphyrin ring with its protein environment and the role of interchain amino-acid side-chain contacts. Such differences have been stated not to be detectable at the present level of resolution of X-ray studies of mammalian haemoglobin[89].

2.1.5 Structural control of porphyrin configuration and iron electronic structure

(1) Hydrogen-bonded contacts of porphyrin propionic groups
Part of the structural basis through which the protein environment of the haem group can control porphyrin orientation within the molecule is suggested by a comparison of the haem environments of the subunits of mammalian haemoglobin to that of sperm-whale myoglobin. A significant structural difference concerns the propionic acid groups of the porphyrin ring, which protrude to the surface of the protein, as illustrated in figure 13. In sperm-whale metmyoglobin the propionic carboxylate group at the porphyrin-6 position is hydrogen bonded to the N_3-position of histidine-97 and is directed towards the proximal side of the porphyrin ring[11,148]. The corresponding position of the α-subunit of methaemoglobin is occupied by a leucine residue, and no hydrogen-bonded structure can be similarly formed[99,147]. In the β-subunit the corresponding propionic carboxylate is hydrogen bonded to lysine-66[16,99]. The propionic carboxylate group from the porphyrin-7 position, on the other hand, forms an analogous polar contact in sperm-whale metmyoglobin with arginine-45[11,148] and is, therefore, directed towards the distal side of the

porphyrin ring. The corresponding carboxylate in haemoglobin is hydrogen bonded in the α-subunit by histidine-45 and possibly by serine-45 in the β-subunit[99]. While the propionic carboxylates both form hydrogen-bonded contacts with surface residues in sperm-whale metmyoglobin, only one forms a similar contact in the α-subunit of methaemoglobin. The contact with serine-45 in the β-subunit might be expected not to be as firm as one with a histidine residue.

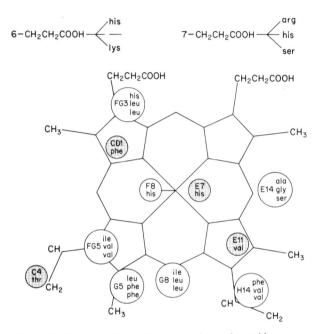

Figure 13 Schematic illustration comparing amino-acid residues that have nonpolar van der Waals contacts with the porphyrin skeleton in sperm-whale myoglobin (top row) and the α and β subunits of horse haemoglobin (second and third rows respectively). Residues on the distal side of the haem plane are indicated by a shaded circle and those on the proximal side by an open circle. The residues that form hydrogen-bonded contacts with the propionic carboxylate groups are similarly indicated above for sperm-whale myoglobin[11] (left) and horse haemoglobin[99] (right). Based on diagrams of Kendrew[148] and Perutz[147].

It is probable that the propionic carboxylate contacts with surface residues in the sperm-whale myoglobin molecule would provide a significantly more rigid anchoring force simultaneously from both proximal and distal sides of the haem group against changes in porphyrin orientation. This could in part constitute the structural basis in sperm whale myoglobin for the more restricted porphyrin movement on change in oxidation state and ligand state. The somewhat greater rotational freedom of the porphyrin ring for corresponding spin-state changes in seal myoglobin (figure 11) is explainable on a similar basis. In this protein the arginine-45 residue is substituted by a lysine group[149]. This constitutes a shortening of the amino-acid side chain providing an amine group for hydrogen-bond formation, a condition which would be expected to result in poorer stabilisation of the propionic carboxylate contact than that with the arginine residue.

The hydrogen-bonded contacts of propionic groups with surface residues supply only a readily identifiable structural basis for the control of porphyrin orientation. There are, in addition, numerous aromatic residues in the haem environments of both proteins[11,99] which could enter into interactions of π-electron systems with the porphyrin ring. These interactions are difficult to define quantitatively and to assess for comparative purposes. The hydrogen-bonded contacts of the propionic residues, however, may have importance in other haem proteins. In yeast cytochrome c peroxidase at least one of the propionic acid groups appears to be buried within the hydrophobic interior of the protein[150], and in cytochrome c both propionic groups form hydrogen-bonded contacts with residues within the hydrophobic interior of the protein[19,20]. This different structural relationship between the protein and the propionic groups may serve to provide a different mode of protein-porphyrin interaction in the control of porphyrin configurational changes.

(2) Steric constraints by nonpolar contacts

There are numerous examples from inorganic chemistry that suggest that small changes in steric constraints of co-ordinating residues promote a shift of spin-state equilibria of metal co-ordination complexes. The most pertinent example to haem proteins may be the tris(o-phenanthroline)Fe(II) complex, known to be a low-spin diamagnetic species. However, substitution of a methyl group into the 2-position of the phenanthroline ring provides sufficient steric hindrance within the octahedrally co-ordinated complex to favour maximum spin multiplicity of the Fe(II) cation[128,152] with subsequent formation of longer Fe—N bonds.

In a comparable manner, the interaction of the protein with the porphyrin ring by residues making van der Waals contact provides a means not only of controlling the orientation of the porphyrin but also of controlling the spin multiplicity of the central cation. A comparison of magnetic properties of ferric haemoglobin and myoglobin derivatives indicates that all the latter have a higher percentage of high-spin character than corresponding haemoglobin derivatives at room temperature[111,112]. As has been pointed out earlier, the metal cation in myoglobin exhibits a more pronounced tendency to remain displaced from the porphyrin ring. This feature undoubtedly reflects protein-imposed steric constraints, which favour maximum spin multiplicity. Its structural origin provides some insight into the manner in which structural perturbations of the polypeptide chain, derived from alteration of surface residues, can be transmitted to the porphyrin centre to control the spin state of the haem iron.

The interaction of the surrounding polypeptide chain with the porphyrin skeleton, influential in controlling the spin state of the haem iron, is demonstrated by studies in which the temperature dependence of the magnetic susceptibility of

several haem proteins has been examined[150,153-6]. Haem proteins in the ferric
state have long been known to exhibit temperature-dependent equilibria between
high- and low-spin states dependent on the nature of the ligand bound at the sixth
co-ordination position[107-12]. Analysis of the thermodynamic parameters of the
spin-state equilibria[112,153-6] has shown that while the energy differences between
the two spin states are dependent on the nature of the co-ordinated sixth ligand, that
is, dependent on the orbital splitting effects of a given ligand, the compensation
temperature at which the concentrations of high-spin and low-spin species are equal
is dependent on the nature of the apoprotein to which the iron–protoporphyrin
IX group is attached. Furthermore, the temperature-dependent equilibria are
altered on substitution of haem derivatives other than iron–protoporphyrin IX
into the corresponding apoproteins[150,156]. Such changes are dependent on the
nonpolar contacts of the porphyrin ring with protein residues. These results have
been further analysed on a theoretical basis[157] to describe temperature-dependent
spin-state equilibria according to co-operative formation and breaking of van der
Waals contacts between the porphyrin and globin. It is probable that alteration in
spin state on the basis of a temperature-dependent spin-state equilibrium is
accompanied by changes in porphyrin orientation analogous to that observed
between high- and low-spin states of methaemoglobin and metmyoglobin[133-6].
A comparable structural basis is most probably responsible for both phenomena
through nonpolar contacts of the porphyrin skeleton with nearby amino-acid side
chains.

(3) Effect of protein–porphyrin interaction on the iron electronic structure

Magnetic-resonance studies of high-spin ferric haem proteins of different
ligand states indicate that the symmetry of the haem with respect to the electronic
structure of the iron depends on the structure of the protein moiety and protein-
porphyrin interactions. High-spin ferric haem proteins as well as high-spin ferric
porphyrins in general display an e.p.r. absorption spectrum characteristic of a high-
spin Fe(III) ion in a strong tetragonal field[158]. The direction of axial symmetry of
the tetragonal field is generally assumed to coincide with the normal to the
porphyrin plane. This axial symmetry implies that the electronic environment of
the d-electron system is equivalent in two perpendicular directions (x and y) lying
in the haem plane and is very different from that in the z direction perpendicular
to the xy-plane. The e.p.r. absorption spectrum of high-spin metmyoglobin[159,160]
and of high-spin methaemoglobin[144,161] are the two best known examples of
tetragonal symmetry of the haem iron in ferric haem proteins. The structural
origin of the tetragonal field is schematically illustrated in figure 14(a). The
co-ordinating imidazole nitrogen in the fifth co-ordination position and the oxygen
atom of the bound water molecule in the sixth position do not provide contributions
to the ligand field equivalent to that of the pyrrole nitrogen atoms in the porphyrin
plane. Tetragonal symmetry is retained under conditions of iron out-of-plane displacement provided the iron cation remains equidistant from each of the four pyrrole
nitrogen atoms.

High-spin ferric haem proteins, however, display, varying degrees of departure from precise tetragonal symmetry[161, 162]. The departure from tetragonality is observed spectroscopically as a broadening or splitting of the e.p.r. absorption band at $g = 6$. This change is caused by introduction of a rhombic component in the ligand field of the donor atoms in the xy-plane making the x and y directions inequivalent. The change in the ligand field from precise tetragonal symmetry is illustrated in figure 14(b). The departure from tetragonality, termed the rhombicity[161, 162], is caused by perturbations of the iron environment. These perturbations may arise from mechanical distortions of the haem, perturbations

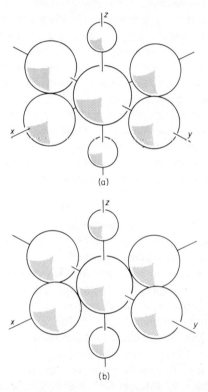

Figure 14 Schematic representation of the ligand field of the haem iron with and without a rhombic component. The central sphere represents the iron atom and the other spheres the donor-ligand atoms. The diameters of the ligand atoms are scaled to reflect the symmetry of the field and their relative contributions to the ligand field. In (a) the tetragonal symmetry of the field is represented by the iron equidistant from the four pyrrole nitrogen donor atoms. The four pyrrole nitrogens are pictured as larger than the ligand atoms along the z-direction since they provide the dominant part of the ligand field. In (b) a rhombic component has been introduced in which the components along the x axis are slightly larger than those along the y axis. From Blumberg *et al.*[164], with permission.

of the π-electron distribution of the haem system or from perturbations of π-electron bonding of ligands to the haem iron at the fifth or sixth co-ordination positions.

The structural basis for the introduction of a rhombic component into the tetragonal symmetry of the iron–porphyrin group is related to the Fe–ligand bonding interactions of the co-ordination centre. For a paramagnetic ion the second-order spin Hamiltonian

$$\mathcal{H} = g\beta H_0 + D(S_z^2 - S(S + 1)/3) + E(S_x^2 - S_y^2) \qquad (2.3)$$

where E and D are the coefficients of the second-rank rhombic and axial-spin operators, defines the proper co-ordinate system[163] for the principal ligand-field directions if E/D has its smallest possible value and $E \geqslant 0$. The axis of the D term defines the axis of greatest covalent bonding or electric-field gradient and therefore defines the axis of greatest metal–ligand interaction. Since the predominant portion of the ligand field is supplied by the porphyrin pyrrole-nitrogen donor atoms, the axis of D, indicating the direction of greatest metal–ligand interaction, must correspond to the direction defined by a pair of pyrrole nitrogen atoms co-ordinating the metal cation on opposite sides. With respect to the side-chain substituents of the porphyrin ring, this direction need not remain the same for all high-spin ferric haem proteins. It could change within the same protein depending on the nature of the mechanical distortion of the porphyrin ring by nearby amino-acid residues and the stereochemistry of the ligands bound in the fifth and sixth co-ordination positions of the iron with respect to the porphyrin plane.

In high-spin ferric proteins the degree of rhombicity introduced into the tetragonal ligand field can be expressed[161,162] to a first approximation as

$$R = (\Delta g/16) \times 100\% \qquad (2.4)$$

where R is the per cent rhombicity and Δg represents the splitting of the absorption component observed near $g = 6$. Figure 15 gives a comparison of the computed percentage of rhombicity of high-spin ferric haem proteins. The wide variation in departure from tetragonality clearly indicates that perturbations of the tetragonal structural and electronic symmetry of the environment of the haem-iron cation differ widely according to individual proteins as well as according to different high-spin ferric derivatives of the same protein. As indicated in figure 15 these proteins can be categorised into separate groups depending on the ligands in the fifth and sixth co-ordination positions. The variation in rhombicity must arise, therefore, from structural perturbations by interaction of nearby amino-acid residues with the porphyrin skeleton or with the ligand at the sixth co-ordination position. The protein thus places structural constraints on the porphyrin, which destroy the electronic equivalence of metal–porphyrin interaction in all directions in the porphyrin ring.

Although the nature of protein–porphyrin interaction and structural changes necessary for electronic distortion of the iron cation cannot be quantitatively assessed by the per cent rhombicity, the comparison of the rhombic and tetragonal

Metal Ions in Proteins

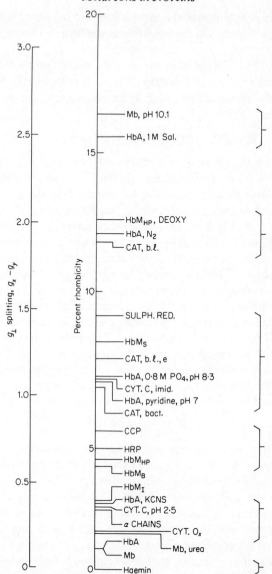

Figure 15 Values of the *g* splitting near *g* = 6 and computed percentage rhombicity of high-spin ferric haem proteins. Mb (metmyoglobin); HbA (methaemoglobin A); Mb, urea (metmyoglobin treated in 5 M urea for $1\frac{1}{2}$ hours); CYT.O_x (partially reduced cytochrome c oxidase); α CHAINS (ferric α-chains prepared from haemoglobin A); CYT.C, (horse-heart cytochrome c, pH 2.5); HbM$_I$ (haemoglobin M Iwate); HbM$_B$ (haemoglobin M Boston); HbM$_{HP}$ (haemoglobin M Hyde Park); HRP (horse radish peroxidase); CCP (cytochrome c peroxidase); CAT, bact. (bacterial *Micrococcus lysodeikticus* catalase); Cyt.C imid. (imidazole-treated cytochrome c); CAT b.l. (bovine liver

components in the haem symmetry of high-spin ferric haem proteins[161,162] does emphasise the importance of contacts between the porphyrin skeleton and surrounding amino-acid side-chain environment. The e.p.r. results of Peisach and coworkers[161, 162] demonstrate that structural changes affecting the porphyrin ring or the axial ligands can be transmitted to the central co-ordinated metal cation. On this basis it is possible to envisage a structural interaction mechanism for the porphyrin skeleton with nearby amino-acid side chains in controlling spin-state transitions and the electron configuration of the haem iron.

While comparable studies cannot be carried out on the paramagnetic high-spin ferrous haem complexes since their e.p.r. absorption has not been detectable, it is probable that similar considerations would apply. The results for the high-spin ferric proteins indicate that small modifications of the porphyrin in its interaction with the protein alter the electronic structure of the iron and co-ordinating pyrrole nitrogen atoms. It has been long known (see, for instance, reference 165) that the crossover point between high- and low-spin states of co-ordination complexes is sensitive to changes of electron density of the donor ligand atoms'. While a more quantitative description of the influence of such changes cannot be made with regard to the high-spin to low-spin conversion of ferrous haemoglobin and myoglobin on oxygen binding, these comparative results of high-spin ferric systems demonstrate a basis through which similar structural and electronic effects can be operative in the normal physiological states of these haem proteins in inducing the reversal changes in iron spin state.

2.1.6 Haem-iron electronic structure and ligand stereochemistry

In the physiological action of haemoglobin and myoglobin the functional species involve a five-co-ordinate ferrous haem complex in the deoxygenated state[17,18,96] and a ferrous derivative in which the sixth co-ordination site is occupied by an oxygen molecule. While the unco-ordinated protein for both haemoglobin and myoglobin has been characterised by X-ray crystallographic studies (see table 2), the oxygen-co-ordinated form has, in general, defied precise and convincing structural characterisation which would serve to define co-ordination stereochemistry and the iron–oxygen electronic structure unambiguously. The importance of precise definition of the structure and electronic bonding of the functional species of both these proteins is obvious, for a satisfactory, quantitative explanation of the relative inertness of the ferrous proteins to oxidation and their ease of oxygenation cannot be achieved otherwise. In addition, the dependence of haem–oxygen interaction on the geometry of the complex containing co-ordinated oxygen may

catalase); HbM$_S$ (haemoglobin M Saskatoon): SULPH RED. (*E. coli* sulphite reductase); HbA N$_2$ (methaemoglobin A over which nitrogen was blown for 1 hour); HbM$_{HP}$ DEOXY (haemoglobin M Hyde Park in deoxygenated form); HbA, 1 M Sal. (Methaemoglobin A incubated for 1 hour in 1 M salicylate); Mb, pH 10.1 (minority constituent of base-treated metmyoglobin). From Peisach *et al.*[161], with permission.

determine the specificity of the biochemical role of haem prosthetic groups of the various types of peroxidases, oxidases, catalases and oxygenases. That the geometry of this interaction must depend on the steric accessibility of an oxygen molecule to the haem group is clearly anticipated and emphasises the need for structural characterisation of the haem environments of other types of haem proteins. Sufficient experimental data to correlate the detailed haem stereochemistry and molecular structures with the electronic basis of haem–oxygen complex formation in oxygen-binding haem proteins is unfortunately lacking. A structural and spectroscopic examination of co-ordinated derivatives of haemoglobin and myoglobin may, however, indicate those factors likely to be of importance in oxygen binding.

(1) Rhombic-distorted low-spin complexes

The difference Fourier synthesis of metmyoglobin azide versus metmyoglobin at 200 pm resolution[61] revealed that the azide ion was bound to the iron atom at the sixth co-ordination position, making an angle of 111° with the perpendicular to the porphyrin plane. This configuration implies that the N_α atom of the azide ligand was hydrogen bonded to the N_3-position of the distal histidine residue (figure 16). An important feature in the study of Stryer and coworkers[70] is that steric constraints on the binding of the azide ligand were revealed from the determination of the configurations of nearby amino-acid residues in the haem

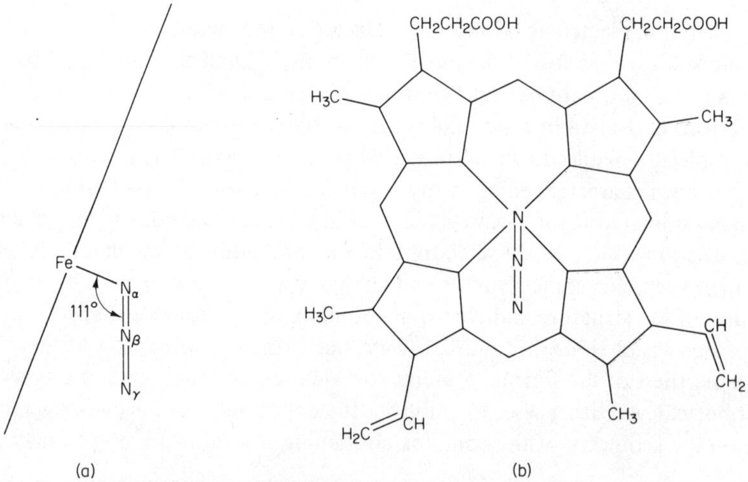

(a)   (b)

Figure 16 Illustrations showing the location of the azide ligand in relation to the haem group in sperm-whale myoglobin. (a) The angle between the Fe–N_α–N_β atoms is 111°. The shortest distance between the N_β and N_γ atoms and the plane of the porphyrin ring system is about 300 pm. (b) The projection of the azide ion on the haem plane is along the line from the iron to the carbon atom of one of the methane bridges between pyrrole rings. From Stryer et al.[70], with permission.

environment. With the iron atom probably slightly displaced from the mean porphyrin plane, the haem should have approximate C_{4v} symmetry, implying that four different co-ordinating positions of the azide ion could be theoretically expected. That the electron-density peaks were restricted to one position with respect to the haem group indicated that only one of the four positions was accessible to the ligand for co-ordinating the iron atom. Referring to figure 16, the sterically restricting residues in the other three positions are phenylalanine-43, valine-68 and histidine-64—the distal histidine[†].

The stereochemical geometry of the azide ligand bound at the sixth co-ordination position is expected to result in significant distortion of the symmetry properties of the iron d orbitals from that associated with the high-spin Fe(III)cation. This distortion results primarily in a change in the relative levels of the t_{2g} set, of which three low-spin configurations are theoretically possible

$$d_{xz}^2 \ d_{yz}^2 \ d_{xy}; \ d_{xz}^2 \ d_{yz} \ d_{xy}^2; \text{ and } d_{xz} \ d_{yz}^2 \ d_{xy}^2$$

E.P.R. studies indicate that the symmetry is lowered with a strong rhombic component and of these orbitals d_{xy} remains lowest in energy and must, therefore, contain two paired electrons[166-8]. The oblique ligand stereochemistry of the azide ion does not permit equivalent interaction of azide nitrogen orbitals with the d_{xz} and d_{yz} orbitals. Stabilisation of the system as a result of inequivalent interaction of the ligand with d_{xz} and d_{yz} orbitals occurs by lowering the symmetry due to introduction of rhombicity[169]. This spontaneous distortion is analogous to the Jahn–Teller effect by perturbing the square-symmetrical arrangement of pyrrole nitrogen atoms around the Fe(III)cation. Structural evidence[93] suggesting that small shifts of the porphyrin pyrrole nitrogen atoms occur on binding the azide anion has already been discussed (section 2.1.3(2)).

A comparable mechanism may account for the rhombic component observed in the e.p.r. absorption spectrum of nitric oxide-co-ordinated haemoglobin and myoglobin. Analysis of the g-tensor anisotropy of the e.p.r. absorption of oriented single crystals of nitric oxide-co-ordinated sperm-whale myoglobin suggests that the NO molecule is co-ordinated at the sixth co-ordination position of the iron in a manner comparable to that of the azide ion, making an angle of $110°$ with a perpendicular to the porphyrin plane[170]. This derivative has a d^7 electronic configuration in which the unpaired electron responsible for paramagnetism resides predominantly in an axially directed d_{z^2} orbital of the metal[171]. With an oblique co-ordinating geometry the bonding arrangement is most probably between an sp^2 nitrogen orbital donating into the iron d_{z^2} orbital with a contribution between the $p\pi$ nitrogen orbital and the d_{yz} orbital of the iron[170]. Again, the inequivalent interaction of the ligand with the d_{xz} and d_{yz} orbitals should result in a spontaneous rhombic distortion effect to stabilise the system as in the case of metmyoglobin azide. No X-ray crystallographic investigation has been undertaken yet on this derivative of myoglobin, but slight shifts of the pyrrole nitrogen atoms are expected

[†] The proximity of the distal histidine to an azide molecule placed near the propionic-acid residue would preclude hydrogen-bond formation with the N_α atom of the ligand.

in analogy to the azide derivative. Because of the steric effects of nearby amino-acid side chains in the haem environment, the binding of the nitric oxide molecule most probably occurs in the same position relative to the methyl and vinyl groups as the azide ligand (figure 16).

The instability of oxyhaemoglobin and oxymyoglobin in the X-ray beam and the marked reluctance of the nitric oxide ligand to dissociate from the haem[172] indicate that the nitric oxide-co-ordinated derivative deserves attention in crystallographic studies for at least two important reasons: (i) Kon[171] has pointed out that the nitric oxide derivative of haemoglobin or myoglobin may have similarities to the electronic basis of oxygen binding; and (ii) the d^7 configuration with an unpaired electron residing predominantly in the axially directed d_{z^2} orbital allows an experimental system for observing the effect of a singly occupied d_{z^2} orbital on all Fe—N bond lengths. Only slight displacement of the iron from the pyrrole nitrogen plane under these conditions is predicted[119]. The effect of the unpaired electron on the bond length of the metal cation with the axially directed proximal histidine residue might therefore be more readily observed by difference Fourier methods.

(2) Other low-spin co-ordinated complexes

X-ray diffraction studies[15,93] suggest that the cyanide anion is bound obliquely with respect to the porphyrin skeleton. Both the cyanide and carbon monoxide ligands were predicted by Pauling[173,174] to bind in a linear fashion with the iron atom on the basis of the electroneutrality principle. In the case of metmyoglobin cyanide an angle of 130° is suggested for the Fe—$\hat{C}$—N angle. However, because of poor resolution of the cyanide ligand in the difference electron-density map, this interpretation[93] assumes that the carbon atom binds at the sixth co-ordination position of the iron atom at the same site as that occupied by the water molecule. The electronic bonding structure[173,174] implies, however, a somewhat shorter Fe—C bond length. The results of Hendrickson and Love[15] indicate that the co-ordinating carbon atom is displaced ≈ 100 pm from the perpendicular to the porphyrin plane through the iron atom (see figure 9) in sea-lamprey methaemoglobin cyanide. While the results of both studies indicate that the stereochemistry predicted by the electroneutrality principle does not hold precisely, probably because of steric hindrance by neighbouring amino-acid residues, quantitative description of sufficient precision is not yet available to define the ligand geometry more accurately. In both cases it was concluded that the nitrogen atom of the cyanide ligand was hydrogen bonded to the imidazole ring of the distal histidine residue. It is not possible to assess firmly whether the Fe-(distal)histidine distance is sufficiently different in sperm-whale metmyoglobin cyanide and sea-lamprey methaemoglobin cyanide to account for the possible difference in stereochemistry of the bound cyanide anion.

The low-spin carboxy-co-ordinated haems in erythrocruorin[96] and *glycera* haemoglobin[95,97] indicate an Fe—$\hat{C}$—O angle of $\approx 145°$. No refinement studies have been completed to define precise bond lengths and stereochemical relationships to nearby amino-acid side chains.

(3) Stereochemistry and electronic structure of the iron–oxygen bond

The stereochemistry of the oxygen molecule with respect to the porphyrin plane and the corresponding d-electron configuration of the iron atom have not been firmly established for the oxygen-co-ordinated complexes of haemoglobin or myoglobin. Pauling[173,174] postulated on the basis of the electroneutrality principle that oxygen co-ordination occurred with an oblique stereochemical arrangement (Fe–Ô–O angle $\simeq 120°$) and that the iron remained in the Fe(II) state. On the other hand, Griffith[175] concluded on the basis of theoretical calculations that bonding could occur with a formally seven-co-ordinate structure in which the interatomic axis of the oxygen molecule was parallel to the porphyrin plane and was bisected by a perpendicular to the Fe(II) cation†. These structures are schematically compared in figure 17.

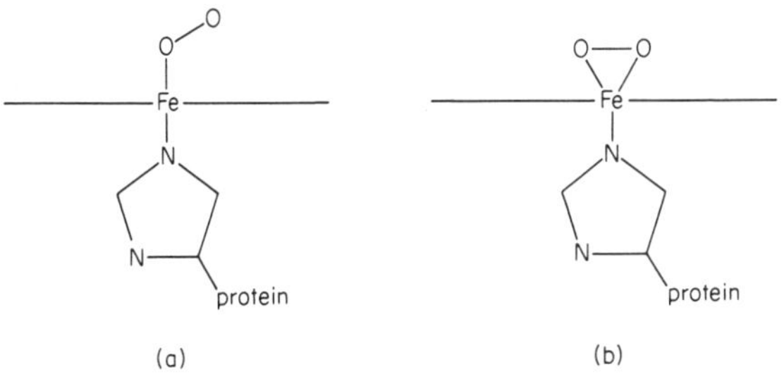

(a)   (b)

Figure 17 Diagrammatic illustration of the stereochemistry of oxygen binding to the haem iron atom according to the Pauling[174] (a) and the Griffith[175] (b) geometry. The illustration is intended to serve as a comparison of the six-co-ordinate and seven-co-ordinate structures only, indicating stereochemical relationships characteristic of each. Electronic bonding structures are not indicated.

An attempt to determine the stereochemistry of the bound oxygen molecule in sperm-whale oxymyoglobin by X-ray diffraction studies has been carried out by Watson and Nobbs[142]. These investigators utilised oxymyoglobin prepared by a method that allowed formation of suitable crystals within 12 hours containing high amounts of the oxygen-co-ordinated species[176]. A difference Fourier electron-density map of oxymyoglobin versus metmyoglobin at 280 pm resolution was then calculated in three dimensions. The resulting difference electron-density map suggested no change in iron–porphyrin stereochemistry. An oblique stereochemical arrangement of the oxygen molecule was interpreted as consistent with the difference electron-density peaks in the haem vicinity with the oxygen molecule hydrogen bonded to the imidazole ring of the distal histidine according to the Pauling hypothesis[173,174]. These results can be considered, however, only as suggestive of the occurrence of the Pauling geometry for oxygen binding in myoglobin. Watson and Nobbs[142] have pointed out that oxidation of the oxygenated

† The seven-co-ordinate haem-oxygen geometry has been shown by Hoard[115] to require displacement of the low-spin metal cation towards the side of the co-ordinating oxygen molecule for sufficient overlap of electron orbitals in binding to occur.

species was promoted by incident X-ray beams during data collection. This oxidation would be expected to result in formation of the high-spin (acid) metmyoglobin species as well as the metmyoglobin hydroxide species because of the solvent conditions (pH $\approx$ 8) utilised to retard oxidation. No indication was given[142] of the relative initial and final concentrations of oxidised myoglobin species to that of oxymyoglobin in crystals utilised for data collection.

In interpretation of the difference electron-density map, it was assumed[142] that the position of the atom of the oxygen molecule co-ordinating the iron atom corresponded to that occupied by the water molecule in metmyoglobin. Since the electronic basis for bonding according to the Pauling model requires the iron atom forming a double bond with the co-ordinating oxygen atom in the predominant resonance form, some decrease in the iron–oxygen bond length would be expected. Furthermore, the Pauling model (figure 17) requires lengthening of the inter-oxygen bond distance on account of the inter-oxygen single bond formed on co-ordination. The difference Fourier synthesis of oxymyoglobin versus metmyoglobin[142] cannot supply sufficient precision to determine these detailed stereochemical relationships. In addition, while hydrogen bonding by the distal histidine residue may provide stabilisation of the oxygen molecule in this myoglobin, it need not be considered an essential requirement. *Chironomus* erythrocruorin[177] and *Aplysia* myoglobin[178] both lack a distal histidine residue corresponding to that in spermwhale myoglobin or mammalian haemoglobins.

Weiss[179] and Wittenberg and coworkers[180] have proposed that the oxygen-co-ordinated haem complex in haem proteins contains a formally low-spin Fe(III) cation with the oxygen molecule bound as a superoxide anion: $Fe(III)-O_2^-$. These suggestions have been made largely on the basis of comparison of the visible absorption spectra of the oxygen-co-ordinated species to that of the low-spin methaemoglobin hydroxide derivative[†]. The postulation[179,180] of a low-spin d^5 Fe(III)ion in oxyhaemoglobin or oxymyoglobin would require coupling of the resultant paramagnetic low-spin Fe(III) cation with the paramagnetic superoxide anion O_2^-, since oxyhaemoglobin is known to be diamagnetic[105]. Such coupling, although possible, has not been demonstrated. Both the Pauling hypothesis[173,174] and the superoxide anion model[179,180] would result in lengthening of the O–O bond distance on binding to the haem iron from that (121 pm) characteristic of the oxygen molecule[58]. The observed O–O internuclear distance for the superoxide anion is 128 pm[58]. Sufficient precision of X-ray diffraction studies of crystalline oxygen-co-ordinated haem proteins is unlikely to be attained to demonstrate this lengthening.

Structural details of low molecular weight oxygen-binding complexes indicate that the electronic structure of oxygen binding might not be inferred from stereochemical details alone. The Pauling[173,174] and the superoxide anion[179,180] models may have comparable stereochemistries with an O–O internuclear distance greater

---

[†] The transitions in the visible absorption spectra in both cases are associated predominantly with the porphyrin (π,π^*) system and are not characteristic of metal–ligand charge-transfer states described[134] for the oxygen-co-ordinated haem complex.

than that of the free oxygen molecule. That lengthening of the O–O distance is to
be expected in the case of the superoxide model has been recently shown by
determination of the structure of Co(N,N'-ethylene-bis(benzoylacetoniminide))-
(pyridine)O_2[181]. In this monomeric oxygen-binding complex, the angle observed
for Co–Ô–O is 126°, and the O–O internuclear distance of 126 pm is comparable
to that observed for the superoxide anion. Furthermore, X-ray diffraction studies[182]
of Vaska's salt, Ir(O_2)Cl(CO)(PPh$_3$)$_2$, a synthetic molecular-oxygen carrier[183], have
demonstrated that while the oxygen atoms are both equidistant from the iridium(II)
cation–that is, oxygen binds according to the Griffith[175] model–there is a lengthening of the O–O internuclear distance to 130 pm. This complex remains diamagnetic
on oxygen binding[183]. Therefore, it is doubtful that stereochemical details alone
will permit an unambiguous assignment of the electronic bonding structure
associated with reversible ligation of the haem iron by the oxygen molecule.

It is possible, however, that some progress toward elucidating the stereochemistry
and electronic bonding of the haem–oxygen complex may be achieved by rigorous
theoretical analysis of the absorption spectroscopic properties of the oxygenated
haem proteins. On the basis of single-crystal polarised-absorption spectroscopic
studies, Makinen and Eaton[134] have recently characterised the intensities,
polarisation and spectral positions of three iron–oxygen charge-transfer bands of
oxyhaemoglobin. One of these, in particular, occurs in the near infrared region
at 925 nm (see figure 10) and has been long recognised as characteristic of
oxyhaemoglobin alone. Correlation of the results of these spectroscopic studies
with those predicted by new molecular-orbital calculations may distinguish
between the various models proposed for the iron–oxygen stereochemistry of the
oxygenated complex of haemoglobin.

The elucidation of the stereochemistry and electronic structure of the haem–
oxygen complex in the oxygenated form of haemoglobin and myoglobin remains
an important unsolved structural problem in haem-protein chemistry. While the
presence of a hydrophobic haem environment is undoubtedly important, as first
suggested by Wang[184], and is inferred[185] to be supplied largely through aliphatic
amino-acid side chains in the haem environment, quantitative data are needed to
explain the exothermicity of formation of the oxygen complex and the unfavourable endothermicity of the oxidation of its Fe(II)cation by molecular oxygen[186].
The marked thermodynamic stability of the oxygenated complex[186] thus
remains imprecisely explained on a detailed structural level. Since the oxygenated
derivatives remain the physiologically functional species in oxygen transport to
tissues and in oxygen storage, and since intricate stereochemical relationships
of haem–oxygen complex formation may primarily determine the biological
specificity of haem–oxygen interaction, renewed efforts in the elucidation of
ligand stereochemistry and electronic structure of the iron–oxygen bond in haemoglobin and myoglobin are required.

3 ZINC-CONTAINING METALLOENZYMES

3.1 CARBOXYPEPTIDASE

Carboxypeptidase A and B are formed by the hydrolytic action of trypsin on their respective procarboxypeptidase precursors synthesised in the pancreas[187]. Of these two enzymes, carboxypeptidase A has been the more widely studied and has received detailed investigation on the basis of X-ray crystallographic studies[29,188,189]. Bovine carboxypeptidase A_α (CPA) is an enzyme comprised of 307 amino acids in a single polypeptide chain which binds tightly one gram-ion of Zn(II) per mole of enzyme. The requirement of Zn(II) for enzymatic activity was first demonstrated by noting that CPA freed of the metal ion remained inactive but activity could be reconstituted on addition of Zn(II)[190,191]. The metal-free enzyme under these conditions appeared to retain the same gross structural properties of active CPA[191]. Later the nature of the role of the Zn(II) ion by binding substrate in peptide hydrolysis was firmly established from crystallographic studies[29]. The enzyme exhibits specific stereochemical requirements in proteolysis by cleaving the C-terminal amino acid from a peptide chain if the C-terminal carboxylate is free and if the amino acid has an L-configuration[192,193]. Higher activity is usually observed if the C-terminal amino acid contains an aromatic or branched-chain residue[194].

3.1.1 Molecular structure and Zn(II) co-ordination geometry

Determination of the structure of CPA to atomic resolution by Lipscomb and co-workers[29,188,189] has enabled a detailed description of the tertiary and secondary structure of the enzyme, and X-ray diffraction studies, particularly through the application of the difference Fourier method in the study of substrates and inhibitors bound to the enzyme[29,195,196], have provided detailed structural information on the probable pathway of enzyme catalysis. The folding of the polypeptide chain of CPA is illustrated in figure 18 to indicate the relationship of the active-site region to the rest of the molecule. As illustrated in figure 19, the potential catalytic groups of the enzyme, deduced on the basis of X-ray diffraction studies, are (i) tyrosine-248, the probable donor to the NH group of the susceptible peptide bond; (ii) the Zn(II) ion, which binds the carbonyl oxygen atom of the susceptible peptide bond, and (iii) glutamate-270, which is the probable base for attack of the carbon atom of the susceptible peptide group either through the carboxylate group of glutamate-270 or possibly through a water molecule. This diagram illustrates that stabilisation of the substrate for hydrolysis is affected by binding to certain amino-acid chains in the active-site cleft.

The anatomical details of the structure of CPA are known at present to 200 pm resolution, and atomic parameters for the positions of the individual amino-acid nuclei[189] have been determined on the basis of the crystallographic co-ordinate refinement procedure of Diamond[61]. The active centre of the enzyme comprises about a quarter of the molecule and includes the Zn(II) ion and its ligands, the active-site pocket, the surface groove for binding of substrates and the immediate boundaries of the region. The donor atoms of the amino-acid residues bonded to the Zn(II) ion within this region are the N_1-positions of histidine-69 and histidine-196, an oxygen atom of the carboxylate group of glutamate-72 and an oxygen

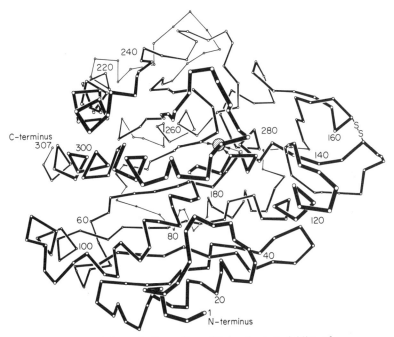

Figure 18 Schematic illustration of the main chain folding of the polypeptide segments of CPA. Solid lines connect the C_α atoms (circles) of each amino-acid residue. The Zn(II) ion is shown near the centre of the molecule as a stippled sphere. The three protein ligands are indicated by arrows. The N-terminus is at the bottom of the drawing, and the C-terminus at the left. From Quiocho and Lipscomb[189], with permission.

atom of a water molecule[29,197]. A second water molecule ≈ 350 pm distant from the metal ion in the active-site cleft has also been noted[67]. In addition, the co-ordinating amino-acid residues form hydrogen-bonded structures with other residues. The N_3-position of histidine-196 is hydrogen bonded to a water molecule, and the N_3-position of histidine-69 is hydrogen bonded to a carboxylate oxygen of aspartate-142.

Table 7 indicates the metal–ligand bond angles of the Zn(II)-containing active site of CPA. The distorted tetrahedral co-ordination of the metal binding site is readily evident from the marked deviation of the angles from that of 109.5°

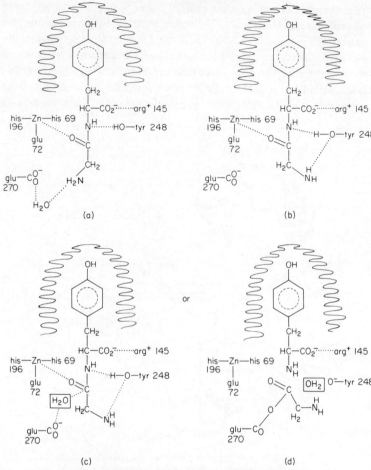

Figure 19 Specific binding interactions of glycyl-L-tyrosine
with CPA, as deduced by difference Fourier studies. The substrate residue probably progresses from a position such as
shown in (a) to one of probable productive binding at the
start of catalysis as in (b). Diagram (c) illustrates the general
base path in which glutamate-270 promotes attack of a lone
pair from H_2O at the carbonyl carbon, probably preceded by,
or concurrent with, proton transfer with tyrosine-248 to the
NH group. Diagram (d) illustrates an anhydride intermediate
pathway in which H_2O later attacks the acylenzyme intermediate. Further studies are necessary to resolve this ambiguity.
From Quiocho and Lipscomb[189], with permission.

expected for perfect tetrahedral co-ordination. Few tetrahedrally co-ordinated
Zn(II) complexes exhibit perfect tetrahedral symmetry, however, in the group-
theoretical sense. A comparison of appropriate bond angles for various Zn(II)
complexes of tetrahedral co-ordination is therefore made in table 8. It is evident

TABLE 7 METAL–LIGAND BOND ANGLES IN CARBOXYPEPTIDASE[†]

L_1–Zn–L_2	Degrees[‡]
$N_1(69)$–Zn–$N_1(196)$	86
$N_1(69)$–Zn–$O(72)$	99
$N_1(69)$–Zn–$O(H_2O)$	120
$O(72)$–Zn–$N_1(196)$	143
$O(72)$–Zn–$O(H_2O)$	99
$N_1(196)$–Zn–$O(H_2O)$	111

[†] from Quiocho and Lipscomb[189].
[‡] bond-angle uncertainties are $\pm 5°$ for protein ligands and $\pm 10°$ for those involving the water molecule.

TABLE 8 COMPARISON OF VALENCE-BOND ANGLES OF Co(II) AND Zn(II) COMPLEXES OF APPROXIMATE TETRAHEDRAL CO-ORDINATION GEOMETRY

Compound	L_1–Co–L_2	Valence angle
Cs_2CoCl_4[†, ‡]	Cl–Co–Cl	$107°20'$
		$108°50'$
		$109°20'$
		$116°20'$
$N(CH_3)_4CoCl_4$[‡, §]	Cl–Co–Cl	$108°18'$
		$109°6'$
		$110°24'$
		$111°42'$
bis(dipivaloylmethanido)Zn(II) [‡, ‖]	O–Zn–O	$94°42'$
		$114°12'$
bis(p-toluidine)CoCl$_2$[¶]	Cl–Co–N	$109°$
	Cl–Co–Cl	$111°$
	N–Co–N	$111°$
bis(L-histidinato)Zn(II) · $2H_2O$[††]	N_{his_1}–Zn–N_{his_2}	$112°$
	N_{his_1}–Zn–N_{amino_1}	$96°27'$
	N_{his_1}–Zn–N_{amino_2}	$116°$
	N_{amino_1}–Zn–N_{amino_2}	$121°$

[†] see reference 208.
[‡] Co(II) and Zn(II) compounds are isostructural.
[§] see reference 209.
[‖] see reference 210.
[¶] see reference 81.
[††] see reference 211.

that the metal–ligand bond angles of Zn(II)CPA deviate, in general, from those observed in low molecular-weight, model tetrahedral co-ordination complexes. The metal–ligand bond lengths of the co-ordination centre in CPA probably compare favourably to those observed for zinc–amino acid complexes (table 4) although they have not been reported. To be sure, the precision of the Zn–ligand bond lengths that could be calculated for the enzyme is considerably less than that for small molecules, and an uncertainty of at least ± 10–20 pm would be probably associated with their estimation.

Although X-ray diffraction studies of Zn(II)CPA are not of sufficient resolution to determine whether small distortions of the imidazole rings have been induced by co-ordinating to the metal cation, it is probable that their geometry has not been altered significantly. A survey[77] of the structural properties of a wide variety of complexes with imidazole derivatives co-ordinated to a metal at the N_1-position has failed to detect marked geometrical alterations in ring structure from that observed for the free ligand despite marked differences of the cations according to ionic radius, nuclear charge and tendency towards covalent-bond formation. These results make it unlikely that metal-ion co-ordination in CPA perturbs the geometry of the imidazole rings of the co-ordinating histidine residues. No estimation has been reported of the displacement of the metal ion in CPA from the plane of the imidazole rings or of the angle that the Zn–N bond makes with the imidazole ring of each co-ordinating histidine residue. Freeman[77] has noted that in no metal–histidine complex does the metal ion lie in the plane of the imidazole ring. As noted in table 4 these geometrical factors depend on the metal ion of the complex, and the data for tetrahedral Zn(II) complexes most probably correspond closely to geometrical relationships of Zn(II) co-ordination in CPA.

Furthermore, Lipscomb and coworkers[198] have concluded on the basis of low-resolution studies of apoCPA that the Zn(II) ion is not critical for maintaining the overall three-dimensional structure of the polypeptide backbone. High-resolution studies[189], demonstrating that the N_3-position of histidine-69 is hydrogen bonded to the carboxylate group of aspartate-142, so maintaining structural integrity in the active-site region, support this conclusion. The N_3-position of histidine-196 is hydrogen bonded to a water molecule. The rigidity which these hydrogen bonds could supply to the active-site region is readily indicated by the high pK (≈ 11.5) for ionisation of the N_3–H bond of a histidine residue when the N_1-position is metal co-ordinated[199]. These hydrogen bonds would be expected to be retained even under strong alkaline conditions.

3.1.2 Co-ordination environment of substituted metal ions

Because of the essential role of Zn(II)[200] for enzymatic activity the effect of metal-cation substitution on peptidase activity would be expected to provide some insight into the nature of the functional role of the metal. Indeed, Vallee and co-workers[190,191] showed that if CPA treated first with *o*-phenanthroline is dialysed against salts of various transition metals significant restoration of enzymatic activity resulted. Since the Zn(II) ion is involved in binding the carbonyl oxygen of the susceptible peptide group as a ligand during proteolysis (see figure 19) and since numerous metal-substituted CPA analogues exhibit retention of catalytic activity[84,85,201,202], it is possible to examine the dependence of enzymatic activity on electronic configurations and co-ordination geometries of substituted metal ions.

On metal-ion substitution in CPA [84,85] the retention of peptidase activity was noted to follow the order

$$Co(II) > Ni(II) \approx Zn(II) \gg Mn(II) > Cu(II) = (zero) \qquad (3.1)$$

No peptidase activity was observed with Hg(II), Pb(II) and Cd(II) substitution. The relevant catalytic rate constants for several substrates are compared in table 9. Since the Hg(II)-substituted enzyme is the only metal-substituted derivative for which structural information at high resolution is available about the metal co-ordination site, structural details and co-ordination geometries of other substituted metal ions must be inferred from spectroscopic investigations and stereochemical considerations.

TABLE 9 STABILITY AND KINETIC CONSTANTS OF PEPTIDE HYDROLYSIS FOR VARIOUS METALLOCARBOXYPEPTIDASES[†]

Metal ion[‡]	$\log K$ [§]	Substrate[‖, ¶]							
		BGGP		CGGP		BGP		CGP	
		K_m	k_{cat}	K_m	k_{cat}	K_m	k_{cat}	K_m	k_{cat}
Zn(II)	10.5	8.0	1.2	2.5	8.0	8.1	5.6	19.5	5.5
Ni(II)	8.2	7.4	1.1	2.5	8.6	–	–	–	–
Co(II)	7.0	6.0	5.9	2.0	17.0	4.8	7.4	11.7	12.3
Mn(II)	5.6	2.9	0.23	1.3	3.6	1.1	0.45	22.9	2.3
Cd(II)	10.8	–	–	–	–	–	–	–	–
Cu(II)	10.6	–	–	–	–	–	–	–	–

† from Coleman and Vallee[84,85], Davies *et al.*[234] and Auld and Vallee[235].
‡ Cr(III) and Fe(II) have been reported to restore activity to apoCPA[190,191].
§ K is corrected for competition by 1 M Cl⁻ and 0.05 M Tris buffer pH 8.
‖ K_m values are M $\times 10^{-4}$; abbreviations are BGGP, benzoylglycylglycyl-L-phenylalanine; BGP, benzoylglycyl-L-phenylalanine; CGP, carbobenzoxyglycyl-L-phenylalanine; CGGP, carbobenzoxyglycylglycyl-L-phenylalanine.
¶ k_{cat} values are min⁻¹ $\times 10^3$.

(1) Co(II)-substituted carboxypeptidase

Substitution of Co(II) into apoCPA has been interpreted as resulting in distorted tetrahedral co-ordination of the metal on the basis of the visible and near infrared absorption properties of the enzyme[203]. Although this distorted co-ordination geometry has been assumed previously on the basis of the visible spectrum alone for the Co(II) enzyme[204], the interpretation of tetrahedral symmetry must be made in part on the energy separation between the spin-allowed ligand-field transitions. Tetrahedral Co(II) exhibits two spin-allowed ligand-field transitions in electric-dipole radiation arising, according to group theory notation for T_d symmetry, from ${}^4A_2(F) \rightarrow {}^4T_1(F)$ and ${}^4A_2(F) \rightarrow {}^4T_1(P)$ electron promotions[205]. On the basis of ligand-field theory the second of these transitions is expected to appear at about 620 nm and is known to exhibit unexpectedly high intensity[206,207]. The transition resulting in a $^4T_1(F)$ excited state is generally observed in the near infrared region at about 1600 nm. A diagrammatic scheme indicating the relative orbital energy levels involved is indicated in figure 20. On the basis of the positions of absorption maxima of Co(II)CPA at 555 nm and 1570 nm with extinction coefficients characteristic of Co(II) in a tetrahedral ligand field[203], it is reasonable to conclude as a first-order approximation that tetrahedral co-ordination is maintained on Co(II) substitution.

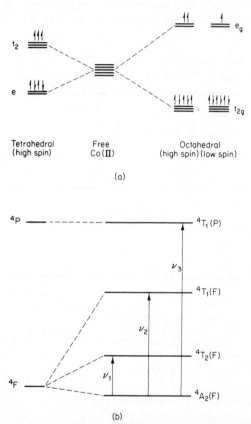

Figure 20 (a) Simplified illustration of orbital-splitting effects of tetrahedral and octahedral ligand fields of the Co(II) ion with a d^7 electron configuration. (b) Simplified energy-level diagram for a metal ion of d^7 electron configuration in a tetrahedral ligand field. The transition labeled ν_3 corresponds to the absorption band for the $^4A_2(F) \rightarrow {}^4T_1(P)$ promotion in the visible region and the transition labeled ν_2 corresponds to the $^4A_2(F) \rightarrow {}^4T_1(F)$ promotion for the absorption band observed in the near infrared region. The transition ν_1 is observed in the infrared region in a lower energy region than that of ν_2. The energy position of its absorption band is a direct measure of Dq.

It is not possible, however, to conclude that the visible absorption spectrum of Co(II)CPA is consistent with distorted tetrahedral co-ordination. The co-ordinating ligands, under the assumption that they remain identical for Co(II) substitution, consist of residues yielding two nitrogen and two oxygen donor atoms, each type with different ligand field and electrostatic contributions. The relative contributions of structural distortion and ligand donor-atom asymmetry cannot be readily determined in view of the absence of suitable model complexes with comparable ligands and perfect tetrahedral geometry.

Cotton and Soderberg[212] have shown for the bis(dipivaloylmethanido)Co(II) complex, a compound exhibiting marked distortion from tetrahedral geometry and having only oxygen donor atoms (see table 8), that the absorption spectrum is markedly altered from those generally observed for tetrahedral complexes of Co(II). The perturbation results primarily in an increase in the absorption intensity of the near infrared transition (ν_2 of figure 20) relative to that of the visible transition (ν_3). For most considerably less distorted tetrahedral complexes, the ratio[212] of absorption intensities $f(\nu_3)/f(\nu_2)$ is in the range $\approx$ 5–10. Although the oscillator strength of the near infrared transition of Co(II)CPA has not been reported[203], the ratio of maximum extinction coefficients of the corresponding transitions is $\approx$ 7. On this basis it would seem that the absorption spectrum of Co(II)CPA is not altered by structural distortion in a manner comparable to that of bis(dipivaloylmethanido)Co(II).

Ferguson[213–15] has investigated and analysed in detail the plane-polarised absorption spectra of single crystals of tetrahedral Co(II) complexes, the results of which appear more readily applicable to our discussion of Co(II)CPA. In these studies he utilised a series of Co(II) complexes of the type [Co(N-donor ligand)$_2$-(halide)$_2$], which on the basis of X-ray studies exhibit deviations from precise tetrahedral co-ordination geometry. Ferguson[213,215] noted for these complexes that the visible absorption spectrum corresponding to the $^4A_2(F) \rightarrow {}^4T_1(P)$ electron promotion appears to be largely *insensitive* to structural distortion. Lowering of temperature, however, reveals simplifying features of the near infrared transition, which are related to structural perturbations of the ligand-field symmetry.

The $^4T_1(F)$ state is orbitally threefold degenerate, and, therefore, three infrared transitions are expected for tetrahedrally co-ordinated Co(II) according to the distortion of the ligand field (see figure 20). The splitting of this transition has been observed in Co(II) complexes of distorted tetrahedral symmetry, and temperature lowering allows simplifying features to be observed[213]. Latt and Vallee[203] have reported that the near infrared absorption spectrum of Co(II)CPA at low temperatures exhibits two bands centred at 940 and 1570 nm. The effect of the lower than tetrahedral symmetry then is thus similar to that observed in [Co(p-toluidine)$_2$Cl$_2$] in which one of the components has been shifted to higher energy and the other two remain mixed[213]. The direction of the change in energy of one component to about 940 nm in Co(II)CPA[203] compared to that at 1100 nm for the [Co(p-toluidine)$_2$Cl$_2$] complex having two nitrogen donor ligands in distorted tetrahedral symmetry[213] may arise from a more pronounced structural perturbation as well as substitution of two oxygen donor ligands for the chloride ligands according to the average environment rule[216].

Because of insufficiently precise spectral data to evaluate the relative effects of various donor ligands, the shift in energy cannot be analysed at present in a more quantitative manner according to co-ordination environment. While the splitting of the near infrared absorption band of Co(II)CPA[203] can be considered to reflect the distorted tetrahedral co-ordination geometry of the metal ion, there is no spectroscopic evidence suggesting that the co-ordination geometry of the Co(II) and Zn(II) ions are precisely identical. Comparison of relevant geometrical relationships of metal–amino-acid complexes in table 4 suggests that co-ordination geometries and configurations of the co-ordinated amino-acid side chains probably differ only slightly if at all.

Co(II)CPA exhibits, furthermore, numerous spectral similarities to tetrahedral complexes of Co(II) with histidine derivatives[217,218]. Comparison of their spectral properties allows some correlation of ligand environment with electronic structure. Although analysis of the fine structure of the visible absorption band corresponding to the $^4A_2(F) \rightarrow {}^4T_1(P)$ transition of model tetrahedrally co-ordinated Co(II) complexes indicates that ligand-field parameters cannot be determined precisely because of mixing of nearby ^2G and ^4P states with the excited $^4T_1(P)$ state[213,214], comparable ligand-field parameters are to be expected for Co(II)CPA and model tetrahedral Co-histidinate complexes. The visible absorption maximum of bis(L-histidinato)Co(II) lies at 558 nm[218] compared to that of 555 nm for Co(II)CPA[203]. Comparison of the valence angles of the bis(L-histidinato)Zn(II) complex in table 8, which presumably also apply to the corresponding Co(II) complex, indicates marked deviation from perfect tetrahedral co-ordination. The high absorption intensity of the visible band of the Co(II) complex[218] compared to that of the infrared band, therefore, suggests that the lower visible absorption intensity of Co(II)CPA does not arise purely from structural distortion. However, the lower absorptivity of the Co(II)-substituted enzyme (with two oxygen and two nitrogen donor ligands) compared to that of bis(L-histidinato)Co(II) having four nitrogen donor ligands[219] is consistent with the general decrease in extinction of $[Co(OH)_4]^{2-}$ compared to the corresponding bands of other $[CoX_4]^{2-}$ species[220].

The total oscillator strength of the visible band of Co(II)CPA is 3×10^{-3} [203] and that for low molecular weight Co(II) complexes is generally approximately 5×10^{-3} [220]. The difference cannot be readily analysed in view of theoretical problems inherent in accounting for the intensities of tetrahedral Co(II) complexes[206,207]. However, since comparison of the oscillator strengths of tetrahedral complexes of Co(II) indicates a correspondence[221] between band intensity and degree of covalence, the lower oscillator strength of the visible transition in Co(II)CPA may indicate a proportionately lower degree of covalent metal–ligand bond formation. This interpretation is in accord with the theoretical analysis of Ballhausen and Liehr[206] of tetrahedral Co(II). It is probable that the visible absorption band of Co(II)CPA remains otherwise similar in electronic origin to that observed for model complexes, on the basis of its low natural circular dichroism and absence of markedly increased fine structure at liquid-helium temperatures[203]. Low optical activity is observed[218] for this transition in model Co–histidinate complexes in the visible region, and lowering of temperature fails to reveal simplifying fine structure of the $^4A_2(F) \rightarrow {}^4T_1(P)$ transition of Co(II) complexes of approximate tetrahedral symmetry with nitrogen donor ligands[213].

Furthermore, Cotton and Bergman[222] have pointed out that in certain Co(II) complexes of high co-ordination numbers, although an approximately tetrahedral ligand field is observed on the basis of spectral and magnetic susceptibility properties, the actual disposition of donor atoms is not tetrahedral. In particular this circumstance occurs with polyatomic ligands in which two chemically equivalent atoms are present, as for instance, in $[Co(NO_3)_4]^{2-}$ and the analogous $[Co(O_2CCF_3)_4]^{2-}$. In such cases the mean positions of the pair of co-ordinating atoms lie roughly at the vertices of one of the usual octahedral or tetrahedral co-ordination polyhedra. In such complexes it appears[222] that the contribution of each oxygen atom to the ligand field is approximately inversely proportional to its distance from the metal ion. On this basis the influence of the second carboxylate oxygen of glutamate-72 and the second water molecule, described[67] as ≈ 350 pm from the metal site, should perhaps be taken into consideration in the interpretation of the origins of the ligand field of Co(II)CPA. Latt and Vallee[203] have correlated a pH-dependent increase in absorptivity of Co(II)CPA at 625 nm with loss of enzymatic activity. Although this increase in absorptivity with increase in pH remains small (as compared to other Co(II)-substituted enzymes), it is of interest to note that crystallographic studies of CPA have been carried out with crystals grown at pH 7.5[198], conditions near that for maximum peptidase activity[202]. The spectral changes imply some change in co-ordination environment. The possible origin of spectral perturbations in this wavelength region from weakly interacting oxygen donor ligands is discussed later in more detail with reference to Co(II)-substituted carbonic anhydrase (section 3.2.4). The decrease in enzymatic activity reflected by the absorptivity at 625 nm may result from a distortion in ligand-field symmetry. Such changes might be more quantitatively assessed in the region of the near infrared transition.

(2) Cu(II), Cd(II) and Hg(II)-substituted carboxypeptidase

Co-ordination geometries comparable to that in the Co(II)-substituted or in the native enzyme cannot be assumed to be retained on introduction of other metal ions. The general co-ordination tendencies of other metal cations and results of spectroscopic investigations support this statement. E.P.R. investigations on single crystals of Cu(II)carboxypeptidase A_γ have suggested[223,224] that the ligand field of the Cu(II) ion is essentially planar in contrast to that of tetrahedrally distorted complexes. Since the ligand hyperfine structure of the spectrum indicates two nitrogen donor ligands[223], it is probable that the Cu(II) ion is bound near the same site as the Zn(II) ion. Crystallographic studies of Cu(II)CPA at 600 pm resolution[195] are consistent with this conclusion.

Although the Cu(II) enzyme is known to bind substrate in solution[85], the Cu(II) analogue is enzymatically inactive. X-ray diffraction studies at low resolution of Cu(II)CPA with bound glycyl-L-tyrosine,, however, indicate that the conformational change associated with glutamate-270 (see figure 19) is absent on substrate binding. This observation suggests that the steric relationships of amino-acid residues in the active-site region required for substrate binding and hydrolysis have been altered by Cu(II) substitution.

Configurational alteration of amino-acid side chains on Cu(II) substitution in the active-site region may occur as a result of flattening of the normal approximate tetrahedral geometry of the co-ordinating residues. The marked tendency of Cu(II) towards square-planar or tetragonally distorted octahedral co-ordination is consistent with the results of Brill and coworkers[223]. Furthermore, numerous examples of Cu-amino-acid complexes with these configurations are known[77]. If the Cu(II) is co-ordinated by the same three ligand donor atoms of protein residues as the Zn(II) ion, then loss of enzymatic activity must result from either prevention of productive binding of the substrate by structural distortion, or from the tighter Cu—O equatorial bond with the water molecule or the carbonyl oxygen atom of the substrate as a result of Jahn–Teller effects.

However, it is probable that two other sources of structural distortion by Cu(II) binding must be considered. In sperm-whale metmyoglobin a Zn(II) ion is specifically bound, in probable tetrahedral co-ordination, by amino-acid side chains exposed to the solvent[66]. The co-ordinating residues are lysine-16, asparagine-116 and the N_1-position of histidine-113. A Cu(II) ion, however, is bound 700 pm distant from the Zn(II) site by the N_3-position of histidine-12, sharing the same lysine and asparagine ligands. There is no obvious explanation for the different binding site of Cu(II) on a steric basis. The binding of the Cu(II) with d^9 electron configuration by the N_3-position of histidine-12 may be favoured because of the apparently greater ligand field of the imidazole N_3-position over that of the N_1-position[77]. A similar situation may obtain in CPA with respect to one of the metal-ion co-ordinating histidine side chains of the active-site region. Freeman[77] has also pointed out that under certain combinations of oxygen and nitrogen donor ligands with amino acids and peptides, the Cu(II) ion favours square-pyramidal co-ordination. These combinations include two or three nitrogen and two or one oxygen donor atoms. In this case the fifth ligand will come from a water molecule. Since in general the e.p.r. characteristics of square-pyramidal Cu(II) complexes are incompletely characterised, it is not readily apparent whether this co-ordination geometry is necessarily inconsistent with the results of Brill and coworkers[223].

The Cd(II)-substituted enzyme exhibits no peptidase activity similar to the Cu(II) enzyme. Unlike the Cu(II) enzyme, however, it retains esterase activity[85]. Comparison of pertinent metal–N_1(histidine), metal–O(carboxylate) and metal–$O(H_2O)$ bond lengths in table 4 indicates that substitution of Cd(II) for Zn(II) should result in a general (approximate) 20 pm increase in bond lengths. In addition, the well-known tendency of Cd(II) for octahedral co-ordination[86] suggests the retention of tetrahedral geometry to be unlikely. On this basis, it is expected that accommodation of the Cd(II) ion with its greater radius would not be possible without configurational distortion of co-ordinating residues and nearby amino-acid side chains in the active-site region (see section 1.3). This conclusion finds support in the observations of Folk and coworkers[225] on carboxypeptidase B. Cd(II)-carb-oxypeptidase B effectively binds (but does not hydrolyse) hippuryl-D-arginine, a substrate analogue, which is not bound by either the Zn(II) or the Co(II) enzyme. Comparable effects are probable in the case of Cd(II)CPA.

Comparable distortion effects are to be expected for the Hg(II)-substituted enzyme, which exhibits no peptidase and only weak esterase activity in solution[85]. In this case, it has been demonstrated by X-ray studies[29] that the Hg(II) ion, co-ordinated by the N_1-positions of histidine-196, is displaced 50-100 pm from the site occupied by the Zn(II) ion. Displacement of the Hg(II) ion has been stated[67] to be accompanied by small configurational shifts of the co-ordinating amino-acid residues. The general tendency of Hg(II) to form linear complexes[86,226] makes retention of tetrahedral co-ordination less likely in the active-site region for this metal ion also.

(3) Ni(II)- and Mn(II)-substituted carboxypeptidase

No spectral or magnetic susceptibility data have been published for Ni(II)CPA which could serve as an unambiguous index of metal co-ordination geometry. It is possible, however, that the Ni(II) cation does not experience a tetrahedral ligand field on substitution in apoCPA. The absence of known Ni(II) complexes of peptides or amino acids (including imidazole derivatives) with tetrahedral co-ordination and the observation that two N(amino)-donor ligands do not exert sufficiently strong orbital splitting effects to result in square-planar co-ordination[77] suggests that the Ni(II) ion in CPA may be co-ordinated within a tetragonally distorted octahedral ligand field. The order of ligand-field splitting effects for nitrogen donor ligands[226]

$$N(amino) > N(peptide) > N_1(imidazole) \qquad (3.2)$$

supports this hypothesis. In addition, as will be discussed later (section 3.2.2), substitution of Ni(II) in carbonic anhydrase with three co-ordinating histidine residues results in an absorption spectrum suggestive of octahedral Ni(II)[228]. However, the possibility of a high-spin five-co-ordinate Ni(II) ion cannot be excluded at present in the case of CPA.

Figure 19 indicates that an essential stereochemical requirement for catalysis is that the position of the water molecule displaced by the carbonyl oxygen of the substrate is not significantly altered with regard to the relationships of the bound substrate molecule to other residues involved in binding and catalysis. There is no *a priori* reason to suggest that this requirement cannot be met by a five- or six-co-ordinate metal cation, depending on the direction of axial symmetry of the metal ion. Although the possibility of small structural distortions affecting the configurations of the three amino-acid ligands cannot be discounted on Ni(II) substitution, it is evident that despite their probable presence they are not incompatible with enzymatic activity.

Mn(II)CPA, on the other hand, appears to exhibit greater than fourfold co-ordination on the basis of nuclear magnetic resonance studies. ^{19}F$^-$ has been demonstrated to bind to the Mn(II) cation in the active-site region with a Mn—F distance of no more than 230 pm, and the binding of ^{19}F$^-$ to the metal cation is not influenced by binding of the substrate inhibitor β-phenylpropionate[229]. On the basis of X-ray crystallographic studies on native CPA [195] it is known that the carbonyl oxygen of the inhibitor binds to the Zn(II) ion and, therefore, presumably also binds to the Mn(II) ion. That the inhibitor β-phenylpropionate displaces a water molecule from the Mn(II) ion within the active-site cleft is suggested, furthermore, by measurements of the relaxation times of Mn–H$_2$O interaction by n.m.r. techniques[230]. Since peptidase activity of Mn(II)CPA is not altered by substitution of Cl$^-$ by F$^-$ in the assay mixture[229], these results indicate that the co-ordination number of the Mn(II) ion can be at least five. Hartsuck and Lipscomb[67] have pointed out that the second water molecule observed in the active-site cleft 350 pm removed from the Zn(II) ion may be responsible for binding to the Mn(II) ion. A six-co-ordinate structure, as pointed out by Navon *et al.*[229] is a more likely possibility in view of the co-ordination tendencies of Mn(II). However, only one *exchangable* water molecule is detectable for the Mn(II)-substituted enzyme[231]. Despite the unusual possibility of fivefold co-ordination, the binding of ^{19}F$^-$ to the Mn(II) ion in the active site during catalysis clearly indicates, however, that four-co-ordinate Mn(II) is not essential for peptidase activity. Despite the retention of enzymatic activity by the Mn(II)-substituted enzyme, comparison of metal–ligand bond lengths in table 4 suggests that some distortion of the co-ordinating ligands and active-site cleft should be expected in Mn(II)CPA.

3.1.3 Correlation of metal-ion co-ordination and peptidase activity in carboxypeptidase

In consideration of the chemical forces that are operative during peptide hydrolysis by CPA, it is essential to establish first the nature of the role of the Zn(II) ion. Initially, X-ray investigations at 200 pm resolution[29] appeared to be consistent with two possible mechanisms of Zn(II)-aided peptide hydrolysis. These are known as the Zn–hydroxide mechanism and the Zn–carbonyl mechanism, respectively illustrated in schemes A and B in figure 21. While the studies of Buckingham and coworkers[232] suggest that with model Co(III) complexes amide and ester hydrolysis is more efficiently catalysed via the hydroxide mechanism, difference Fourier studies at 280 pm resolution of the complex of CPA with the tetrapeptide phenylalanylglycylphenylalanylglycine (phe-gly-phe-gly) indicate that the difference electron-density peaks are not compatible with a water molecule (or hydroxide ion) remaining bound to the Zn(II) ion on substrate binding[196]. These results can be therefore regarded as structural evidence supporting the Zn–carbonyl mechanism.

Metal-substituted CPA derivatives also exhibit considerable esterase activity. Unfortunately the lack of suitable ester derivatives that can form stable complexes with crystalline CPA has prevented X-ray diffraction studies for determining the animo-acid residues involved in esterase activity[67]. Stereochemical requirements for the catalytic pathway have therefore not been established as for the hydrolysis of peptides. The relative order of esterase activity is[84,85]

$$Cd(II) > Mn(II) > Co(II) > Zn(II) \approx Ni(II) > Hg(II) > Cu(II) = (zero) \quad (3.3)$$

The probable distortion of the co-ordinating residues in the active-site region on substitution of Cd(II), Hg(II) and Mn(II) may influence the capacity for esterase activity. In view of the lack of detailed structural information for ester hydrolysis, we cannot therefore meaningfully discuss the dependence of esterase activity on metal-ion substitution. Moreover, Vallee and coworkers[233] have suggested that the binding sites for ester and peptide hydrolysis, although overlapping, may not be identical.

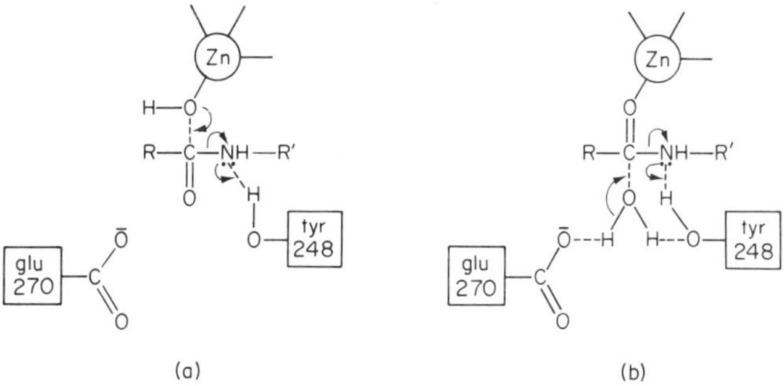

Figure 21 Diagrammatic illustration of the mechanism of carboxypeptidase-catalysed peptide hydrolysis: (a) Zn–hydroxide mechanism and (b) Zn–carbonyl mechanism.

Thus far in our discussion, results of structural or spectroscopic investigations have not suggested the essential qualities of the metal ion necessary for catalytic activity. Structural studies on the crystalline enzyme demonstrate that the Zn(II) ion is co-ordinated by the oxygen atom of the carbonyl group of the susceptible peptide bond on displacement of a co-ordinating water molecule (figure 21). Although no spectroscopic evidence directly indicates that the substituted Co(II) and Ni(II) ions behave in an analogous manner, binding of the carbonyl oxygen atoms by the substituted metal ion is probably essential for peptidase activity. Furthermore n.m.r. studies[229] on the Mn(II)-substituted enzyme suggest that this binding occurs in a similar way to that in the Zn(II)-containing molecule[29,196]. It is therefore probable, especially in view of low-resolution difference Fourier studies[195], that the substrate carbonyl oxygen cannot be bound productively in the Cu(II) enzyme and that the lack of peptidase activity in this case rests on a structural basis. Similar considerations apply in the case of the Cd(II)- and Hg(II)-substituted enzymes, for which evidence of disturbance of structural relationships required for substrate binding in the active-site cleft has been discussed (section 3.1.2(2)). It shall therefore be necessary to exclude the Cu(II)-, Cd(II)- and Hg(II)-substituted enzymes from any further discussion on the correlation of metal-ion co-ordination and peptidase activity.

A rigorous correlation of co-ordination environment, metal d-electron configuration and peptidase activity is at present rendered difficult because of the uncertainty of the co-ordination stereochemistry of the Ni(II)-substituted enzyme. In addition, since the efficiency of peptide hydrolysis reflects numerous kinetic and structural factors, including binding of the carbonyl oxygen atom to the metal ion, binding of the substrate to amino-acid side chains in the active-site region, accessibility of solvent molecules, substrate and product inhibition, all of which may be in part differently affected by distortions induced by metal-ion substitution, only certain general features of peptide hydrolysis, which must remain common to metal–substrate interaction for all metallo-CPA analogues, can be considered. One of these features must involve the influence of the metal ion on the electronic structure of the susceptible carbonyl bond of the peptide substrate. Since the co-ordination environments of the Zn(II)- and Co(II)-substituted enzymes are probably closely similar, correlation of metal-ion induced electronic perturbations of the carbonyl bond with metal d-electron configuration would be expected to obtain in a rigorous manner for these two enzyme derivatives.

Co-ordination of the metal ion in CPA by the carbonyl oxygen atom of the substrate has been postulated to result in polarisation of the C=O bond of the carbonyl group, rendering the carbon atom more susceptible to nucleophilic attack[188,189]. This polarisation effect is expected to be enhanced by the carboxylate group of glutamate-270 in the sense

$$M^{2+} \ldots O^{\delta -} = C^{\delta +} \ldots O^{-} \text{ (glutamate-270)} \tag{3.4}$$

On this basis the relative efficiency of inducing polarisation of the C=O bond by metallo-CPA analogues would be expected to be influenced primarily by factors determining the relative order of metal–oxygen bond strengths under the condition that other requirements of substrate binding, namely stereochemical relationships of amino-acid side chains, solvent accessibility to the active site, etc., remain relatively constant. While a discussion of enzymatic activity of metallo–CPA analogues towards one substrate cannot necessarily be applied towards all substrates, it remains unusual that readily measurable peptidase activity is observed on metal substitution for several substrates (table 9). Because of the possible alterations in co-ordination environment and co-ordination number for Ni(II) and Mn(II) substitution, indicating that four-co-ordinate structures are not an absolute requirement for enzymatic activity, it is of interest to compare electronic properties of metal–oxygen interaction in assessing the capacity of metal ions to induce C=O bond polarisation.

A large body of experimental evidence and theoretical analysis of the far infrared absorption spectra of co-ordination complexes show that metal–ligand bond strengths and changes in bonding structure can be assessed on the basis of bond-stretching frequencies and the calculated bond-stretching force constants[236]. Haigh et al.[237] have pointed out that for divalent metal ions of the first transition series metal–ligand bond-stretching frequencies measured in the far infrared region and the calculated force constants as a measure of metal–ligand bond strength are expected on the basis of theoretical calculations to be influenced by crystal-field stabilisation energies. The relative order of metal–ligand bond strengths would be expected to be

$$Zn(II) < Cu(II) < Ni(II) > Co(II) > Mn(II) \qquad (3.5)$$

This finds support in the conclusions of George and McLure[238] that formation of co-ordination complexes is favoured in terms of enthalpies for transition-metal cations compared to closed-shell metal cations and that the order of added stability of the co-ordination complex is dependent on the number of d orbital electrons as well as the nature of the donor ligand. For structurally similar octahedral complexes with nitrogen donor ligands, however, the order of metal–ligand bond-stretching force constants is observed to be

$$Zn(II) < Cu(II) > Ni(II) > Co(II) > Mn(II) \qquad (3.6)$$

and is, furthermore, reflected by the order[151] of affinity constants of these metals for a single ligand. The change in the relative order of Cu(II) and Ni(II) occurs as a result of the tighter in-plane metal–ligand bonds associated with the Cu(II)ion in a d^9 electron configuration (Jahn–Teller effect).

Comparison[237] of the metal–nitrogen bond-stretching frequencies and bond strengths of octahedrally co-ordinated Mn(II) and Zn(II), in which ligand-field stabilisation effects are absent, indicates that nearly identical metal–ligand stretching frequencies are observed. Furthermore, the Zn—N bond strength in tetrahedral co-ordination, as reflected by metal–ligand bond-stretching frequencies, is greater than in octahedral co-ordination. This increase in bond strength results from the redistribution of bonding capacity of the metal ion with six ligands to four ligands. Therefore, the Zn–ligand bond strengths for tetrahedral co-ordination would also be expected to remain greater than Mn–ligand bond strengths under conditions of fivefold or sixfold co-ordination. Similarly, in tetrahedral complexes Co—N bond strengths are increased because of the redistribution of bonding capacity from octahedral to tetrahedral co-ordination and are greater than those for octahedrally co-ordinated Ni(II)[237], and presumably would remain greater than for five-co-ordinate Ni(II) complexes. These considerations allow a modification of the above order of metal–ligand bond strengths determined on the basis of far infrared metal–ligand stretching frequencies tabulated by Haigh and coworkers[237] to be the following

$$Cu(II)_o > Co(II)_t > Ni(II)_o > Zn(II)_t > Mn(II)_o \qquad (3.7)$$

where the subscripts o and t refer to octahedral and tetrahedral co-ordination geometries.

While spectroscopic data of comparable precision are not available for metal-oxygen bond strengths of metals co-ordinated by the carbonyl oxygen atom of suitable model ligand molecules, an identical order would be expected on the basis of crystal-field stabilisation energies and co-ordination geometries. This conclusion finds firm support in the studies of Nakamoto and coworkers[239,240] on metalglycinato complexes. They observed that metal–nitrogen and metal–oxygen bond-stretching force constants change in a parallel manner according to the metal ion complexed. Most importantly, the shift in force constants, and hence bond strength, followed the same order as the stability constants of the metal ions with a single ligand[151]. These results clearly indicate that the bonding between the metal ion and the co-ordinating oxygen donor atoms is influenced in a manner dependent on the metal ion and that the change in bond strength and subsequently covalency follows an order closely related to the electronic configuration and co-ordination geometry of the metal ion.

The relative order (3.7) of metal–ligand, that is metal–oxygen, bond strengths modified according to co-ordination geometry follows closely the relative order[84,234] of peptidase activity† of metallocarboxypeptidases (table 9)

$$Co(II) > Ni(II) \approx Zn(II) \gg Mn(II) \tag{3.8}$$

Because of the uncertainty introduced by the unknown co-ordination environment of the Ni(II) ion, the seemingly parallel behaviour with regard to change in metal–oxygen bond strength and peptidase activity can be most rigorously considered only for the Co(II)-, Zn(II)- and Mn(II)-substituted enzymes. On this basis it none-theless appears that the relative order of peptidase activity can be related to the relative order (3.7) of metal–oxygen (carbonyl) bond strength.

Since metal–oxygen interaction is essential to peptide hydrolysis, it would appear then that the influence of metal ions on the electronic structure of the susceptible carbonyl bond to promote hydrolysis can be best determined spectro-scopically. Far infrared absorption techniques comparable to those utilised for model compounds may allow assignments of metal–ligand bond-stretching frequencies for a more quantitative assessment of the postulated mechanism[188,189] of CPA proteolytic action.

Furthermore, that polarisation of the carbonyl bond is indeed promoted by metal-ion co-ordination, is amply supported by numerous infrared spectroscopic studies on metal–amino-acid complexes. Various workers[239] have shown that on co-ordination of a metal ion by a carboxylate group, the C–O and C=O bond-stretching force constants change in such a manner as to suggest formation of a more symmetrical carboxylate group

and that the symmetrical carboxylate electronic structure is favoured by a *more ionic* metal–oxygen bond. Such a structure is presumably formed during peptide hydrolysis by metal-CPA derivatives after binding of the carbonyl oxygen atom to the metal ion. It has been pointed out earlier (section 3.1.2(1)) that the visible

† We exclude Cu(II), Cd(II) and Hg(II) from this discussion on the basis of structural distortions (see section 3.1.2) resulting in nonproductive substrate binding in peptide hydrolysis.

absorption spectrum of Co(II)CPA suggests a lesser degree of covalent metal-ligand bond character than found in model Co-histidinate complexes. The lesser degree of metal-ligand covalency in CPA may be a property of the metal-ligand active site directed by all of the co-ordinating amino-acid side chain donor ligands. It is possible that this function is preserved by the co-ordinating amino-acid residues by efficiently promoting polarisation of a carbonyl group and subsequent formation of a carboxylate group during hydrolysis of the bound substrate.

A hypothesis formulated by Pauling[241] on the nature of transition-state enzyme-substrate complexes states that enzymes function by virtue of their ability to bind the transition-state species more tightly than the substrate molecule and that this property is responsible for lowering the free energy of activation of the reaction. The structural studies of Lipscomb and coworkers[29,188,189], demonstrating the binding of the carbonyl oxygen of the substrate by the metal cation, imply that this co-ordinating arrangement exists in the transition-state enzyme-substrate complexes of metal-CPA analogues. The strength of the metal-oxygen bond formed on substrate binding may therefore be an important component in stabilising the transition-state enzyme-substrate complex. On this basis the similar orders of relative metal-oxygen bond strengths and peptidase activity for metal-CPA analogues suggest a chemical basis for the dependence of peptidase activity on the co-ordination geometry and electron configuration of the substituted metal ion.

The correlation of relative order of peptidase activity with metal-ion co-ordination geometry and electron configuration allows a rational description of the electronic rearrangements of the substrate molecule that are consistent with the mechanism of enzyme action formulated on the basis of X-ray studies. It supplys no explanation of why Zn(II) is found in the native protein rather than the more active Co(II) and gives no all-inclusive explanation for the basis of catalytic action of carboxypeptidase. It is likely, however, that the mechanism postulated by Lipscomb and coworkers reflects the essential correctness of the various substrate-binding sites including the metal ion. The correlation of relative order of peptidase activity with the expected relative order of metal-ligand bond strengths ((3.7) and (3.8)) may therefore indicate the importance of the electronic structure and co-ordination geometry of the metal cation in the proteolytic action of metallocarboxypeptidases.

3.2 CARBONIC ANHYDRASE

Carbonic anhydrase is a metalloenzyme containing one gram-ion of Zn(II) per molecule[242], essential for enzymatic activity[243]. The enzyme has a molecular weight of approximately 30 000 and is widely distributed in plants, animals and bacteria. The physiological role of carbonic anhydrase centres on rapid hydration of metabolic CO_2 formed in tissues, dehydration of HCO_3^- in the lungs and transfer and accumulation of H^+ or HCO_3^- in organs of secretion[244]. Polymorphism of the enzyme in a number of animals has been observed[245], and the two forms of human carbonic anhydrase B (HCAB) and human carbonic anhydrase C (HCAC), differing in amino-acid composition and activity[245-8], are perhaps the best known and most extensively studied. The high turnover rate of 10^6 s^{-1} for CO_2 hydration[249] makes this enzyme one of the fastest known. A recent comprehensive review of the biochemistry and structure of carbonic anhydrase has been written by Lindskog *et al.*[250]

3.2.1 Molecular structure and active-site region

X-ray diffraction studies initiated on HCAC[251] have been completed to 200 pm resolution. A Kendrew skeleton model of the enzyme, which has been constructed by interpretation of the electron-density map, serves to characterise the general

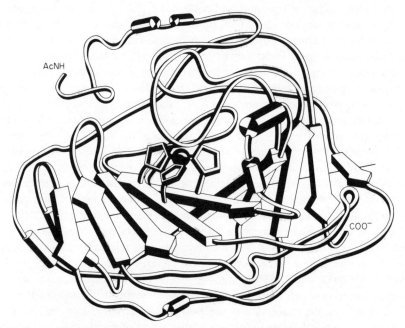

Figure 22 Schematic illustration of the main chain folding of HCAC. The helices are represented by cylinders and the pleated sheet strands by arrows in the direction amino to carboxyl. The ball supported by three histidine residues represents the Zn(II) ion within the active-site cleft. From Kannan *et al.*[37], with permission.

structural properties of the enzyme[37,252]. X-ray diffraction studies have revealed that the enzyme is comprised of an extensive structure of the polypeptide chain in a β-pleated sheet conformation which divides the molecule into two halves. A schematic representation of the main chain folding of HCAC is illustrated in figure 22. The active site containing the co-ordinated Zn(II) ion is found in a large hydrophobic cavity located between the β-structure in one half of the molecule and six polypeptide sequences remaining essentially perpendicular to the chain direction of the β-structure. The Zn(II) ion is thus almost in the centre of the molecule and is located approximately 1500 pm from the ellipsoidal surface of the protein.

There are three ligands to the Zn(II) from the β-structure: the N_3-position of each of the imidazole rings of histidine-93 and histidine-95 and the N_1-position of histidine-117. The fourth ligand in the tetrahedral complex is supplied by a water molecule. The greatest departure from precise tetrahedral co-ordination for metal–ligand bond angles is approximately estimated[252] as 20°. A comparison of the distances from the Zn(II) cation to each donor ligand atom and to various residues of the active-site region is given in table 10. These distances, obtained by

TABLE 10 THE DISTANCE FROM THE Zn(II) ION TO VARIOUS GROUPS IN THE ACTIVE-SITE CLEFT OF HCAC[†]

Residue no.	Group	Atom	Distance (pm)
93	his	N_3	200
95	his	N_3	200
117	his	N_1	200
	H_2O (1)	O	200
	H_2O (4)	O	350
104	glx	$X^{\varepsilon 1}$	400
104	glx	$X^{\delta 1}$	500
198	leu	$C^{\delta 1}$	400
197	thr	O^{γ}	450
	H_2O (A)	O	500

† from reference 37.

measurements of the atomic skeleton model[37,252], indicate that only four donor-ligand groups can be considered sufficiently close to the metal ion to form bonds in the usual sense with the Zn(II) ion, and the metal cation is probably formally four-co-ordinate. Since the atomic parameters of the polypeptide chain have not yet been subjected to refinement methods, a comparison of structural details of the active site further to those of low molecular weight metal–amino acid complexes is not warranted.

The results of numerous studies supply firm evidence that the metal ion in carbonic anhydrase participates in catalytic activity. The Zn(II) cation is necessary for activity since apocarbonic anhydrase does not catalyse the hydration of CO_2[243], and the enzyme is known to be inhibited by anions that bind the metal cation[253]. The chloride anion, which has been shown to inhibit enzymatic activity on binding of a single anion per molecule of enzyme[253], has been demonstrated on the basis of ^{35}Cl n.m.r. studies[254] to bind directly to the Zn(II) ion, presumably on replacement of the co-ordinated solvent molecule. However, the exceptionally fast turnover of the enzyme-catalysed hydration reaction

$$CO_2 + H_2O \leftrightarrows H_2CO_3 \leftrightarrows H^+ + HCO_3^- \qquad (3.9)$$

has thus far prevented direct study by spectroscopic methods of transient enzyme–substrate complexes. Working at low pH values (< 6.5) where the enzyme does not appreciably catalyse the hydration reaction as compared to alkaline conditions, Riepe and Wang[255] observed on the basis of infrared absorption maxima of inhibited and CO_2-equilibrated enzyme that (i) azide, a powerful inhibitor of enzymatic activity, is bound to the Zn(II) ion, and (ii) the shift in the absorption

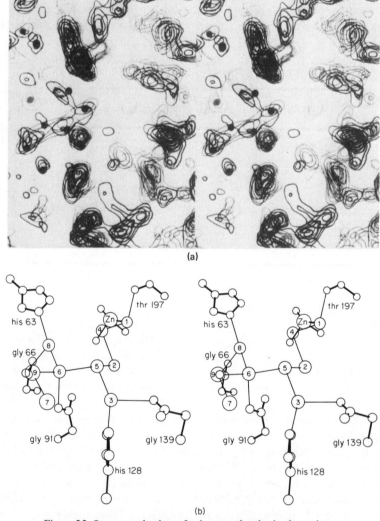

(b)

Figure 23 Stereoscopic view of solvent molecules in the active site of HCAC. Nine peaks of electron density not connected to protein residues were interpreted as water molecules. In this drawing the solvent molecules are connected by a network of possible hydrogen bonds. From Liljas *et al.*[252], with permission.

maximum of CO_2 in the presence of the enzyme suggested asymmetric stretching of the gaseous molecule loosely bound to a hydrophobic region of the enzyme but not directly to the metal ion. It is, however, not certain that active intermediates were observed by Riepe and Wang[255] in view of the difficulty in attempting to confirm these results[249,256]. While the metal ion may not be directly involved in binding CO_2 or bicarbonate, it remains essential for CO_2 hydration–dehydration reactions presumably by interaction with solvent molecules in the active-site region[250,257].

The enzymatic action of carbonic anhydrase involves water in either forward or backward directions (3.9). Of particular interest, therefore, in this enzyme is the unusual extensively ordered solvent structure detected as light peaks of electron density visible in the vicinity of the metal-ion containing active site[37,252]. These peaks are displaced on binding of aromatic sulphonamide inhibitors, which bind near the Zn(II) cation and therefore cannot be part of the protein molecule. Their relationship to the Zn(II) ion in the active site is illustrated in figure 23. Although the presence of ammonia and chloride cannot be entirely excluded as the origin of the electron-density features[252], the shortest distances between their centres is comparable to that of 276 pm, the hydrogen-bond distance in water [258]. The less dense arrangement of solvent molecules outside the active site also strongly suggests nine molecules of water arranged in the active site illustrated in figure 23. The role of these ordered water molecules in catalysis is not known, but in view of the probable mechanism of CO_2 hydration involving the CO_2 molecule, the bicarbonate anion and water[249,259], the ordered solvent structure may be important in interaction with the water ligand of the tetrahedrally co-ordinated Zn(II) cation (see section 3.2.3).

3.2.2 Metal-ion substitution, co-ordination geometry and solvent structure

(1) Structural studies

As for the substitution of Zn(II) in CPA, metal ions can be added to apocarbonic anhydrase, and specific substitution of the Zn(II) ion by added cations is observed[260]. Table 11 gives a summary of the enzyme activities and the binding affinity of the inhibitor acetazolamide to metallocarbonic anhydrases. Cation-dependent enzymatic activity and binding of specific inhibitors of carbonic anhydrase are primarily restricted to the presence of either Zn(II) or Co(II) ions. The results tabulated in table 11 therefore strongly suggest that significant perturbations on the active-site region occur as a result of metal-cation substitutions.

Liljas[69] has compared the co-ordinate positions of Mn(II), Co(II), Cu(II), and Hg(II) to that of Zn(II) in HCAC. These positions were determined on the basis of two-dimensional difference Fourier maps of each metal-substituted enzyme versus the apoenzyme. The distance between the positions of the Zn(II) binding site and that of each substituted cation are tabulated in table 12. For these studies data were collected to 250 pm resolution for the metal-substituted enzyme. The metal-ion parameters were then determined by least-squares refinement. Final difference electron-density maps for each metal–HCAC derivative were computed using

TABLE 11 METAL-ION SPECIFICITY OF METALLO-CARBONIC ANHYDRASES

Metal ion	Relative specific activity in excess of residual activity[†]			Moles of inhibitor acetazolamide bound per mole of enzyme[§]
	BCAB[‡]	HCAB	HCAC	
Zn(II)	100	100	100	1.00
Cu(II)	0	0	–	0.14
Ni(II)	2 ± 1	5 ± 1	–	0.02
Co(II)	50 ± 5	55 ± 5	50 ± 10	1.00
Fe(II)	4 ± 1	5 ± 2	–	–
Mn(II)	8 ± 2	12 ± 2	–	0.40

† from reference 261.
‡ bovine carbonic anhydrase B.
§ HCAB from ref. 262.

phase angles obtained from three heavy-atom derivatives, and the compound under investigation was not allowed to contribute to the phase-angle calculation[69]. From table 12 it can be seen that the positions of the substituted metal cations do not deviate greatly from that of Zn(II) in the native protein and that the largest deviation is observed for Hg(II). In the difference electron-density maps calculated in projection for Hg(II)HCAC versus apoHCAC only small difference electron-density peaks are evident in the active-site region although the position of the Hg(II) ion is significantly closer to the O^γ of threonine-197[69].

TABLE 12 DISTANCES BETWEEN THE REFINED POSITION OF THE Zn(II) ION AND THAT OF SUBSTITUTED CATIONS IN HCAC[†]

Metal ion	Mn(II)	Co(II)	Cu(II)	Hg(II)
Zn–metal distance (pm)	20	20	10	60

† from reference 69.

Although the difference electron-density maps determined for metal–HCAC derivatives have indicated that substituted metal cations bind near the site occupied by Zn(II), they supply no direct structural explanation for the general loss of enzymatic activity on metal substitution. Calculated three-dimensional difference Fourier maps of metal–HCAC analogues versus the apoenzyme will be required to resolve structural changes of the ordered solvent molecules or of co-ordinating amino-acid residues on metal-ion substitution. Structural details of small configurational changes of groups near the co-ordinated metal cation or changes in the ordered solvent molecules in the active-site cleft cannot be expected to be readily deter-mined on the basis of difference Fourier maps calculated in projection only. Furthermore, the absence of large conformational changes in the active-site region does not preclude the possibility of a change in co-ordination geometry on metal-ion substitution involving only small shifts of co-ordinating residues. Although the sites of the Mn(II) and Co(II) ions do not differ greatly from that of the Zn(II) ion, such slight shifts are nonetheless associated with a marked decrease in enzymatic activity. On the basis of comparative structural data of metal–amino acid complexes in table 4, small configurational changes of co-ordinating residues and alterations in co-ordination geometry are expected. Since a water molecule co-ordinated to the tetrahedrally bound Zn(II) forms part of the intricate structure of ordered solvent molecules in the active-site cleft[37,252], small configurational differences in the co-ordinating histidine residues and small shifts in the binding sites of substituted metal cations may profoundly effect the ordered solvent structure. These perturbations in solvent structure would consequently alter conditions for reactions involving water in the active-site cleft.

(2) Spectroscopic studies

The mechanism of catalytic-action of carbonic anhydrase in the hydration of CO_2 that appears most attractive on the basis of kinetic and chemical studies is that proposed by Khalifah[249] and Coleman[259], in which the metal ion and solvent molecules are the most important participants. These proposals receive strong support from the results of X-ray diffraction studies[37,252], demonstrating that ordered solvent molecules, readily perturbed by inhibitors, are found in the active-site cleft, and from the ^{35}Cl n.m.r. studies of Ward[254], which show that chloride inhibition occurs on binding of the anion to the metal cation. A schematic illustration of the mechanism most likely is shown in figure 24. An explanation of

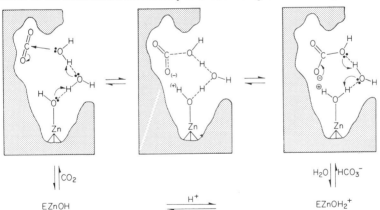

Figure 24 Schematic illustration of a catalytic mechanism for CO_2 hydration–dehydration by carbonic anhydrase in which the Zn-bound water molecule participates in both the electron donor and acceptor requirements of the reaction. It is not intended to imply that CO_2 and HCO_3^- bind competitively to the enzyme. From Khalifah[249], with permission.

the marked decrease in enzymatic activity on metal-ion substitution must be sought from an evaluation of possible changes in co-ordination geometry induced by the substituted metal cations in the active-site region and the concomitant effects induced on the ordered solvent structure. Changes in co-ordination geometry must be deduced on the basis of spectroscopic investigations in view of the absence of direct structural information.

The absorption spectrum of Co(II)HCAC at $\approx$ pH 6 has long been considered to be consistent with tetrahedral co-ordination on the basis of the energy separation of visible and infrared ligand-field transitions and band intensities[261]. Furthermore, magnetic-susceptibility properties of the Co(II) enzyme indicate that the magnetic moment ranges between 4.26 and 4.37 Bohr magnetons in good agreement with the value expected for high-spin Co(II) in a tetrahedral environment[263]. In view of the absence of large configurational changes of metal-co-ordinating residues on Co(II) substitution in HCAC[69], it must be concluded on the basis of the structural data in table 4 and the absorption spectrum of the Co(II) enzyme that the stereochemistry of the co-ordinating ligands of Co(II) is similar but not necessarily precisely identical to that of Zn(II) in the native enzyme. The small displacement of the Co(II) ion of 20 pm from the Zn(II) binding site (table 12) is at first difficult to reconcile as underlying the decrease in enzymatic activity, for Co(II) substitution in CPA is associated with an increase in enzymatic activity[85,202]. However, as illustrated in figure 23, the water molecule co-ordinating the Zn(II) cation is hydrogen bonded to the O^{γ} of threonine-197 and to two other solvent molecules. Small perturbations on the ordered solvent structure may be readily transmitted by a small shift in metal-cation position. On this basis catalytic activity may be altered, since conditions for reactions involving water in the active-site cleft would have been changed. Such perturbations may result from displacement of the substituted metal cation from the site occupied by the Zn(II) ion as well as from small configurational changes in amino-acid residues in the active-site cleft close to the ordered solvent structure.

The basis for lack of enzymatic activity on Cu(II) substitution is more readily appreciated by the results of Taylor and Coleman[264]. Their e.p.r. spectra of Cu(II) carbonic anhydrase are consistent with square-planar or tetragonally distorted octahedral symmetry with nitrogen ligands co-ordinating the Cu(II) ion. Unfortunately the difference Fourier studies in projection[69] are not able to suggest the stereochemica changes associated with the co-ordinating residues on Cu(II) substitution, and e.p.r. dat on orientated crystals of Cu(II)-substituted carbonic anhydrase are not available to determine the direction of axial symmetry of the Cu(II) co-ordination site with respect to fixed axes in the molecule. If it is assumed, as is likely, from figure 23 that the three Cu—N bonds define the co-ordination plane of the Cu(II) ion formed by flattening of the usual tetrahedral arrangement, the solvent molecule occupying the fourth equatorial site would not necessarily correspond to the same site as that in the tetrahedral complex of Zn(II). Furthermore, the change in co-ordination geometry on substitution of the Cu(II) ion would result in a reordering of water structure in the active-site region since the added co-ordination sites would probably be occupied by solvent molecules. The reordered solvent structure may not be favourable for enzyme catalysis with respect to other nearby amino-acid side chains essential for CO_2 hydration. This perturbation on the ordered solvent molecules in addition to the effects of probable configurational changes of co-ordinating residues can thus readily explain the lack of enzymatic activity.

The visible absorption spectrum of the Ni(II)-substituted enzyme[228] can be considered as consistent with octahedral or possibly square-planar co-ordination on the basis of the low intensity of its visible absorption band near 600 nm. It appears more suggestive of octahedral co-ordination by comparison to the spectrum of $[Ni(NH_3)_6]^{2+}$[265] and Ni-histidine complexes[266]. Since square-planar co-ordination of Ni(II) requires ligands with strong ligand-field properties, a circumstance observed only in Ni–peptide complexes with four peptide-nitrogen donor ligands[77], it is probable that the Ni(II) ion experiences only a tetragonally distorted octahedral ligand field. As with the Cu(II) ion, this situation similarly dictates a change in the position of the co-ordinated water molecule from that associated with the Zn(II) ion and an increase in the number of co-ordinating solvent molecules. The decrease in enzymatic activity is thus expected to occur on a similar basis to that concluded for the Cu(II)-substituted enzyme. For comparable co-ordination geometries Cu–ligand bond lengths are generally shorter by approximately 10 pm than corresponding Ni–ligand bond lengths (see table 4). The increased distortion of co-ordinating residues expected for Cu(II) as compared to Ni(II) may account for the small amount of enzymatic activity retained on Ni(II) substitution (table 11).

No spectroscopic data are available to suggest the co-ordination geometry of the substituted Mn(II) ion. Despite its general tendency towards octahedral co-ordination[86], the unusual possibility of five co-ordination in Mn(II)CPA (section 3.1.2(3)) allows no reliable assessment of its co-ordination environment in carbonic anhydrase. The displacement of the Mn(II) ion from the Zn(II) binding site (table 12) and the generally longer metal–ligand bond lengths associated with Mn(II) suggest a basis for significant distortion, which would readily alter structural requirements for retention of ordered solvent structure in the active-site cleft.

It is probable that the general decrease or loss of enzymatic activity on metal-ion substitution is primarily the result of reordering or disruption of solvent structure in the active-site cleft as a result of change in co-ordination geometry rather than through gross alterations in protein structure. Since the preliminary difference Fourier data[69] for metal-substituted derivatives awaits re-evaluation on the basis of three-dimensional difference electron-density maps, the conclusion relating loss of enzymatic activity to disruption of solvent structure remains supported only by inference of the effects of change in co-ordination geometry on solvent molecules within the active-site cleft. Tetrahedral co-ordination in the case of Co(II) substitution, as suggested by the absorption spectrum[261], is associated with retention of approximately 50 per cent of the native enzymatic activity (table 11). The slight displacement of the Co(II) ion (table 12) and possibly small differences in geometries of metal–histidine interaction between Zn(II) and Co(II) in tetrahedral environments indicate the profound effect of small perturbations on the ordered solvent structure.

The series of metallocarbonic anhydrases studied by chemical and crystallographic methods suggests the importance of delicate steric requirements in the interaction of ordered solvent molecules with the substrate molecule. That catalytic action cannot be a simple function of the nature of the metal–oxygen bond formed in the active-site cleft is demonstrated by comparison of the activities on the Co(II)- and Zn(II)-containing enzymes (table 11). On the basis of the suggestion of Orgel[267] and Jencks[268] that the role of a metal ion in a protein is to increase the acidity of a bound water molecule, polarisation of an O–H bond of the bound water molecule, as illustrated in figure 24, and of probable importance in the catalytic mechanism[249,259], would be expected, nevertheless, to be promoted more readily by co-ordination to Co(II) than to Zn(II). The overwhelming influence of slight steric perturbations on the ordered solvent structure in the active-site cleft underlines the delicate nature of steric requirements which must be fulfilled in the catalysis of CO_2 hydration–dehydration by this enzyme.

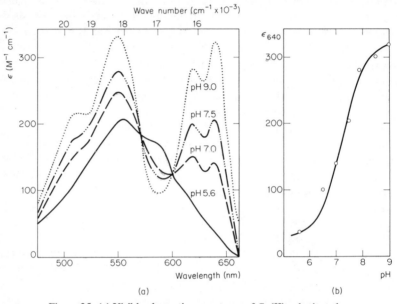

Figure 25 (a) Visible absorption spectrum of Co(II)-substituted HCAC as a function of pH. Unbuffered enzyme solution $(3.5 \times 10^{-4}$ M) was titrated with small aliquots of HCl or NaOH. (b) The change of molar extinction coefficient at 640 nm as a function of pH. The curve has been drawn to correspond to that theoretically expected with a pK_a of 7.3. From Lindskog and Nyman[261], with permission.

3.2.3 Metal-solvent interaction related to enzyme activity

Although Zn(II) and Co(II) ions are co-ordinated within approximate tetrahedral environments by nitrogen and oxygen donor ligands in CPA and carbonic anhydrase, marked differences in the visible absorption properties of the two Co(II)-substituted enzymes are observed. The basis of these differences, despite seemingly comparable ligand environments, appears not to have been explained satisfactorily. In view of the wide application of Co(II) substitution in this enzyme, it is of importance to define the structural and electronic origin of the spectral changes in order to study structure–function relationships of carbonic anhydrase catalytic activity[257]. The visible absorption spectrum of Co(II)HCAC is illustrated in figure 25. As observed first by Lindskog and Nyman[261] there is a pronounced increase in absorption intensity centred near 625 nm and 640 nm on increase in pH. The visible absorption spectrum of Co(II)CPA[203] is similar to that of Co(II)HCAC at pH 6 according to general band contour and positions of maximum absorption. Co(II)-substituted HCAC exhibits greater absorption intensity, however. An increase in alkalinity results only in a small increase in absorption near 625 nm with somewhat increased resolution of the bands centred near 555 and 530 nm for Co(II)-substituted CPA[269]. As pointed out earlier (section 3.1.2(1)) the visible absorption spectrum of Co(II)CPA (and also of Co(II)HCAC at pH 6) is similar according to general band contour to the absorption spectrum of bis(L-histidinato)-Co(II)[218] and other model tetrahedral complexes of Co(II) with imidazole derivatives[220]. Therefore, the origin of the pronounced change in absorption properties of Co(II)HCAC and the absence of comparable changes in Co(II)CPA are not readily apparent. Since tetrahedral bis(L-histidinato)Co(II), as shown by n.m.r. studies[219], is formed only under conditions of high alkalinity, and the Co(II) ion is then surrounded by four nitrogen donor ligands in a strongly alkaline medium (pH ≈ 12), the origin of the pH-dependent spectral changes of Co(II)HCAC must be looked for from within the active site of the molecule.

The X-ray diffraction studies[37,69,252] on HCAC are particularly pertinent to the spectra illustrated in figure 25 since data were collected on crystals of Co(II)- and Zn(II)-containing HCAC at pH 8.5[69]. These conditions correspond to titration-monitoring of the spectral changes of the Co(II) enzyme to at least 90 per cent completion. The absence of prominent difference electron-density peaks for the Co(II) enzyme at pH 8.5, suggestive of large configurational changes in the active-site cleft, the low pH absorption spectrum of Co(II)HCAC consistent with approximate tetrahedral symmetry and the four co-ordinate geometry of ligands for the Zn(II) ion as determined by X-ray studies, appear therefore to exclude the possibility of donation of a new ligand to the Co(II) ion from an amino-acid residue as the origin of the spectral change. The Co(II) ion is probably, therefore, not formally five-co-ordinate. Furthermore, the low sensitivity of the visible absorption spectrum of tetrahedral Co(II) to structural distortions[213-15] indicates that the spectral changes observed in Co(II)-substituted carbonic anhydrases[261,270] probably do not arise primarily from structural alterations of the metal-co-ordinating ligands on increase in pH. That the spectral changes are not caused by co-ordination of HCO_3^- or CO_2 to the metal ion has also been established by Lindskog[270].

Numerous hypotheses have been made that the active species of carbonic anhydrase contains a hydroxyl ion bound to the Zn(II) cation[255,271-3]. It is generally presumed on this basis that the ionisation of the Zn(II)-bound water molecule has a $pK_a \approx 7$ although the corresponding value for aquo Zn(II) complexes[274,275] is $\geqslant 9$. According to this hypothesis, it could be concluded that the ionisation of a water molecule bound to the Co(II) ion in the active-site cleft could occur with a similarly low pK_a, and be responsible for the spectral changes indicated in figure 25. The obvious accessibility of solvent molecules to the metal cation in CPA and the absence of comparable spectral changes in Co(II)CPA on increase in pH[269] makes the hypothesis of a hydroxyl ion bound to the Co(II) ion as the only species responsible for the spectral changes difficult to rationalise.

The spectral changes illustrated in figure 25 must be considered as reflecting an alteration in the effective site symmetry of the metal cation. Since the symmetry of the excited states associated with the transitions at 625 nm and 640 nm is not known, it is not possible to designate the orbital origin of the pH-dependent spectral changes with respect to the electronic configurations of the Co(II) ion and its co-ordinating ligands. The polarised single-crystal absorption spectra of bisacetobis(ethylenethiourea)Co(II)[276], however, suggest a possible origin of the spectral changes in view of the ordered solvent structure in the active site of HCAC. This complex crystallises with a distorted tetrahedral configuration of four ligands with normal metal–ligand bond lengths but with two oxygen atoms of the aceto groups approximately 290 pm distant from the Co(II) ion. While the long Co—O distance indicates only long-distance, weak bonding interactions, they have a significant effect on the Co(II) electronic spectrum, indicating that pseudo-tetrahedral symmetry obtains. Significantly, transitions induced by the distant carbonyl oxygen atoms are observed at comparable energy positions to those of Co(II)HCAC at high pH and reflect electron promotions either to low symmetry states[276] or to the spin–orbit coupled ^2G states[207]. Table 10 indicates that in HCAC there are at least three oxygen-containing groups as candidates which could participate in long-distance, weak interactions with the Co(II) ion under the assumption that similar stereochemical relationships obtain as for the Zn(II) enzyme: a H_2O molecule approximately 350 pm distant and possibly the O^γ of threonine-197 or another water molecule 500 pm distant from the Co(II) ion. Of these the closer water molecule is the most likely.

The conclusions that distant oxygen-containing groups perturb the electronic structure of the Co(II) ion implies that the spectrum of Co(II)HCAC observed at high pH reflects formation of the ordered structure of solvent molecules. Spectral studies[270,277] on the Co(II)-substituted enzyme in the presence of anionic and acetazolamide inhibitors in conjunction with crystallographic studies[37,252,278] support this conclusion. X-ray studies indicate that binding of acetazolamide and sulphonamide inhibitors displace solvent molecules in the active-site region. Increasing concentrations of acetazolamide under alkaline conditions as well as azide and cyanide anions abolish the spectral characteristics of the high pH Co(II) enzyme illustrated in figure 25. Lindskog[270] has pointed out, furthermore, that inhibitory anions such as I^-, Br^- and Cl^- shift the pK_a of the spectral change to higher values. ^{35}Cl n.m.r. studies[254] show that Cl^- binding to the Zn(II) ion occurs at pH ≈ 6 and is gradually abolished on increase in alkalinity, with complete loss of Zn–chloride interaction at pH 9. Difference Fourier studies of halide-inhibited HCAC crystals[69,252,278] demonstrate that the halide anions bind near the Zn(II) ion at low pH (≈ 7.2) and that the Zn–halide bond distance increases with increase in pH. Binding of halide anions near the Zn(II) ion and the O^γ of threonine-197[278] prevents the formation of the ordered solvent structure. Under these conditions the spectrum of the Co(II)-substituted enzyme is indicative of a tetrahedrally co-ordinated Co(II) ion[270,277] with loss of the prominent bands in the 618–640 nm region.

The spectral properties in the visible region of the Co(II)-substituted enzyme thus suggest that the ordered solvent structure in the active-site cleft exists in a pH-dependent equilibrium, presumably in exchange with unordered solvent molecules at the protein surface. The structural basis with regard to pH-induced alterations of amino-acid side chains in the active-site cleft responsible for inducing the ordered solvent structure is not evident. Indeed, since crystals of HCAC appear to be unstable below pH 7.2[69], it may be difficult to determine the nature of the pH-dependent structural alterations involved. These considerations do not imply, however, that no solvent molecule is co-ordinated to the metal ion at low pH. Indeed the low-pH spectrum of Co(II)HCAC is consistent with tetrahedrally co-ordinated Co(II).

The ordered solvent structure, as reflected by the visible absorption properties of the Co(II)-substituted enzyme, is necessary for catalytic function and can be related to the catalytic efficiency of various isoenzymes of carbonic anhydrase. The value of k_{cat} for CO_2 hydration by the bovine enzyme[279,280] increases with pH, indicating that a catalytic group necessary for activity has a $pK_a \approx 6.9$. Khalifah[249] has observed that while HCAC exhibits an increase in k_{cat} dependent on pH such that the apparent pK_a of the variation was close to 7.0, the value of k_{cat} for the B isoenzyme continued to increase with pH, and activity did not level off at high pH. Significantly, the pH of the associated spectral changes in the visible region for the Co(II)-substituted enzymes are 6.4 for the bovine enzyme[257], 7.2 for HCAC[261] and 8.1 for HCAB[281]. Whitney[282] has pointed out a parallel course of increase in (esterase) activity with increase in the formation of the species absorbing at 618–640 nm for Co(II)-substituted HCAB, and Lindskog[270] has concluded on the basis of inhibition studies that the active form of the enzyme in CO_2 hydration is to be associated with the spectral species having the characteristic double absorption peak in the 618–640 nm range. This correlation of solvent structure, spectral absorption properties and catalytic efficiency of isoenzymes of carbonic anhydrase suggests that the catalytic power of the enzyme is dependent on the presence of ordered water molecules in the active-site cleft.

It is evident that structural, spectroscopic and chemical studies have not determined the amino-acid residues responsible for stabilising the CO_2 molecule for hydration, or the HCO_3^- anion in the dehydration reaction. This still remains an important task in further investigations of this enzyme, as indicated by the nature of electronic rearrangements that occur during carbonic anhydrase activity. Edsall[245] has pointed out that a reason why the nonenzymatic reaction of CO_2 hydration is considerably slower than the enzymatically catalysed reaction is that the CO_2 molecule must undergo extensive electronic rearrangement in passing from a symmetrically linear molecule with C=O bond lengths of 116 pm to the triangular HCO_3^- anion with inequivalent C—O bonds and with different O–$\hat{C}$–O bond angles. Structural and spectroscopic studies have indicated the chemical nature of certain groups in the active-site region involved in enzymatic catalysis. The importance of determining the structural basis by which the substrate molecules CO_2 and HCO_3^- are stabilised and through which these electronic rearrangements are efficiently promoted remains a demanding task for future investigations.

4 CALCIUM-BINDING PROTEINS

4.1 LOW MOLECULAR WEIGHT CARP-MUSCLE ALBUMINS

Numerous low molecular weight albumins occurring in the white muscle of fish have been characterised according to biochemical properties[283,284]. While the physiological function of these proteins of approximately 11 000 molecular weight is not known, they manifest unusual characteristics in having amino-acid compositions of approximately 10 per cent phenylalanine and 20 per cent alanine content with little or no tryptophan, tyrosine, methionine, histidine, cysteine or arginine. In addition, they exhibit high calcium affinity. These two characteristics suggest that the carp-albumin protein may be analogous to the troponin-A protein of mammalian and avian muscle[285], and may be involved, therefore, in mediation of the effect of calcium in muscle contraction[286].

Crystallographic determination of the three-dimensional structure of one carp albumin has been carried out to 200 pm resolution[287,288]. The protein consists of a single polypeptide chain of 108 amino acids with four main regions consisting of α-helix†. These lie within the residues 40–51, 60–70, 79–88 and 99–107. From initial X-ray diffraction studies[287] a spherical region of high electron density centrally located approximately 250 pm from the carboxylate oxygen atoms of aspartate-94, aspartate-90, aspartate-92 and glutamate-101, and the carbonyl group of lysine-96 has been interpreted as a bound Ca(II) ion. More recent investigations[288,289] have revealed that a second binding site is located between the carboxylate groups of aspartate-51, aspartate-53, glutamate-62 and carbonyl groups of serine-55 and phenylalanine-57. The second binding site has not been characterised in sufficient detail at the time of writing to be included in discussion.

Williams[290] has pointed out that the high co-ordination number usually associated with Ca(II) and its ability to act as a 'bridge' between anions make it highly suitable for integration as a structural component in biological systems. The stereochemical basis of this structural characteristic is demonstrated on the basis of the co-ordination geometries of Ca(II) ions in simple, low molecular weight complexes. Ca(II) ions readily form complexes which are frequently characterised by distorted octahedral, pentagonal bipyramidal and even eight-co-ordinate environments with a large variation in bond angles and, especially, in Ca–ligand bond lengths[291]. In a model Ca–peptide complex[292] Ca–O bond distances ranging from 220 to 260 pm are observed. These structural characteristics of Ca(II) co-ordination explain the preference of Ca(II) ions within an environment of carbonyl and carboxylate oxygen donor ligands in carp albumin and indicate that Ca(II)

---

† The numbering of amino-acid residues has been modified from that in the first publication on the structure of carp albumin[287] according to a personal communication from Dr R. H. Kretsinger. On this basis the fourth α-helical region between residues 60 and 70 has also been included[288].

ions can be easily accommodated into irregular co-ordination environments. In view of the suggestion of Williams[290] of Ca(II) as a 'bridge', its role in binding to carp albumin may be structural stabilisation linking carboxylate groups of amino-acid side chains between α-helical regions of the protein.

Although crystallographic co-ordinate refinement of the structure of carp albumin has not been completed and the co-ordination geometries for both binding sites cannot yet be compared, the results of Kretsinger and coworkers[287,288] do establish the structural basis of Ca(II) co-ordination in this protein. On the basis of the stereochemical characteristics of low molecular weight complexes, Ca-ligand environments are expected to exhibit considerably distorted co-ordination geometries. The relative contribution of the ligands to charge neutralisation of the Ca(II) ion will depend consequently on Ca-ligand bond distances. While no direct functional role of Ca(II) in binding carp albumin can yet be assigned, the Ca(II) co-ordinating properties of this protein indicate that similar ligand environments may be expected in other enzymes and proteins. Urry[293] has pointed out that the co-ordination of Ca(II) ions in the structural proteins elastin and collagen probably occurs primarily through carbonyl oxygen atoms. Such oxygen-co-ordinated environments of Ca(II) ions may thus serve as important integral units of these proteins in providing rigidity and stability of structure.

4.2 STAPHYLOCOCCAL NUCLEASE

4.2.1 Ca(II) requirement for enzymatic activity

Staphylococcal nuclease is a single-chain polypeptide enzyme of 149 amino-acid residues, which exhibits hydrolytic activity towards both deoxyribonucleic acid (DNA) as well as ribonucleic acid (RNA) polymers[294]. Of particular interest is that enzymatic activity appears to be completely dependent on the presence of Ca(II) ions in comparable concentrations whether RNA or DNA is hydrolysed. Ca(II) is also required for binding of inhibitors to the enzyme active site[295]. Chemical studies with the use of synthetic substrates have also demonstrated that Ca(II) is required not only for binding of substrates, but also for the subsequent hydrolytic process. These deductions have been made on the basis of the inhibition of substrate hydrolysis by substituted metal cations[296]. Certain metal ions, however, do retain the ability to promote nucleotide binding without subsequent hydrolysis.

It is unusual in the case of staphylococcal nuclease that in solution only one mole of Ca(II) per mole of enzyme is bound in the presence of monophospho-nucleotides, but two moles are bound with the inhibitor thymidine-3′,5′-diphosphate (pdTp) despite the high Ca(II) concentrations necessary for optimal activity (as high as 0.01 M depending on pH)[296]. However, in contrast to solution studies, only one Ca(II) binding site has been identified by crystallographic studies of the ternary nuclease–Ca(II)–pdTp complex at 200 pm resolution[33,297]. This differ-ence, as pointed out by Cotton and Hazen[298], may reflect the different solvent conditions of crystallographic and chemical studies. It is presumed that the second

Ca(II) is bound between the phosphate groups of the ribose ring and that binding is prevented because of a nearby side chain of a neighbouring molecule[297]. No Ca(II) ion is bound to the enzyme in the absence of substrate[295,296].

4.2.2 Co-ordination site of Ca(II)

The role of Ca(II) in the enzyme mechanism of staphylococcal nuclease has been primarily considered as stabilising a conformation of the protein necessary for enzymatic action and binding of the substrate during phosphate hydrolysis[299]. Further analysis of the electron-density map of the ternary complex at high resolution[297] has suggested that the Ca(II) ion may interact with the 5′-phosphate group of the ribose ring during hydrolysis either through a co-ordinated water molecule or a bound hydroxide ion. Figure 26 illustrates the active-site region

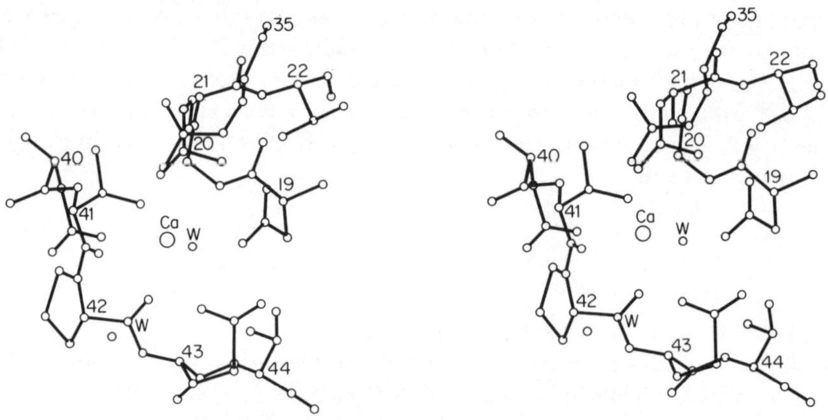

Figure 26 Stereoview of the structural array around the Ca(II) ion in staphylococcal nuclease. The carboxylate groups of aspartate-21, aspartate-40 and glutamate-43 are clearly co-ordinated to the Ca(II) ion with the carboxylate of aspartate-19 somewhat more distant from a roughly square array around the metal ion. The peptide carbonyl oxygen of threonine-41 is within the primary co-ordination sphere but in a position displaced from regular octahedral geometry. Residues are labelled by sequence number at the position of their α-carbon atoms. W indicates a water molecule (or hydroxide ion). From Cotton *et al.*[297], with permission.

within which the Ca(II) ion is co-ordinated. The Ca(II) ion on the basis of the interpretation of the electron-density map at 200 pm resolution appears to be six-co-ordinate with respect to a roughly planar array of oxygen donor atoms from the carboxylate groups of aspartate-19, -21 and -40 and glutamate-43[297,299]. The peptide carbonyl oxygen atom of threonine-41 appears to be within co-ordination distance of the central Ca(II) opposite the site indicated by the water molecule. A more complex secondary co-ordination sphere formed by an array of hydrogen-bonded threonine residues, water molecules and other structural elements[297] has not yet been completely interpreted on the basis of the 200 pm electron-density map.

As indicated first by Arnone *et al.*[33], the electron-density peaks in the region of the Ca(II) ion are difficult to interpret as a result of loss of some isomorphism by use of the Ba(II)-inhibited enzyme for phase determination. While a detailed stereochemical description of the Ca(II) co-ordination site will require further studies, it is evident that the Ca(II) site is characterised by a wide variation in Ca—O bond lengths and bond angles[299]. Smearing of the electron density in the Ca(II) regions has prevented an unambiguous determination of co-ordination of the metal ion by a water molecule (or hydroxide ion)[297]. While it appears that a water molecule may be co-ordinating the Ca(II) ion in a position opposite the carbonyl oxygen of threonine-41, a firmer interpretation of the presence of this ligand awaits further crystallographic studies.

4.2.3 Tryptic digests of nuclease with enzymatic activity

In the presence of Ca(II) and the inhibitor dpTp, digestion of the enzyme with trypsin results in limited cleavage to yield fragments, termed Nuclease-T, which are separately inactive but interact noncovalently to generate enzymatically active complexes[294]. The relative portions of the enzyme involved in such enzymatically productive complementation studies are illustrated in figure 27. The active

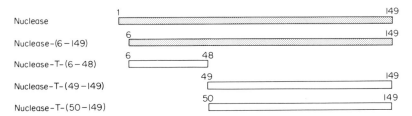

Figure 27 Schematic illustration of fragments of staphylococcal nuclease obtained by tryptic digestion. The combination of the fragment containing residues 6–48 with those with residues 49–149 and 50–149 results in a noncovalently bound complex having approximately 8 per cent of the activity of the native enzyme. Based on diagrams of Anfinsen *et al.*[294].

complexes are formed between Nuclease-T-(6–48) and Nuclease-T-(49–149) and between Nuclease-T(6–48) and Nuclease-T-(50–149). These complexes have optical rotatory dispersion and circular-dichroism properties similar to those of the native enzyme but separately have only random structure[300,301]. In addition, since the complexes can be crystallised in the identical space group of the native enzyme with similar unit-cell parameters[302], they are expected to retain consider-able structural similarity to the native enzyme.

Solid-phase peptide synthesis[303] has been employed[304,305] to make synthetic peptide analogues of the shorter Nuclease-T fragment, designed to give information about amino-acid side-chain requirements for binding. These studies were based

on the observation that the synthetically derived Nuclease-T-(6–47) fragment
combines with Nuclease-T-(48,49–150) to yield an active complex. Strict stereochemical requirements of cation binding necessary for substrate binding and
hydrolysis can be demonstrated on the basis of the combined results of X-ray
crystallographic studies[33] and the results obtained with synthetic peptide analogues
of Nuclease-T-(6–47), in which certain residues in the active site have been changed.
Figure 28 illustrates a schematic drawing of the binding relationships of the

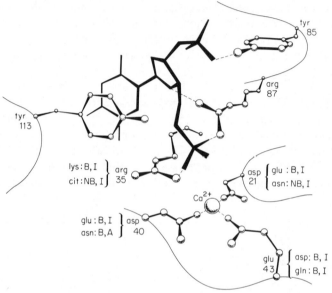

Figure 28 Correlation of active-site synthetic analogue results
for Nuclease-(49, 50–149) with the X-ray crystallographic
model for the binding-site region for nuclease. B indicates
binding of a synthetic (6–47) analogue to Nuclease-T (49, 50–149)
with noted change; NB, ineffective binding; A, ability of the
synthetic analogue to generate at least partial enzymatic
activity on addition to Nuclease-T; I, little or no ability to
generate such activity. From Chaiken and Anfinsen[305], with
permission.

nucleotide inhibitor dpTp to the amino-acid carboxylate side chains responsible
for binding Ca(II) in the active-site region of staphylococcal nuclease. Replacement of aspartate-40 by asparagine allows retention of nuclease activity, suggesting
that structural requirements for activity can be satisfied by co-ordination of the
Ca(II) with the carbonyl oxygen atom. Replacement of glutamate-40 by glutamine
or aspartic acid results in complete loss of enzymatic activity. Similarly, if
aspartate-21 is replaced by the shorter glutamic acid, no enzymatic activity is
observed. In all of the substitutions described, however, binding of the synthetic
analogue peptide of Nuclease-T-(6–47) to the larger tryptic fragment Nuclease-T-

(48,49–149) is promoted. These results clearly indicate the sensitivity of essential amino-acid residues in the metal binding-site region to small stereochemical perturbations. Other substitutions of amino-acid residues resulting in no binding of the synthetic analogue peptide to Nuclease-T are also indicated in figure 28.

4.2.4 The possible functional role of Ca(II) in nucleotide hydrolysis

Cotton *et al.*[297] have implicated a co-ordinated water molecule (or possibly a hydroxide ion in view of the pH range of 8–10 for optimal activity) to be involved in the catalytic mechanism of phosphate ester hydrolysis. This hypothesis receives some indirect support from chemical studies.

As evident in figure 28, an approach to the tetrahedral 5'-phosphate group by an oxygen atom of a water molecule bound to the Ca(II) could result in nucleophilic attack on the phosphate group[297]. The phosphate group is stabilised by the positively charged residues arginine-35 and arginine-87 as suggested by X-ray studies[33] and active-site analogue peptide-complementation studies[305]. This could promote polarisation of the 5'-phosphate group with resultant increased susceptibility for nucleophilic attack. Furthermore, cleavage of the 5'-C–O–P group occurs between oxygen and phosphorus[306] making nucleophilic attack a structurally favourable situation.

This situation may be analogous to the mechanism of action of ribonuclease. It has been concluded[307] on the basis of synthetic transition-state substrate analogues that the action of ribonuclease in the hydrolysis of a ribosyl phosphate ester involves a substrate transition-state species in which the phosphate group resembles a pentacovalent trigonal bipyramid. The entering and leaving groups are axial. A transition-state species of comparable phosphate bipyramidal stereochemistry can be readily imagined with the help of figure 28 for staphylococcal nuclease action and could probably obtain in nucleophilic attack by the oxygen donor ligand co-ordinating the Ca(II) ion.

4.2.5 Metal-substitution studies

Table 13 gives a summary of nuclease enzyme activity with RNA and DNA as substrates in the presence of various metal ions. This comparison shows that no metal ion can be effectively substituted for Ca(II). That the metal ions were bound by the enzyme was indicated by the release of their inhibitory effects by addition of Ca(II) to the assay mixture. This inhibition was most marked for Zn(II), Hg(II) and Cd(II). Furthermore, enhanced inhibition of nuclease activity did not result from preincubation of the enzyme with the appropriate metal salts[296]. These results indicate that the inhibitory effects of other metal cations resulted from binding of metals at or near the site involved in Ca(II) co-ordination and not as a result of nonspecific binding to other regions of the protein.

Of especial interest in comparing the effects of metal-cation substitution is the effect of Sr(II) and Ba(II) on staphylococcal nuclease. Like Ca(II), both cations favour eight-co-ordinate structures and oxygen donor ligands[290]. X-ray crystallographic studies[33] of the Ba(II)-inhibited enzyme indicate that the Ba(II) ion is

displaced approximately 75 pm from the site occupied by Ca(II). It is probable
that the Sr(II) ion is somewhat less displaced from the Ca(II) site because of its
smaller ionic radius (table 3). These results thus suggest that stereochemical
requirements of metal-cation co-ordination for RNA hydrolysis must differ slightly
from those for DNA hydrolysis since Sr(II) promotes DNA but not RNA hydrolysis.
The structural basis for the diminished DNA hydrolytic activity in the presence of
Sr(II) and its absence in the presence of Ba(II) cannot be straightforwardly deduced
on the basis of crystallographic studies. It is possible that geometrical distortions
of the Ca(II) binding site dependent on ionic radius are transmitted to the nucleotide
binding regions. Such effects could result in poorer steric fit of the substrate in the
active site, resulting in diminished enzymatic activity. Moreover, the effect of an
increased ionic radius on the possible catalytic role of a water molecule, serving
as a bridge between the metal ion and the phosphorus atom in the ribosyl
5'-position, must also be considered.

TABLE 13 COMPARISON OF EFFECT OF SUBSTITUTED
METAL IONS ON STAPHYLOCOCCAL NUCLEASE
ACTIVITY[†]

Metal ion	Conc. (mM)	% Activity of Ca(II)[‡] substrate	
		DNA	RNA
Fe(II)	0.1	6	0
Cu(II)	0.1	16	0
Sr(II)	10	40	0
	1	60	0
	0.1	14	0

† from reference 296.
‡ no activity observed with Mg(II), Mn(II), Ni(II), Zn(II),
  Cd(II), Ba(II), Hg(II), La(III), Y(III) or Eu(III).

Numerous studies[290,308-10] of magnetic resonance and absorption-spectroscopic
properties of cations of the lanthanide series have demonstrated that these metal
ions closely mimic the properties of Ca(II) in biological systems because of comparable ionic radii, weak ligand-field stabilisation energies and high electropositivities.
On the basis of these properties it would be expected, as for Ca(II), that the
lanthanide metals would tend to form complexes primarily with oxygen-donor
ligands. This conclusion receives support from numerous crystallographic
studies[26,311-13], in which lanthanide ions utilised to form isomorphous heavy-metal
derivatives are bound by carboxylate groups.

It is perhaps somewhat surprising therefore that staphylococcal nuclease exhibits
no hydrolytic activity in the presence of La(III), Y(III) or Eu(III)[296]. However, a
comparative study[309] of the effect of lanthanide cations on the activation of
α-amylase, a Ca(II) metalloenzyme, has clearly shown that enzyme systems discriminate on the basis of charge and ionic radius and that not all lanthanide cations

are equally effective as substitutes for Ca(II). As pointed out by Smolka *et al.*[309], a direct comparison of ionic radii of the lanthanide elements to that of Ca(II) is made difficult because of the different crystalline salts and oxides utilised to determine metal-cation radii by X-ray diffraction methods. Furthermore, the EDTA complexes of lanthanide-ions[314,315], which may be considered analogous to a polyfunctional binding site on a protein, indicate that co-ordination numbers as high as ten are characteristic. The increased number of ligand-binding sites as compared to those of most Ca(II) complexes are generally occupied by solvent molecules. This factor may be of importance in the structural requirements of lanthanide binding to staphylococcal nuclease.

Nd(III) is a good Ca(II) substitute in activating α-amylase[309] and in the conversion of trypsinogen to trypsin[308]. Preliminary studies[316,317] show that the specific activity of staphylococcal nuclease with thymidine-3'-phosphate-5'-(*p*-nitrophenylphosphate) as substrate is approximately one and one-half-fold higher in the presence of Nd(III) than in the presence of Ca(II). Binding of the Nd(III) ion, however, appears on the basis of n.m.r. studies to occur near histidine-46 as well as with a glutamate residue[316,318]. A detailed analysis of the stereochemical relationships of the lanthanide ions to nearby amino-acid residues is necessary to determine differences in structural requirements for binding of Ca(II) and lanthanide ions to staphylococcal nuclease. In contrast to the case of substitution of lanthanide ions for Ca(II) in thermolysin[311], attempts to obtain lanthanide-substituted staphylococcal nuclease in crystalline form have not thus far been successful[299]. These results thus suggest that the structural requirements of metal-ion binding by staphylococcal nuclease are exquisitely sensitive to ionic radius, charge, resultant electropositivity and co-ordination requirements. These requirements appear to be satisfied best only by Ca(II) for effective enzymatic activity. More detailed chemical studies in conjunction with further crystallographic analysis are clearly still necessary to determine the mechanistic importance of the co-ordination properties of the bound metal cation.

5 SUMMARY

This review has emphasised the importance of precise stereochemical data in the determination of the structural basis of metal-ion function and reactivity in proteins and enzymes. The resolving power of protein X-ray crystallography has been discussed in general to indicate the origins of the approximate stereochemical data characteristic of protein structures at present, in comparison to the results of small-molecule X-ray crystallography. Possible directions through which improved stereochemical detail of proteins may be obtained in the future have been briefly discussed, especially with respect to their metal–ligand co-ordination centres.

A comparison of stereochemical origins of three categories of metal–protein interaction has been made on the basis of high-resolution X-ray diffraction studies of a number of proteins and enzymes. With reference to metal–ligand co-ordination centres in proteins, detailed correlations of molecular structure, stereochemistry and electronic structure in assessing the biological role of metal-ion function have been made to outline those structural and electronic factors responsible for control of metal-ion reactivity. The interaction of the porphyrin ring of haemoglobin and myoglobin with the nearby amino-acid environment through hydrophobic contacts is examined with regard to the results of magnetic susceptibility, paramagnetic resonance and polarised single-crystal absorption spectroscopic investigations. The interaction of substituted metal ions in carboxypeptidase A with the carbonyl group of substrates in peptide hydrolysis is discussed with reference to their d orbital electron configuration and co-ordination geometry, and the formation of an ordered-solvent structure near the substituted Co(II) ion catalytically active in carbonic anhydrase is suggested to be responsible for the pH-dependent spectral perturbations observed in the electronic absorption spectrum. Correlations of molecular structure, as determined by X-ray diffraction methods, with the electronic structure of the metal–ligand co-ordination centre, as assessed from spectroscopic investigations, thus illustrate the origins of structural control of metal-ion reactivity in proteins and enzymes.

Several problems pertinent to each metal-ion-requiring protein or enzyme have been outlined which, while still relatively poorly understood, are none the less important in understanding the chemical and structural basis of metal function. Some aspects of the theoretical considerations, spectroscopic investigations and high-precision structural data necessary to bring further insight into these problems have been outlined. These problems may be of particular interest to theoretical, co-ordination and structural chemists alike in the investigation of biological roles of metal-ion function.

REFERENCES

1. Dixon, M., and Webb, E. C., *Enzymes*, Longmans, London (2nd edn 1964).
2. Williams, R. J. P., *Biol. Rev.*, **28** (1953), 381.
3. Malmström, B. G., and Rosenberg, A., *Adv. Enzymol.*, **21** (1959), 131.
4. Vallee, B. L., and Wacker, W. E. C., in *The Proteins* (ed. H. Neurath), vol. 5, Academic Press, New York (1970).
5. Malkin, R., and Malmström, B. G., *Adv. Enzymol.*, **33** (1970), 177.
6. Bray, R. C., and Swann, J. C., *Struct. Bond.*, **11** (1972), 107.
7. Wood, J. M., and Brown, D. G., *Struct. Bond.*, **11** (1972), 47.
8. Mildvan, A. S., in *The Enzymes* (ed. P. D. Boyer), vol. 2, Academic Press, New York (1970), p. 445.
9. Mildvan, A. S., and Cohn, M., *Adv. Enzymol.*, **33** (1970), 1.
10. Kendrew, J. C., Dickerson, R. E., Strandberg, B. E., Hart, R. G., Davies, D. R., Phillips, D. C., and Shore, V. C., *Nature, Lond.*, **185** (1960), 422.
11. Watson, H. C., *Prog. Stereochem.*, **4** (1968), 299.
12. Scouloudi, H., *J. molec. Biol.*, **40** (1969), 353.
13. Lattman, E. E., Nockolds, C. E., Kretsinger, R. H., and Love, W. E., *J. molec. Biol.*, **60** (1972), 271.
14. Huber, R., Epp, O., and Formanek, H., *Naturwissenschaften*, **56** (1969), 362.
15. Hendrickson, W. A., and Love, W. E., *Nature, Lond.*, **232** (1971), 197.
16. Perutz, M. F., Muirhead, H., Cox, J. M., and Goaman, L. G. C., *Nature, Lond.*, **219** (1968), 131.
17. Bolton, W., and Perutz, M. F., *Nature, Lond.*, **228** (1970), 551.
18. Muirhead, H., and Greer, J., *Nature, Lond.*, **228** (1970), 516.
19. Dickerson, R. E., Takano, T., Eisenberg, D., Kallai, O. B., Samson, L., Cooper, A., and Margoliash, E., *J. biol. Chem.*, **246** (1971), 1511.
20. Takano, T., Swanson, R., Kallai, O. B., and Dickerson, R. E., *Cold Spring Harb. Symp. quant. Biol.*, **36** (1971), 397.
21. Mathews, F. S., Argos, P., and Levine, M., *Cold Spring Harb. Symp. quant. Biol.*, **36** (1971), 387.
22. Kraut, J., Singh, S., and Alden, R. A., in *Structure and Function of Cytochromes* (eds K. Okunuki, M. D. Kamen and I. Sekuzu), Univ. of Toko Press, Tokyo (1968). p. 252.
23. Larsson, L. O., Hagman, L.-O., Kierkegaard, P., and Yonetani, T., *J. biol. Chem.*, **245** (1970), 902.
24. Hagman, L.-O., Larsson, L. O., and Kierkegaard, P., in *Probes of Structure and Function of Macromolecules and Membranes*, vol. 2 (*Probes of Enzymes and Hemoproteins*), (eds B. Chance, T. Yonetani and A. S. Mildvan), Academic Press, New York (1971), p. 519.
25. Longley, W., *J. molec. Biol.*, **30** (1967), 323.
26. Sieker, L. C., Adman, E., and Jensen, L. H., *Nature, Lond.*, **235** (1972), 40.
27. Carter, C. W., Jr, Freer, S. T., Xuong, Ng H., Alden, R. A., and Kraut, J., *Cold Spring Harb. Symp. quant. Biol.*, **36** (1971), 381.
28. Watenpaugh, K. D., Sieker, L. C., Herriott, J. R., and Jensen, L. H., *Cold Spring Harb. Symp. quant. Biol.*, **36** (1971), 359.
29. Lipscomb, W. N., Hartsuck, J. A., Reeke, G. N., Jr, Quiocho, F. A., Bethge, P. H., Ludwig, M. L., Steitz, T. A., Muirhead, H., and Coppola, J. C., *Brookhaven Symp. Biol.*, **21** (1968), 24.
30. Matthews, B. W., Jansonius, J. N., Colman, P. M., Schoenborn, B. P., and Duporque, D., *Nature, new Biol.*, **238** (1972), 37.
31. Matthews, B. W., Colman, P. M., Jansonius, J. N., Titani, K., Walsh, K. A., and Neurath, H., *Nature, new Biol.*, **238** (1972), 41.
32. Swan, I. D. A., *J. molec. Biol.*, **60** (1971), 405.

33. Arnone, A., Bier, C. J., Cotton, F. A., Day, V. W., Hazen, E. E., Jr, Richardson, D. C., Richardson, J. S., and Yonath, A., *J. biol. Chem.*, **246** (1971), 2302.
34. Hanson, A. W., Applebury, M. L., Coleman, J. E., and Wyckoff, H. W., *J. biol. Chem.*, **246** (1971), 2302.
35. Bränden, C.-I., Zeppezauer, E., Söderberg, B.-O., Boiwe, T., Nordström, B., Söderlund, G., Zeppezauer, M., Werner, P.-E., and Äkeson, A., in *Wenner-Gren Symposium on Structure and Function of Oxidation-Reduction Enzymes* (eds A. Äkeson and A. Ehrenberg), Pergamon, Oxford (1972), p. 93.
36. Wiley, D. C., Evans, D. R., Warren, S. G., McMurray, C. H., Edwards, B. F. P., Franks, W. A., and Lipscomb, W. N., *Cold Spring Harb. Symp. quant. Biol.*, **36** (1971), 285.
37. Kannan, K. K., Liljas, A., Waara, I., Bergstén, P.-C., Lövgren, S., Strandberg, B., Bengtsson, U., Carlbom, U., Fridborg, K., Järup, L., and Petef, M., *Cold Spring Harb. Symp. quant. Biol.*, **36** (1971), 221.
38. Kannan, K. K., Fridborg, K., Bergstén, P.-C., Liljas, A., Lövgren, S., Petef, M., Strandberg, B., Waara, I., Adler, L., Falkbring, S. O., Göthe, P. O., and Nyman, P. O., *J. molec. Biol.*, **63** (1972), 601.
39. Blundell, T. L., Cutfield, J. F., Dodson, E. J., Dodson, G. G., Hodgkin, D. C., and Mercola, D. A., *Cold Spring Harb. Symp. quant. Biol.*, **36** (1971), 233.
40. Blundell, T. L., Dodson, G. G., Hodgkin, D. C., and Mercola, D. A., *Adv. Protein Chem.*, **26** (1973), 279.
41. Hodgkin, D. C., and Mercola, D. A., personal communication.
42. Eisenberg, D., Heidner, E. G., Goodkin, P., Dastoor, M. N., Weber, B. H., Wedler, F. and Bell, J. D., *Cold Spring Harb. Symp. quant. Biol.*, **36** (1971), 291.
43. Blake, C. C. F., Evans, P. R., and Scopes, R. K., *Nature, new Biol.*, **235** (1972), 195.
44. Hardman, K. D., Wood, M. K., Schiffer, M., Edmundson, A. B., and Ainsworth, C. F., *Cold Spring Harb. Symp. quant. Biol.*, **36** (1971), 271.
45. Reeke, G. N., Jr, Becker, J. W., and Quiocho, F. A., *Cold Spring Harb. Symp. quant. Biol.*, **36** (1971), 277.
46. Woolfson, M. M., *An Introduction to X-ray Crystallography*, University Press, Cambridge (1970).
47. Lipson, H., and Cochran, W., *The Determination of Crystal Structures*, Bell, London (1966).
48. Phillips, D. C., in *Advances in Research by Diffraction Methods* (eds R. Brill and R. Mason), vol. II (1966), p. 75.
49. North, A. C. T., and Phillips, D. C., *Prog. Biophys. molec. Biol.*, **19** (1969), 5.
50. Blundell, T. L., and Johnson, L. N., in *M.T.P. Intern. Rev. Sci., Physical Chemistry, Series 1*, vol. 11, *(Chemical Crystallography)*, (ed. J. M. Robertson), Butterworths, London (1972), 199.
51. Blake, C. C. F., *Adv. Protein Chem.*, **23** (1967), 59.
52. Eisenberg, D., in *The Enzymes* (ed. P. Boyer), Vol. 1, Academic Press, New York (1970), p. 1.
53. Davies, D. R., and Segal, D. M., *Meth. Enzym.*, **22** (1971), 266.
54. Green, D. W., Ingram, V. M., and Perutz, M. F., *Proc. R. Soc.*, **A225** (1954), 287.
55. Blow, D. M., and Crick, F. H. C., *Acta crystallogr.*, **12** (1959), 794.
56. Srinivasan, R., *Proc. Indian Acad. Sci.*, **A53** (1961), 252.
57. James, R. W., *Acta crystallogr.*, **1** (1948), 132.
58. Pauling, L., *The Nature of the Chemical Bond*, Cornell University Press, Ithaca, N.Y. (3rd edn 1960).
59. Crick, F. H. C., and Magdoff, B. S., *Acta crystallogr.*, **9** (1956), 901.
60. Richards, F. M., *J. molec. Biol.*, **37** (1968), 225.
61. Diamond, R., *Acta crystallogr.*, **A27** (1971) 436.
62. Leavitt, M., and Lifson, S., *J. molec. Biol.*, **46** (1969), 269.
63. Crick, F. H. C., and Kendrew, J. C., *Adv. Protein Chem.*, **12** (1957), 134.
64. Matthews, B. W., *J. molec. Biol.*, **33** (1968), 491.
65. Dickerson, R. E., Kendrew, J. C., and Strandberg, E., *Acta crystallogr.*, **14** (1961), 1188.
66. Banaszak, L. J., Watson, H. C., and Kendrew, J. C., *J. molec. Biol.*, **12** (1965), 130.
67. Hartsuck, J. A., and Lipscomb, W. N., in *The Enzymes* (ed. P. D. Boyer), Vol. 3, Academic Press, New York (1970), p. 1.

68. Fridborg, K., Kannan, K. K., Liljas, A., Lundin, J., Strandberg, B., Strandberg, R., Tilander, B., and Wiren, G., *J. molec. Biol.*, **25** (1967), 505.

69. Liljas, A., Dissertation, University of Uppsala, *Acta Univ. Uppsal.*, Weilands Tryckeri, Uppsala, Sweden (1971).

70. Stryer, L., Kendrew, J. C., and Watson, H. C., *J. molec. Biol.*, **8** (1964), 96.

71. Henderson, R., and Moffat, J. K., *Acta crystallogr.*, **B27** (1971), 1414.

72. Hendrickson, W. A., Love, W. E., and Murray, G. C., *J. molec. Biol.*, **33** (1968), 829.

73. Banner, D. W., Bloomer, A. C., Petsko, G. A., Phillips, D. C., and Pogson, C. I., *Cold Spring Harb. Symp. Quant. Biol.*, **36** (1971), 151.

74. Kendrew, J. C., *Science, N.Y.*, **139** (1963), 1259.

75. Watenpaugh, K. D., Sieker, L. C., Herriott, J. R., and Jensen, L. H., *Acta crystallogr.*, **B29** (1972), 943.

76. Herriott, J. R., Sieker, L. C., Jensen, L. H., and Lovenberg, W., *J. molec. Biol.*, **50** (1970), 391.

77. Freeman, H. C., *Adv. Protein Chem.*, **22** (1967), 257.

77b. McAuliffe, C. A., Quagliano, J. V. and Vallarino, L. M. *Inorg. Chem.* **5** (1966) 1996.

78. Weast, R. C., *Handbook of Chemistry and Physics*, Chemical Rubber Co., Cleveland, Ohio, (50th edn, 1969), p. F152.

79. Templeton, D. H., and Dauben, C. H., *J. Am. Chem. Soc.*, **76** (1954), 5237.

80. Willstadter, E., Hamor, T. A., and Hoard, J. L., *J. Am. Chem. Soc.*, **85** (1963), 1205.

81. Bokii, G. B., Malinovskii, T. I., and Ablov, A. V., *Kristallografiya*, **1** (1956), 49.

82. Bowman, K., Gaughan, A. P., and Dori, Z., *J. Am. Chem. Soc.*, **74** (1972), 727.

83. Prout, C. K., and Wiseman, T. J., *J. chem. Soc.* (1964), 497.

84. Coleman, J. E., and Vallee, B. L., *J. biol. Chem.*, **235** (1960), 390.

85. Coleman, J. E., and Vallee, B. L., *J. biol. Chem.*, **236** (1961), 2244.

86. Gillespie, R. J., and Nyholm, R. S., *Q. Rev. chem. Soc.*, **11** (1957), 359.

87. Caughey, W. S., *A. Rev. Biochem.*, **36** (1967), 611.

88. Schoenborn, B. P., *Cold Spring Harb. Symp. quant. Biol.*, **36** (1971), 569.

89. Perutz, M. F., and TenEyck, L. F., *Cold Spring Harb. Symp. quant. Biol.*, **36** (1971), 295.

90. Matthews, F. S., Levine, M., and Argos, P., *J. molec. Biol.*, **64** (1972), 449.

91. Nobbs, C. L., Watson, H. C., and Kendrew, J. C., *Nature, Lond.*, **209** (1966), 339.

92. Watson, H. C., and Chance, B., in *Hemes and Hemoproteins* (eds B. Chance, R. W. Estabrook and T. Yonetani), Academic Press, New York (1966), p. 149.

93. Bretscher, P., Ph. D. Dissertation, University of Cambridge (1968).

94. Perutz, M. F., and Mathews, F. S., *J. molec. Biol.*, **21** (1966), 199.

95. Padlan, E., Ph. D. Dissertation, The Johns Hopkins University, Baltimore (1968).

96. Huber, R., Epp, O., and Formanek, H., *J. molec. Biol.*, **52** (1970), 349.

97. Padlan, E., and Love, W. E., in *Probes of Structure and Function of Macromolecules and Membranes*, vol. 2 (*Probes of Enzymes and Hemoproteins*), (eds B. Chance, T. Yonetani and A. S. Mildvan), Academic Press, New York (1971), p. 187.

98. Antonini, E., and Brunori, M., *Hemoglobin and Myoglobin in Their Reactions with Ligands*, North Holland, Amsterdam (1971).

99. Perutz, M. F., *Proc. R. Soc.*, **B173** (1969), 113.

100. Hill, A. V., *J. Physiol., Lond.*, **40** (1910), iv.

101. Muirhead, H., and Perutz, M. F., *Nature, Lond.*, **199** (1963), 633.

102. Muirhead, H., Cox, J. M., Mazzarella, L., and Perutz, M. F., *J. molec. Biol.*, **28** (1967), 177.

103. Perutz, M. F., *Nature, Lond.*, **228** (1970), 726.

104. Hoard, J. L., in *Hemes and Hemoproteins* (eds B. Chance, R. W. Estabrook and T. Yonetani), Academic Press, New York (1966), p. 9.

105. Pauling, L. and Coryell, C. D., *Proc. natn. Acad. Sci. U.S.A.*, **22** (1936), 210.

106. Perutz, M. F., Rossmann, M. G., Cullis, A. F., Muirhead, H., Will, G., and North, A. C. T., *Nature, Lond.*, **185** (1960), 416.

107. Coryell, C. D., Stitt, F., and Pauling, L., *J. Am. chem. Soc.*, **59** (1937), 633.

108. Theorell, H., and Ehrenberg, A., *Acta chem. scand.*, **5** (1951), 823.

109. Williams, R. J. P., *Chem. Rev.*, **56** (1956), 299.

110. Scheler, W., Schoffa, G., and Jung, F., *Biochem. Z.*, **329** (1957), 232.

111. George, P., Beetlestone, J., and Griffith, J. S., in *Haematin Enzymes* (eds J. E. Falk, R. Lemberg and R. K. Morton), Pergamon, Oxford (1961), p. 105.

112. Beetlestone, J., and George, P., *Biochemistry*, 3 (1964), 707.
113. Perutz, M. F., *J. Cryst. Growth*, 2 (1968), 54.
114. Ballhausen, C. J., *Introduction to Ligand Field Theory*, McGraw-Hill, New York (1962).
115. Hoard, J. L., in *Structural Chemistry and Molecular Biology* (eds A. Rich and N. Davidson), Freeman, San Francisco (1968), p. 573.
116. Hoard, J. L., *Science, N.Y.*, 174 (1971), 1295.
117. Hoard, J. L., *Ann. N.Y. Acad. Sci.*, 206 (1973), 3.
118. Collins, D. M., Countryman, R., and Hoard, J. L., *J. Am. chem. Soc.*, 94 (1972), 2066.
119. Radonovich, L. J., Bloom, A., and Hoard, J. L., *J. Am. chem. Soc.*, 94 (1972), 2073.
120. Hoard, J. L., Hamor, M. J., Hamor, A., and Caughey, W. S., *J. Am. chem. Soc.*, 87 (1965), 2312.
121. Koenig, D. F., *Acta crystallogr.*, 18 (1965), 663.
122. Hoard, J. L., Cohen, G. H., and Glick, M. D., *J. Am. chem. Soc.*, 89 (1967), 1992.
123. Crute, M. B., *Acta crystallogr.*, 12 (1959), 24.
124. Koenig, D. F., Ph.D. Dissertation, The Johns Hopkins University, Baltimore (1962).
125. Nobbs, C. L., in *Hemes and Hemoproteins* (eds B. Chance, R. W. Estabrook and T. Yonetani), Academic Press, New York (1966), p. 143.
126. Simon, S. R., Konigsberg, W. H., Bolton, W., and Perutz, M. F., *J. molec. Biol.*, 28 (1967), 451.
127. Williams, R. J. P., *Fedn Proc. Fedn Am. Socs. exp. Biol.* (20 Suppl.) 10 (1961), 5.
128. Williams, R. J. P., *Cold Spring Harb. Symp. quant. Biol.*, 36 (1971), 53.
129. Wiekliem, H. A., and Hoard, J. L., *J. Am. chem. Soc.*, 91 (1969), 549.
130. Hoard, J. L., Kennard, C. H. L., and Smith, G. S., *Inorg. Chem.*, 2 (1963), 131.
131. Lind, M. D., Hamor, M. J., Hamor, T. A., and Hoard, J. L., *Inorg. Chem.*, 3 (1964), 34.
132. Cohen, G. H., and Hoard, J. L., *J. Am. chem. Soc.*, 88 (1966), 3228.
133. Makinen, M. W., and Eaton, W. A., *162nd Annual Meeting Am. chem. Soc.,* Washington, D.C., September 13–17, Abstract No. 39 (1971).
134. Makinen, M. W., and Eaton, W. A., *Ann. N.Y. Acad. Sci.*, (1972).
135. Makinen, M. W. and Eaton, W. A., *Nature, Lond.*, 247 (1974), 62.
136. Makinen, M. W., and Eaton, W. A., manuscript in prepartion.
137. Eaton, W. A., and Hochstrasser, R. M., *J. chem. Phys.*, 46 (1967), 2533.
138. Eaton, W. A., and Hochstrasser, R. M., *J. chem. Phys.*, 49 (1968), 985.
139. Anderson, L., *J. molec. Biol.* 79 (1973), 495.
140. Boyes-Watson, J., Davidson, E., and Perutz, M. F., *Proc. R. Soc.*, A191 (1947), 83.
141. Makinen, M. W., unpublished observations.
142. Watson, H. C., and Nobbs, C. L., in *Biochemie des Sauerstoffs* (eds B. Hess and H. Staudinger), Springer, Berlin (1968), p. 37.
143. Antonini, E., and Brunori, M., *A. Rev. Biochem.*, 39 (1970), 977.
144. Ingram, D. J. E., Gibson, J. F., and Perutz, M. F., *Nature, Lond.*, 178 (1956), 906.
145. George, P., and Hanania, G. I., *Biochem. J.*, 53 (1953), 236.
146. Cullis, A. F., Muirhead, H., Perutz, M. F., Rossman, M. G., and North, A. C. T., *Proc. R. Soc.*, A265 (1962), 161.
147. Perutz, M. F., *J. molec. Biol.*, 13 (1965), 646.
148. Kendrew, J. C., *Brookhaven Symp. Biol.*, 15 (1962), 216.
149. Bradshaw, R. A., and Gurd, F. R. N., *J. biol. Chem.*, 244 (1969), 2167.
150. Yonetani, T., Iizuka, T., Asakura, T., Otsuka, J., and Kotani, M., *J. biol. Chem.*, 247 (1972), 863.
151. Irving, H., and Williams, R. J. P., *J. chem. Soc.* (1953), 3182.
152. Goodwin, H. A., and Sylva, R. N., *Aust. J. Chem.*, 21 (1968), 83.
153. Iizuka, T., and Kotani, M., *Biochim. biophys. Acta*, 154 (1968), 417.
154. Iizuka, T., and Kotani, M., *Biochim. biophys. Acta*, 194 (1969), 351.
155. Iizuka, T., and Kotani, M., *Biochim. biophys. Acta*, 181 (1969), 275.
156. Iizuka, T., Kotani, M., and Yonetani, T., *J. biol. Chem.*, 246 (1971), 4731.
157. Otsuka, J., *Biochim. biophys. Acta*, 214 (1970), 233.
158. Griffith, J. S., in *Molecular Biophysics* (eds B. Pullman and M. Weissbluth), Academic Press, New York (1965), p. 191.
159. Bennett, J. E., Gibson, J. F., and Ingram, D. J. E., *Proc. R. Soc.*, A240 (1957), 67.
160. Eisenberg, P., and Pershan, P. S., *J. chem. Phys.*, 45 (1966), 2832.

161. Peisach, J., Blumberg, W. E., Ogawa, S., Rachmilewitz, E. A., and Oltzik, R., *J. biol. Chem.*, **246** (1971), 3342.
162. Peisach, J., and Blumberg, W. E., in *Probes of Structure and Function of Macromolecules and Membranes*, vol. 2 (*Probes of Enzymes and Hemoproteins*), (eds B. Chance, T. Yonetani and A. S. Mildvan), Academic Press, New York (1971), p. 231.
163. Blumberg, W. E., in *Magnetic Resonance in Biological Systems* (eds A. Ehrenberg, B. G. Malmström and T. Vänngard), Pergamon, Oxford (1967), p. 119.
164. Blumberg, W. E., Peisach, J. Wittenberg, A., and Wittenberg, J. B., *J. biol. Chem.*, **243** (1968), 1854.
165. Martin, R. L., and White, A. H., in *Transit. Metal Chem.*, **4** (1968), 113.
166. Griffith, J. S., *Nature, Lond.*, **180** (1957), 30.
167. Gibson, J. F., and Ingram, D. J. E., *Nature, Lond.*, **180** (1957), 29.
168. Helcké, G. A., Ingram, D. J. E., and Slade, E. F., *Proc. R. Soc.*, **B169** (1968), 275.
169. Kotani, M., *Ann. N. Y. Acad. Sci.*, **158** (1969), 20.
170. Dickinson, L. C., and Chien, J. C., *J. Am. chem. Soc.*, **93** (1971), 5036.
171. Kon, H., *J. biol. Chem.*, **243** (1968), 4350.
172. Gibson, Q. H., and Roughton, F. J. W., *Proc. R. Soc.*, **B163** (1966), 197.
173. Pauling, L., in *Hemoglobin* (eds F. J. W. Roughton and J. C. Kendrew), Butterworths, London (1949), p. 57.
174. Pauling, L., *Nature, Lond.*, **203** (1964), 182.
175. Griffith, J. S., *Proc. R. Soc.*, **A235** (1956), 23.
176. Yamazaki, I., Yokota, K., and Shikama, K., *J. biol. Chem.*, **239** (1964), 4151.
177. Huber, R., Epp, O., Steigemann, W., and Formanek, H., *Eur. J. Biochem.*, **19** (1971), 42.
178. Tentori, L., Vivaldi, G., Carta, S., Marinucci, M., Massa, A., Antonini, E., and Brunori, M., *FEBS Lett.*, **12** (1971), 181.
179. Weiss, J. J., *Nature, Lond.*, **202** (1964), 83.
180. Wittenberg, J. B., Wittenberg, B. A., Peisach, J., and Blumberg, W. E., *Proc. Natn. Acad. Sci. U.S.A.*, **67** (1970), 1846.
181. Rodley, G. A., and Robinson, W. T., *Nature, Lond.*, **235** (1972), 438.
182. LaPlaca, S. J., and Ibers, J. A., *J. Am. chem. Soc.*, **87** (1965), 2581.
183. Vaska, L., *Science, N.Y.*, **140** (1963), 809.
184. Wang, J. H., in *Haematin Enzymes* (eds J. E. Falk, R. Lemberg and R. K. Morton), Pergamon, Oxford (1961), p. 98.
185. Wüthrich, K., Keller, R. M., Brunori, M., Giacometti, G., Huber, R., and Formanek, H., *FEBS Lett.*, **21** (1972), 63.
186. George, P., and Stratmann, C. J., *Biochem. J.*, **57** (1964), 568.
187. Neurath, H., in *The Enzymes* (eds P. D. Boyer, H. Lardy and K. Myrbäck), vol. 4, Academic Press, New York (2nd edn 1960), p. 11.
188. Lipscomb, W. N., Reeke, G. N., Jr, Hartsuck, J. A., Quiocho, F. A., and Bethge, P. H., *Phil. Trans R. Soc.*, **B257** (1970), 177.
189. Quiocho, F. A., and Lipscomb, W. N., *Adv. Protein Chem.*, **25** (1972), 1.
190. Vallee, B. L., Rupley, J. A., Coombs, T. L., and Neurath, H., *J. Am. chem. Soc.*, **80** (1958), 4750.
191. Vallee, B. L., Rupley, J. A., Coombs, T. L., and Neurath, H., *J. biol. Chem.*, **235** (1960), 64.
192. Waldschmidt-Leitz, E., *Physiol. Rev.*, **11** (1931), 358.
193. Hofmann, K. and Bergmann, M., *J. biol. Chem.*, **134** (1940), 225.
194. Stahmann, M., Fruton, J. S., and Bergmann, M., *J. biol. Chem.*, **164** (1946), 753.
195. Steitz, T. A., Ludwig, M. L., Quiocho, F. A., and Lipscomb, W. N., *J. biol. Chem.*, **242** (1967), 4662.
196. Quiocho, F. A., Bethge, P. H., Lipscomb, W. N., Studebaker, J. F., Brown, R. D., and Koenig, S. H., *Cold Spring Harb. Symp. quant. Biol.* 36 (1971), 561.
197. Bradshaw, R. A., Ericsson, L. H., Walsh, K. A., and Neurath, H., *Proc. Natn. Acad. Sci. U.S.A.*, 63 (1969), 1389.
198. Lipscomb, W. N., Coppola, J. C., Hartsuck, J. A., Ludwig, M. L., Muirhead, H., Searl, J., and Steitz, T. A., *J. molec. Biol.*, **19** (1966), 423.
199. Morris, P. J., and Martin, R. B., *J. inorg. nucl. Chem.*, **33** (1971), 2913.
200. Vallee, B. L. and Neurath, H., *J. Am. chem. Soc.*, **76** (1954), 5006.

201. Folk, J. E., and Gladner, J. A., *Biochim. biophys. Acta*, **48** (1961), 139.
202. Folk, J. E., and Gladner, J. A., *J. biol. Chem.*, **235** (1960), 60.
203. Latt, S. A., and Vallee, B. L., *Biochemistry*, **10** (1972), 4263.
204. Vallee, B. L., and Williams, R. J. P., *Proc. Natn. Acad. Sci. U.S.A.*, **59** (1968), 498.
205. Carlin, R. L., *Transit. Metal Chem.*, **1** (1965), 1.
206. Ballhausen, C. J., and Liehr, A. D., *J. molec. Spectrosc.*, **2** (1958), 342. Also *see erratum*, *J. molec. Spectrosc.*, **4** (1960), 190.
207. Weakliem, H. A., *J. chem. Phys.*, **36** (1962), 2117.
208. Porai-Koshitz, M. A., *Kristallografiya*, **1** (1956), 291.
209. Morosin, B., and Lingafelter, E. C., *Acta crystallogr.*, **12** (1959), 611.
210. Cotton, F. A., and Wood, J. S., *Inorg. Chem.*, **3** (1964), 245.
211. Kretsinger, R. H., Cotton, F. A. and Bryan, R. F., *Acta crystallogr.*, **16** (1963), 651.
212. Cotton, F. A., and Soderberg, R. H., *Inorg. Chem.*, **3** (1964), 1.
213. Ferguson, J., *J. chem. Phys.*, **32**, (1960), 528.
214. Ferguson, J., *J. chem. Phys.*, **39** (1963), 116.
215. Ferguson, J., *Prog. Inorg. Chem.*, **12** (1966), 256.
216. Tsuchida, R., *Bull. chem. Soc. Japan*, **13** (1938), 388.
217. Goodgame, M., and Cotton, F. A., *J. Am. chem. Soc.*, **84** (1962), 1543.
218. Morris, P. J., and Martin, R. B., *J. Am. chem. Soc.*, **92** (1970), 1543.
219. McDonald, C. C., and Phillips, W. D., *J. Am. chem. Soc.*, **85** (1963), 3736.
220. Cotton, F. A., Goodgame, D. M. L., and Goodgame, M., *J. Am. chem. Soc.*, **83** (1961), 4690.
221. Cotton, F. A., and Soderberg, R. H., *J. Am. chem. Soc.*, **84** (1962), 872.
222. Cotton, F. A., and Bergman, J. G., *J. Am. chem. Soc.*, **86** (1964), 2941.
223. Brill, A. S., Kirkpatrick, P. R., and Scholes, C. P., in *Probes of Structure and Function of Macromolecules and Membranes*, vol. 1, *Probes and Membrane Function* (eds B. Chance, C. P. Lee and J. K. Blasie), Academic Press, New York (1971), p. 135.
224. Brill, A. S., personal communication.
225. Folk, J. E., Wolff, E. C., and Schirmer, E. W., *J. biol. Chem.*, **237** (1962), 3100.
226. Grdenić, D., *Q. Rev. chem. Soc.*, **19** (1965), 303.
227. Bryce, G. F., and Gurd, F. R. N., *J. biol. Chem.*, **241** (1966), 122.
228. Coleman, J. E., in *Prog. Biorg. Chem.* (eds E. T. Kaiser and F. J. Kezdy), vol. 1, Wiley, New York (1971), p. 159.
229. Navon, G., Shulman, R. G., Wyluda, B. J., and Yamane, T., *J. molec. Biol.*, **51** (1970), 15.
230. Shulman, R. G., Navon, G., Wyluda, B. J., Douglass, D. C., and Yamane, T., *Proc. Natn. Acad. Sci. U.S.A.*, **56** (1966), 39.
231. Navon, G., *Chem. Phys. Lett.*, **1** (1970), 390.
232. Buckingham, D. A., Foster, D. M., and Sargeson, A. M., *J. Am. chem. Soc.*, **92** (1970), 6151.
233. Vallee, B. L., Riordan, J. F., Bethene, J. L., Coombs, T. L., Auld, D. S., and Sokolovsky, M., *Biochemistry*, **7** (1968), 3547.
234. Davies, R. C., Riordan, J. F., Auld, D. S., and Vallee, B. L., *Biochemistry*, **7** (1968), 1090.
235. Auld, D. S., and Vallee, B. L., *Biochemistry*, **9** (1970), 602.
236. Nakamoto, K., *Infrared Spectra of Inorganic and Coordination Compounds*, Wiley-Interscience, New York (2nd edn, 1970).
237. Haigh, J. M., Hancock, R. D., Hulett, L. G., and Thornton, D. A., *J. molec. Struct.*, **4** (1969), 369.
238. George, P., and McClure, D. S., *Prog. Inorg. Chem.*, **1** (1959), 75.
239. Condrate, R. A., and Nakamoto, K., *J. chem. Phys.*, **42** (1965), 2590. McAuliffe, C. A. and Perry, W. D. *J. chem. Soc.* A, (1969) 634.
240. Nakamoto, K., McCarty, P. J., and Miniatas, B., *Spectrochim. Acta*, **21** (1965), 379.
241. Pauling, L., *Chem. Engng News*, **24** (1946), 1375.
242. Keilin, D., and Mann, T., *Nature, Lond.*, **144** (1939), 442.
243. Lindskog, S. and Malmström, B. G., *Biochim. biophys. Res. Commun.*, **2** (1960), 213.
244. Maren, T. H., *Physiol. Rev.*, **47** (1967), 595.
245. Edsall, J. T., *Harvey Lect.*, **62** (1968), 191.
246. Nyman, P. O., *Biochim. biophys. Acta*, **52** (1961), 1.
247. Laurent, G., Marriq, C., Nahon, D., Charrel, M., and Derrien, Y., *Bull. Soc. Chim. biol.*, **44** (1962), 419.

248. Rickli, E. E., and Edsall, J. T., *J. biol. Chem.*, **237** (1962), PC258.
249. Khalifah, R. G., *J. biol. Chem.*, **246** (1971), 2561.
250. Lindskog, S., Henderson, L. E., Kannan, K. K., Liljas, A., Nyman, P. O., and Strandberg, B., in *The Enzymes* (ed. P. D. Boyer), vol. 5, Academic Press, New York (3rd edn, 1972), p. 587.
251. Strandberg, B., Tilander, B., Fridborg, K., Lindskog, S., and Nyman, P. O., *J. molec. Biol.*, **5** (1962), 583.
252. Liljas, A., Kannan, K. K., Bergstén, P.-C., Waara, I., Fridborg, K., Strandberg, B., Carlbom, W., Järup, L., Lövgren, S., and Petef, M., *Nature, new Biol.*, **235** (1972), 131.
253. Roughton, F. J. W., and Booth, V. H., *Biochem. J.*, **40** (1946), 319.
254. Ward, R. L., *Biochemistry*, **8** (1969), 1879.
255. Riepe, M. E., and Wang, J. H., *J. biol. Chem.*, **243** (1968), 2779.
256. Christiansen, E., and Magid, E., *Biochim. biophys. Acta.*, **220** (1970), 630.
257. Lindskog, S., *Struct. Bond.*, **8** (1970), 153.
258. Barnes, W. H., *Proc. R. Soc.*, **125A** (1929), 670.
259. Coleman, J. E., in *Inorganic Biochemistry* (ed. G. L. Eichorn), Elsevier, Amsterdam (1973), p. 488.
260. Lindskog, S., and Malmström, B. G., *J. biol. Chem.*, **237** (1962), 1129.
261. Lindskog, S., and Nyman, P. O., *Biochim. biophys. Acta*, **85** (1964), 462.
262. Coleman, J. E., *Nature, Lond.*, **214** (1967), 193.
263. Lindskog, S., and Ehrenberg, A., *J. molec. Biol.*, **24** (1967), 133.
264. Taylor, J. S., and Coleman, J. E., *J. biol. Chem.*, **246** (1971), 1058.
265. Sacconi, L., *Transit. Metal Chem.*, **4** (1968), 199.
266. Meredith, P. L., and Palmer, R. A., *Inorg. Chem.*, **10** (1971), 1049.
267. Orgel, L. E., in *Metals and Enzyme Activity* (ed. E. M. Crook), University Press, Cambridge (1958), p. 8.
268. Jencks, W. P., *Catalysis in Chemistry and Enzymology*, McGraw-Hill, New York (1969), p. 181.
269. Vallee, B. L., and Latt, S. A., personal communication.
270. Lindskog, S., *Biochemistry*, **5** (1966), 2641.
271. Davis, R. P., *J. Am. chem. Soc.*, **81** (1959), 5674.
272. Riepe, M. E., and Wang, J. H., *J. Am. chem. Soc.*, **89** (1967), 4229.
273. Kaiser, E. T., and Lo, K.-W., *J. Am. chem. Soc.*, **91** (1969), 4912.
274. Chaberek, S., Sr, Courtney, R. C., and Martell, A. E., *J. Am. chem. Soc.*, **74** (1952), 5057.
275. Hunt, J. P., *Metal Ions in Aqueous Solution*, Benjamin, New York (1963), p. 50.
276. Holt, E. M., Holt, S. L., and Watson, K. J., *J. Am. chem. Soc.*, **92** (1970), 2721.
277. Lindskog, S., *J. biol. Chem.*, **238** (1963), 945.
278. Bergstén, P.-C., Waara, I., Lövgren, S., Liljas, A., Kannan, K. K., and Bengtsson, U., in *Oxygen Affinity of Hemoglobin and Red Cell Acid–Base Status. Alfred Benzon Symp. IV*, Munksgaard, Copenhagen, and Academic Press, New York (1971).
279. Kernohan, J. C., *Biochim. biophys. Acta*, **81** (1960), 346.
280. Kernohan, J. C., *Biochim. biophys. Acta*, **96** (1965), 304.
281. Coleman, J. E., *J. biol. Chem.*, **242** (1967), 5212.
282. Whitney, P. L., *Eur. J. Biochem.*, **16** (1970), 126.
283. Konosu, S., Hamoir, G., and Pechere, J. F., *Biochem. J.*, **96** (1965), 98.
284. Bhushana, Bao, K. S. P., Focant, B., Gerday, Ch., and Hamoir, G., *Comp. Biochem. Physiol.*, **30** (1969), 33.
285. Ebashi, S., Endo, M., and Ohtsuki, I., *Q. Rev. Biophys.*, **2** (1969), 351.
286. Pechere, J. F., Capony, J. P., and Ryden, L., *Eur. J. Biochem.*, **23** (1971), 421.
287. Kretsinger, R. H., Nockolds, C. E., Coffee, C. J., and Bradshaw, R. A., *Cold Spring Harb. Symp. quant. Biol.*, **36** (1971), 217.
288. Nockolds, C. E., Kretsinger, R. H., Coffee, C. J., and Bradshaw, R. A., *Proc. Natn. Acad. Sci. U.S.A.*, **69** (1972), 581.
289. Kretsinger, R. H., personal communication.
290. Williams, R. J. P., *Q. Rev. chem. Soc.*, **24** (1970), 331.
291. Truter, M. R., *Chem. Br.*, **7** (1971), 203.
292. Van der Helm, D., and Willoughby, T. V., *Acta crystallogr.*, **B25** (1969), 2317.
293. Urry, D. W., *Proc. Natn. Acad. Sci. U.S.A.*, **68** (1971), 810.

294. Anfinsen, C. B., Cuatrecasas, P., and Taniuchi, H., in *The Enzymes* (ed. P. D. Boyer), vol. 4, Academic Press, New York (1971), p. 177.
295. Cuatrecasas, P., Fuchs, S., and Anfinsen, C. B., *J. biol. Chem.*, **242** (1967), 3063.
296. Cuatrecasas, P., Fuchs, S., and Anfinsen, C. B., *J. biol. Chem.*, **242** (1967), 1541.
297. Cotton, F. A., Bier, C. J., Day, V. W., Hazen, E. E., Jr, and Larsen, S., *Cold Spring Harb. Symp. quant. Biol.*, **36** (1971), 243.
298. Cotton, F. A., and Hazen, E. E., Jr, in *The Enzymes* (ed. P. D. Boyer), vol. 4, Academic Press, New York (1971), p. 153.
299. Cotton, F. A., and Hazen, E. E., Jr, personal communication.
300. Taniuchi, H., and Anfinsen, C. B., *J. biol. Chem.*, **243** (1968), 4778.
301. Taniuchi, H., Moróvek, L., and Anfinsen, C. B., *J. biol. Chem.*, **244** (1969), 4600.
302. Taniuchi, H., Davies, D. R., and Anfinsen, C. B., *J. biol. Chem.*, **247** (1972), 3362.
303. Merrifield, R. B., *J. Am. chem. Soc.*, **85** (1963), 2149.
304. Ontjis, D. A., and Anfinsen, C. B., *J. biol. Chem.*, **244** (1969), 6316.
305. Chaiken, I. M., and Anfinsen, C. B., *J. biol. Chem.*, **246** (1971), 2285.
306. Cuatrecasas, P., Wilchek, M., and Anfinsen, C. B., *Biochemistry*, **8** (1969), 2277.
307. Lienhard, G. E., Secemski, I. I., Koehler, K. A., Lindquist, R. N., *Cold Spring Harb. Symp. quant. Biol.*, **36** (1971), 45.
308. Darnall, D. W., and Birnbaum, E. R., *J. biol. Chem.*, **245** (1970), 6484.
309. Smolka, G. E., Birnbaum, E. R., and Darnall, D. W., *Biochemistry*, **10** (1971), 4556.
310. Dwek, R. A., Richards, R. E., Morallee, K. G., Nieboer, E., Williams, R. J. P., and Xavier, A. V., *Eur. J. Biochem.*, **21** (1971), 204.
311. Colman, P. M., Weaver, L. H., and Matthews, B. W., *Biochem. biophys. Res. Commun.*, **46** (1972), 1999.
312. Ludwig, M. L., Andersen, R. D., Apgar, P. A., Burnett, R. M., LeQuesne, M. E., and Mayhew, S. G., *Cold Spring Harb. Symp. quant. Biol.*, **36** (1971), 369.
313. Ramaseshan, S., *Curr. Sci.*, **35** (1966), 87.
314. Lind, M. D., Lee, B., and Hoard, J. L., *J. Am. chem. Soc.*, **87** (1965), 1611.
315. Lind, M. D., Lee, B., and Hoard, J. L., *J. Am. chem. Soc.*, **87** (1965), 1612.
316. Williams, M. N., personal communication.
317. Williams, M. N., *Fedn Proc. Fedn Am. Socs exp. Biol.*, **30** (1971), 1292.
318. East, D., Nieboer, E., Cohen, J. S., and Schechter, A. N., *Fedn. Proc. Fedn. Am. Socs exp. Biol.*, **31** (1972), 502.
319. Benesch, R. E., Ranney, H. M., Benesch, R., and Smith, G. M., *J. biol. Chem.*, **236** (1961), 2967.

PART 2
Principles of Catalysis by Metallo-Enzymes

J. M. PRATT

Imperial Chemical Industries Ltd, Corporate Laboratory
The Heath, Runcorn, Cheshire
Present address: Department of Chemistry, University of the Witwatersrand,
Jan Smuts Avenue, Johannesburg, South Africa

6 AIMS OF THE REVIEW

The main role of transition metals in organisms depends on their catalytic properties. Many enzymes consist only of protein (that is, polypeptides), but others consist of a protein (called the 'apoenzyme') and one or more smaller molecules or ions ('cofactor', 'coenzyme' or 'prosthetic group'), which together form the complete enzyme or 'holoenzyme'. The cofactor may be an organic molecule such as flavin, pyridoxal, pyridine nucleotide, etc., attached to the protein by covalent bonds, hydrogen bonds or van der Waals interaction; or it may be a simple metal ion (for example, Cu) or a metal complex with one or more ligands (for example, Fe porphyrins, Co corrinoids). Where one or more of the amino-acid residues is co-ordinated to the metal, the protein may be regarded as the ligand, albeit a rather unusual one. Obviously these metallo-enzymes can be considered either as a special group of enzymes or as a special group of metal complexes, and hence compare the catalytic activity either of enzymes with and without transition metals, or of transition-metal complexes with and without a protein. In this review we shall neglect those metallo-enzymes where the metal functions merely as a Lewis acid (as in certain hydrolytic enzymes—see, for example reference 59) and include only those metallo-enzymes and metal-containing proteins that undergo reactions (for example, redox reactions) and deal with ligands (for example, O_2) that are usually considered as more typical of transition-metal complexes.

Co-ordination chemists may be forgiven for thinking that, although the purely organic enzymes are remarkable catalysts in their own way, the presence of a metal considerably enhances both their reactivity and their interest. There are many examples where different enzymes, some with and some without a metal, can catalyse the same reaction, attack the same substrate or form the same product; for example electron-transfer proteins may contain flavins, Fe porphyrins or ferredoxins, and enzymes that catalyse the reduction of H_2O_2 by organic substrates may also be based on flavins or Fe porphyrins (see section 8.1). But there are other reactions that, as far as we know at present, can be catalysed only by enzymes containing transition metals, for example the fixation of nitrogen (see section 9.2), the reduction of NO_3^- to NO_2^- (see, for example, reference 32) and certain isomerisation reactions involving Co corrinoids (see section 10.2)[18,181]. And there must clearly be many reactions that can be catalysed more efficiently by enzymes containing transition metals. These metal-protein complexes or metallo-enzymes are involved in many different areas of metabolism, but there are two that deserve special mention for two reasons. Firstly, they form the gateway by which inorganic nitrogen in the form of either N_2 or NO_3^- enters the metabolic pool. Secondly, they are closely associated with the main schemes of energy production or interconversion: as electron carriers, and possibly also in the evolution of O_2, in chloroplasts; as electron carriers and in the reaction with O_2 in oxidative phosphorylation; and in the evolution of H_2 and CH_4 in anaerobic fermentation.

There are many examples where the same reaction can be observed with the simple complex as with the metal–protein complex, though the rates may be very different. For example, the rate of the catalysed decomposition of H_2O_2 at pH 7 varies in the ratio $1:10^3:10^8$ in the presence of equal molar quantities of the simple aquated Fe(III) ion, simple Fe(III) porphyrins and the enzyme catalase, respectively (see section 8.7). But there are certain other enzymatic reactions, which as yet have no analogues among protein-free complexes; for example the isomerisations catalysed by enzymes containing Co corrinoids, which involve rearrangement of the carbon skeleton (see section 10.2)[18,181]. Nor do chemists appear to have discovered true analogues of the enzymatic fixation of nitrogen (see section 9.1). The presence of the protein (and any additional cofactors) can clearly have a very pronounced effect on the reactivity of transition-metal complexes.

The aim of this review is to examine some of the ways in which the protein affects the reactions of transition-metal complexes and hence to try to pinpoint some of the factors that contribute to the enhanced catalytic activity of enzymes, with particular reference to metallo-enzymes. The protein (or assembly of proteins) must be able to vary one or more of the following:

(i) the thermodynamics of individual steps in the overall reaction, for example the equilibrium constant for the binding of a ligand, or the electrode potential for oxidation and reduction.
(ii) the kinetics of the individual steps; some rates must be increased, but others must be severely inhibited.
(iii) the thermodynamics of the overall reaction, that is coupling together with a second reaction, either to conserve some of the free-energy change of a thermo-dynamically favourable reaction (as in oxidative phosphorylation) or to drive a thermodynamically unfavourable reaction (as in nitrogen fixation).

In order to be able to identify the mechanisms by which the protein enhances the catalytic activity of metal complexes we must firstly be able to identify the individual steps and intermediates in the overall enzymatic reaction and, secondly, to compare the rate constants of the individual steps with those of closely analogous reactions in the absence of protein. This should then establish whether the apparent enhancement of any particular rate constant is primarily due to the effect of the protein on the kinetics or on the thermodynamics (that is on some equilibrium involving the substrate alone, the metal ion alone or some intermediate). A third point is, therefore, that we cannot expect to understand how the protein affects rate constants (which are related to the difference in free energy between a ground state of more or less known structure and a transition state of unknown structure) before we have examined how the protein can affect equilibrium constants (which are related to the difference in free energy between two ground states).

We have chosen to examine one example of each category in depth, rather than to survey a wider field more superficially. Haemoglobins and myoglobins provide the ideal case for studying how the protein controls the reversible co-ordination of a ligand (namely O_2) in the absence of complications due to any overall reaction (chapter 7); because they do not catalyse any reaction they are, of course, not enzymes, but simply proteins. The catalases and peroxidases also provide an excellent opportunity for studying the affect of the protein on the rates of several individual steps (chapter 8). Finally, the case of nitrogen fixation allows us to draw certain conclusions about the general mechanism and the overall thermodynamics, even though we have not yet identified the active sites (chapter 9).

For each case study we set out to consider what the simple protein-free metal complex can and cannot do, and hence what problems nature has to overcome; and then to examine how the protein solves these problems. Sufficient information on the distribution, function and nature of these proteins and enzymes has also been provided to sketch in the physiological and biochemical background, and to show that nature has often developed several completely different ways of, for example, binding O_2 or catalysing the removal of H_2O_2. This approach requires some of the experimental evidence to be looked at from an angle rather different from that adopted by previous reviewers, hence the fair amount of detail in certain sections. The evidence and conclusions for each case study are summarised at the end of chapters 7–9. The more general conclusions regarding the mechanisms by which the protein can affect the thermodynamic and kinetic properties of transition-metal complexes are presented in section 10.1 and used to make a few tentative suggestions about the mechanism of the so-called 'isomerase' reactions catalysed by cobalt corrinoids in section 10.2.

Most of the haemoglobins, myoglobins, catalases and peroxidases are relatively simple and fairly well-characterised proteins. They contain one Fe protoporphyrin IX (see figure 29) per polypeptide chain, and there are either one or four polypeptide chains per protein molecule. The structures of several haemoglobins and myoglobins have been determined by X-ray analysis. But other metalloproteins may contain several polypeptide sub-units (for example, over a hundred in certain haemocyanins;

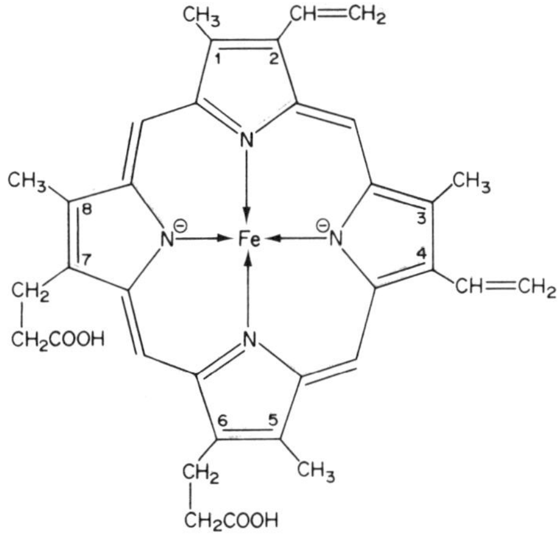

Figure 29 Structure of Fe protoporphyrin IX (valency and axial ligands not indicated).

see section7.1) and each polypeptide may contain large numbers of metal ions (for example, 40 Fe and 2 Mo in some of the nitrogenase proteins; see section 9.2). In addition, enzymes (or enzyme systems) may consist of only one protein (as in peroxidase; see section 8.1), two proteins (as in nitrogenase; see section 9.2) or a whole series of proteins (as in the cytochrome chain). Finally, we do not yet know the nature of the ligands in any of the enzymes containing Cu or Mo. This merely serves to emphasise the complexity of the situation and the limited extent of our present knowledge. Any general conclusions to be drawn in this review can therefore, at best, only represent part of the picture.

The following abbreviations are used in more than one of the chapters:

Hb	haemoglobin
Mb	myoglobin
ATP	adenosine triphosphate

Other abbreviations used in only one chapter are mentioned in the introduction to that chapter. We have not used the biochemist's terms for Fe porphyrin complexes (haem, haemin, haematin, haemochrome, methaemoglobin, etc.) Since these would only be confusing to the co-ordination chemist, and we usually omit the "IX" in protoporphyrin IX, etc.

7 HAEMOGLOBINS AND MYOGLOBINS

The solubility of O_2 in aqueous media is relatively low; one litre of pure water in equilibrium with air at $20°C$ and atmospheric pressure dissolves 6.59 cm^3 of O_2 [93], which corresponds to a 3×10^{-4} M solution. This limits the rate of diffusion of O_2 from the surface of an organism and also the rate at which O_2 can be delivered by a circulatory system. This could in turn place a severe restriction on the rate of energy production and hence on the metabolic level and general activity of an organism. Nature has therefore developed proteins with the aim of:

(i) increasing the rate of delivery of O_2 to the site of consumption, that is transporting O_2, in the circulatory system and even within the cell;
(ii) counteracting fluctuations in the supply (for example, when an air-breathing animal dives or a marine organism is uncovered by the retreating tide) or demand (for example, during sudden muscular activity), by storing O_2.

Such proteins are often called 'respiratory pigments'. As a very broad generalisation, they are found with increasing frequency as we ascend the phyla of the animal kingdom, but are uncommon in plants and micro-organisms (see section 7.1). They act by reversibly co-ordinating O_2 to a transition metal (Fe, Cu or possibly V), so that the concentration of bound O_2 can then be raised by increasing the concentration of the protein. Human blood, for example, dissolves about 200 cm^3 of O_2 per litre when in equilibrium with air at $20°$ [81], which corresponds to a 9×10^{-3} M solution; that is the blood can carry about thirty times as much oxygen as pure water. But oxygen is delivered to the enzyme or transferred across the cell wall in the form of free, unbound O_2. The function of these proteins is therefore to increase the local steady-state concentration of free O_2 above the level that would otherwise result from the balance between removal by the enzymes and supply by unassisted diffusion and circulation.

In their simplest form (represented by mammalian myoglobins) the proteins reversibly bind O_2 according to the straightforward equilibrium

$$\text{protein} + O_2 \rightleftharpoons \text{protein}.O_2 \qquad (7.1)$$

which leads to the typical 'hyperbolic' curve shown in figure 30; the equilibrium constant is not significantly affected by external reagents or other factors. Nature has subsequently developed other proteins, for which the equilibrium constant can be reversibly altered by external reagents (for example, pH, organic phosphates) or, in proteins that bind more than one mole of O_2, by the degree of oxygenation of the whole protein. This not only provides a far more sophisticated mechanism for increasing and controlling the local concentration of free O_2, but also enables the concentration to be raised above that which would be in equilibrium with the environment surrounding the organism, and hence provides a mechanism for the active transport of O_2 against a concentration gradient, for example in inflating the swim bladder of a fish[138,208] or in maintaining a high oxygen tension in the eye[240].

The interaction between the binding of O_2 and the binding of external reagents is sometimes called *heterotropic* interaction, and the specific effect of pH is termed the 'Bohr effect'. When the uptake of O_2 yields a *sigmoid* curve, such as that in figure 31, it indicates that the equilibrium constant is dependent on the degree of oxygenation of the whole protein and that there is some interaction between the

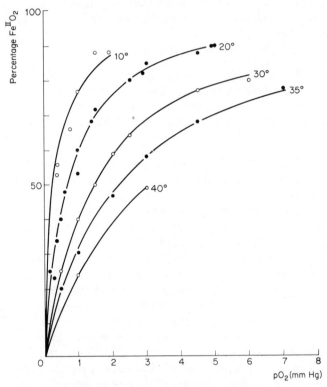

Figure 30 Oxygen equilibrium curves of human Mb at different temperatures, pH 7.5. From reference 189.

separate metal ions (see section 7.6); these are sometimes referred to as *homotropic* or *co-operative* interactions or, in the specific case of the Fe porphyrins in Hb, as *haem–haem* interaction.

Several completely different groups of transition-metal complexes (Fe porphyrins, non-porphyrin Fe, Cu and possibly V) are used for binding O_2 in proteins. These are described in section 7.1 together with the protein-free Co complexes, which show analogies to the Fe porphyrins. Thereafter we concentrate exclusively on the Hbs and Mbs.

Hbs and Mbs form an ideal group of proteins in which to examine the effect of the protein in controlling the equilibrium constant for the binding of one particular ligand, namely O_2. These proteins all contain the same Fe protoporphyrin complex and, in all but a few mutants, the same axial ligand; in addition, all the proteins that we shall discuss have a very similar tertiary structure. Yet they show a wide variation in the equilibrium constant and in the effects of homotropic and heterotropic interaction. Starting with the pioneering work of Kendrew and Perutz and their coworkers

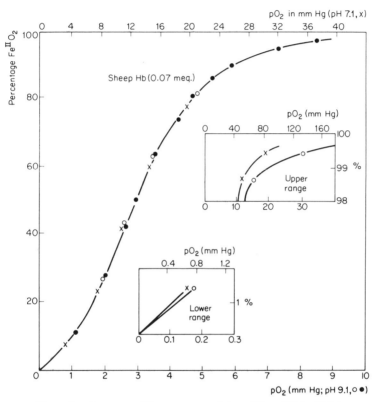

Figure 31 Oxygen equilibrium curves of sheep Hb at 19°C, pH 7.1 (top scale) and pH 9.1 (bottom scale). From reference 194.

on sperm-whale Mb and on horse and human Hb respectively, X-ray analysis has provided us with detailed knowledge of the structure of several Hbs and Mbs. Thanks to their medical interest and importance, many mutant Hbs have been isolated and studied, and we can examine the effect of changing a single amino acid both on the structure of the protein and on its affinity for O_2.

The structures of Hbs and Mbs are described in section 7.2. The problems involved in developing proteins for the reversible co-ordination of O_2 are delineated in section 7.3 and discussed in turn in sections 7.4–7.7. The main points are summarised in section 7.8.

As already mentioned, we shall use the following two abbreviations throughout this section:

Hb   Haemoglobin

Mb   Myoglobin

An excellent source of information and references is provided by Antonini and Brunori's recent book *Haemoglobin and Myoglobin and their Reactions with Ligands*, published in 1971[8]. Lemberg and Legge's book *Hematin Compounds and Bile Pigment*, published in 1949[143], still provides useful background material.

7.1 THE DISTRIBUTION AND NATURE OF O_2-CARRYING PROTEINS

Nature has developed respiratory pigments based on three different transition metals and four completely different types of complex. These are described here in order of decreasing importance.

7.1.1 Fe porphyrins

These proteins bind one molecule of O_2 per Fe(II) ion. Only two different porphyrins have so far been discovered, namely protoporphyrin IX (see figure 29) and chlorocruoroporphyrin or Spirographis porphyrin; these differ only in that the vinyl group at position 2 is replaced by a formyl group (−CHO) in the latter.

Proteins containing the Fe complex of the latter are called *chlorocruorins* (Chl) and are found only in certain polychaete worms of the phylum Annelida. Proteins containing Fe protoporphyrin, on the other hand, are very widespread; for further details and references see reference 82. They occur in all vertebrates with very few exceptions (for example, certain sluggish Antarctic fish). They are present only in the blood (when they are called *haemoglobins*) and in muscle (*myoglobins*), in addition to bone marrow and liver where the red blood cells are formed in mammals and lower vertebrates, respectively; particularly high concentrations of Mb are found in muscles characterised by repetitive activity and long contraction times (for example, heart muscles of large mammals) and in the muscles of diving animals (for example, whales). They also occur sporadically among invertebrates, being particularly widespread in the phyla Annelida, Crustacea and Mollusca, represented in others such as Insecta and Nematoda, but so far unknown in the Coelenterata, Polyzoa, etc; and, in contrast to the situation in vertebrates, the proteins are not confined to corpuscles in the blood (or coelomic fluid) and to muscle, but occur in a variety of other tissues (for example, nerve cells, fat cells, eggs, etc.) and even dissolved in the blood or other fluid. In classifying these proteins, the name *myoglobin* is usually extended to include all the noncirculating proteins, and *haemoglobin* the circulating proteins contained in cells, while the extracellular proteins in the blood and other fluids are called *erythrocruorins*. However, some workers refer to all these invertebrate proteins as haemoglobins, and for the sake of simplicity we shall do the same here. Isolated occurrences have been recorded among yeasts, moulds and ciliate protozoans. They also occur in the nitrogen-fixing bacteria of the genus *Rhizobium*, which live symbiotically in the root-nodules of legumes (*leghaemo-globins*). But, by and large, Hbs and related proteins are absent from the plant king-dom, from the lower animals and from micro-organisms. The role of the leghaemo-globins appears to be to facilitate the diffusion of O_2 within the cells of the root-nodule[239].

It appears that the basic unit of all these O_2-carrying proteins contains one Fe and usually has a M.W. in the region of 16 000 to 20 000, though values up to 40 000 are occasionally found[165]. Most or all vertebrate Mbs are monomers, but Hbs (including invertebrate Mbs and erythrocruorins) and chlorocruorins show a far greater variation. Mammalian Hbs are almost always tetrameric, and usually consist of two units each of two different polypeptide chains (see section 7.2). But a few octamers and higher polymers are known[28] and sickle-cell Hb probably forms a linear polymer in the deoxy form[170]. Lamprey Hb shows equilibria between monomers, dimers and higher aggregates[144]. A few dimers[31,122] and an octamer[165] have also been found among invertebrates. But many invertebrate proteins have much higher M.W.s; the erythro-cruorins from *Arenicola*, *Lumbricus* and *Planorbis* and the chlorocruorin from *Spirographis* have M.W.s of 3.0, 2.9, 1.6 and 2.8 x 10^6, respectively, and must contain well over a hundred sub-units; electron microscopy of *Spirographis* Chl reveals a regular cylindrical structure composed of disc-shaped units[192]. The structures of the mammalian Hbs and Mbs and related compounds are considered in more detail in the next section.

Control of O_2-binding in response to physiological requirements is well exemplified in this group. The binding of O_2 by mammalian Hb, for example, shows co-operativity and is influenced by pH and organic phosphates (sections 7.5–7.6). Polymorphism or heterogeneity, that is the presence of several closely related proteins, is frequently found for both Hbs and Mbs and may be the rule rather than the exception; for good examples see references 31 and 241. The needs of the organism are presumably better fulfilled by a range of proteins with slightly different properties to cover different eventualities. The relative and absolute concentrations of these closely related proteins may vary with age (see, for example, reference 241 and section 7.2) or in response to a change in the environment. Man, for example, can increase the concentration of Hb in his blood by 20 per cent when living at high altitudes, while the crustacean *Daphnia* can vary its concentration of Hb tenfold in response to the concentration of O_2 in the surrounding water[82].

7.1.2 Haemocyanins[82,87]

Haemocyanins contain Cu and bind one molecule of O_2 for every pair of Cu(I) ions; the oxy form is blue, the deoxy form almost colourless. The nature of the ligands co-ordinated to the metal is not known. They are found only in molluscs and arthropods (crustacea and arachnids), and occur dissolved in the blood (or haemo-lymph), where they usually comprise 90–98 per cent of the total protein present. The sub-units contain two atoms of Cu and have a M.W. of 50 000–74 000; they form aggregates with a M.W. up to 9×10^6. These particles are exceeded in size only by viruses. Electron microscopy shows that the sub-units associate to form regular spheres, cubes, rings, etc. (see, for example, reference 78). The equilibrium constant for binding O_2 may show co-operativity and be influenced by pH and the concentration of Ca and Mg ions (see, for example, reference 76).

7.1.3 Haemerythrins

Haemerythrins contain Fe, but no porphyrin; that is, the Fe is co-ordinated only to the protein; but the identity of the ligands is not known. They bind one O_2 per two Fe(II) ions; the oxy form is red, the deoxy pale yellow. They have a very limited and scattered distribution covering four different phyla: several sipunculids and priapulids, one polychaete annelid and one brachiopod. They usually occur in blood corpuscles, some of which may, like the mammalian erythrocytes (red cells), lack nuclei. The proteins are all associated and have M.W.s up to 120 000[87]; that from *Golfingia gouldii* has a M.W. of 107 000 and consists of 8 sub-units, each containing two Fe atoms[127].

7.1.4 Haemovanadins

The vacuoles in the blood cells (vanadocytes) of the ascidians contain a vanadium–protein complex dissolved in 1.5–2 N H_2SO_4! There has been controversy as to whether this compound can act as an O_2-carrier, but recent work with carefully prepared vanadocytes has demonstrated a reversible uptake of O_2 with $P_{1/2}$ (partial pressure of O_2 required for 50 per cent oxygenation) of ≈ 250 N m^{-2} (2 mm Hg)[46].

The apparently haphazard distribution of these groups, even among closely related species, has often been commented upon. For example, some of the polychaete worms have Chl, others have Hb, and at least one species has no respiratory pigment; one genus has Chl in the blood and Hb in the muscles, while another genus has both Chl and Hb in its blood. But at least Chl and Hb are both Fe porphyrin complexes. Certain molluscs, however, contain haemocyanin in their blood and Hb in their muscles[82]. The apparently random distribution between blood plasma and corpuscles, muscles and other tissues, is equally surprising.

The existence of these four completely different groups of proteins provides an excellent example of nature's ability to tackle and solve many problems of co-ordination chemistry (in this case the reversible co-ordination of O_2) in more than one way. Even the mechanism for linking the equilibrium constant to the degree of oxygenation of the whole protein has been developed in both the Hbs and the haemocyanins. But the recent preparation of functionally active Co-containing analogues of Hb and Mb demonstrates that nature has not found and developed all the solutions that are possible even within the range of metals, amino acids and other ligands at her disposal. The Co analogues have a lower affinity for O_2 than the native proteins[207]; but both pH and diphosphoglycerate have virtually the same effect on the binding of O_2 by Hb and by its Co analogues, though the haem–haem or homotropic interaction is less in the latter[108]. The Co analogues are prepared by dissociating Hb and Mb into protein and the Fe porphyrin (see section 7.4), and then recombining the protein with the analogous Co(II) porphyrin complex.

With the possible exception of certain Fe porphyrin (see section 7.3) and phthalo-cyanine tetrasulphonate complexes[224,233], there are no authenticated examples of the reversible co-ordination of O_2 by protein-free complexes of Fe, Cu or V. By contrast, there are now well over 50 such complexes known for Co, though many are unstable and have only been detected at low temperature. The $Co:O_2$ ratio may be either 1 : 1 as in Hb and Mb or 2 : 1 as in haemerythrin and haemocyanin. The mononuclear complexes derived from Co(II) show many analogies to the O_2 com-plexes of Hb and Mb, as emphasised by the properties of their Co-containing analogues; they are, in fact, the only protein-free complexes so far discovered in which the co-ordinated O_2 occupies one co-ordination site in an octahedral complex. They can, therefore, provide useful information on (i) the structure of co-ordinated O_2, (ii) factors that affect the equilibrium constant for binding O_2, and (3) the reactions of co-ordinated O_2 and the autoxidation of the Co(II) ion. They will be referred to several times in part 2 of the book. A brief summary of these complexes is therefore appropriate.

A great variety of Co(II) complexes can react with O_2 to give mononuclear and/or dinuclear complexes, and these reactions may in certain cases be reversible depending on the rate of irreversible formation of simple Co(III) complexes; that is, we can write schematically

$$2Co(II) + O_2 \rightleftharpoons Co.O_2 + Co(II) \rightleftharpoons Co.O_2.Co \rightarrow 2Co(III) \qquad (7.2)$$

We shall here be interested only in the mononuclear adducts. For references to the dinuclear adducts (μ-peroxides), see reference 234. Mononuclear adducts are most frequently found where the Co(II) complex possesses a conjugated equatorial ligand, which occupies the four co-ordination sites in the plane. Typical examples of such ligands are: Schiff bases derived from ethylenediamine, such as salen[79], BAE[69], amben[2], bisdimethylglyoxime[200], corrinoids[22], porphyrins[225] and phthalo-cyanines[1] (only a few representative references are given here, and common abbreviations are used for some of the ligands with unwieldy names).

In the absence of O_2 these complexes may contain 0, 1 or 2 axial ligands—that is, they are four-co-ordinate (square planar), five-co-ordinate (square pyramidal) or six-co-ordinate; the square-planar complexes may also dimerise to achieve Co five-co-ordination. The equilibrium constant for binding one mole of a nitrogenous base such as pyridine is always greater than that for binding the second—that is, $K_1 > K_2$—and the five-co-ordinate complex may be readily obtained, in contrast to Fe(II) (see section 7.3). In many cases K_2 is so small that the bispyridine com-plex is only observed at very low temperature; see, for example, the corrinoids[181]. All these d^7 Co(II) complexes with conjugated equatorial ligands are low spin and hence possess one unpaired electron.

The mononuclear O_2 adducts can be formed with a wide range of axial ligands (B); see, in particular, the corrinoids[22,250]. The ligand atom may be: O as in methanol or dimethylformamide; N as in NH_3, pyridine or other amines; C as in CN^-; P as in

PPh_3; or I^-. But the most commonly used ligand is pyridine. O_2^- adducts, which are probably five-co-ordinate, can also be formed in solvents such as toluene in the absence of any base (see, for example, reference 3). The mononuclear O_2-adducts are all paramagnetic, possess one unpaired electron and show a very characteristic e.s.r. spectrum. The general pattern of equilibria discussed above can be shown schematically as shown in equation 7.3.

$$\text{Dimer} \rightleftharpoons {-}Co{-} \xrightleftharpoons{\text{B}} {-}Co{-}\ \overset{\text{B}}{\underset{\text{B}}{|}} \xrightleftharpoons{\text{B}} {-}Co{-}\ \overset{\text{B}}{\underset{\text{B}}{|}}$$

$$\big\updownarrow O_2 \qquad\qquad \big\updownarrow O_2 \tag{7.3}$$

$$\underset{-Co-}{\overset{O_2}{|}} \xrightleftharpoons{\text{B}} \underset{-Co-}{\overset{O_2}{\underset{\text{B}}{|}}}$$

Qualitative results suggest that the equilibrium constant for the binding of O_2 increases as the donor power of the axial ligand increases in the general order no ligand $<$ ligands co-ordinated through O $<$ ligands co-ordinated through N and, among γ-substituted pyridines, as the substituent changes in the order CN $<$ H $<$ CH_3 $<$ NH_2, but there is no systematic effect from changing the equatorial ligand (see, for example, references 3 and 69). The majority of these O_2 adducts are only formed below room temperature. Their formation appears to be rapid, even at very low temperatures, although no quantitative kinetics have yet been reported.

7.2 THE STRUCTURE OF HAEMOGLOBINS AND MYOGLOBINS

Greatest attention has naturally been devoted to vertebrate Hbs and Mbs, in particular of mammalian origin. The Mbs consist of one polypeptide chain; the mammalian Hbs all contain four sub-units, two of one kind of chain, two of another.

The blood of normal adult humans contains two Hbs. The major component ($\geqslant$95 per cent) is called Hb A_1 and consists of two α chains and two β chains; the protein is written $\alpha_2\beta_2$. The minor component (2–4 per cent) is Hb A_2, written $\alpha_2\gamma_2$. The human foetus contains HbF, written $\alpha_2\delta_2$. A decreased rate of synthesis of the α chain may lead to the formation of Hbs composed of only one chain, for example Hb H, β_4, and Hb Barts, γ_4. Where the protein is not more exactly specified, human Hb means Hb A_1. Such a diversity of closely related proteins ('heterogeneity') may be common to all mammals, but none have yet been examined in such detail; the presence of more than one Hb has, however, been shown for several other mammals.

But this is only the start of the complexity. In solution Hb is in dynamic equilibrium with the asymmetric dimers ($\alpha\beta$) and even monomers[172], but the equilibrium is displaced far to the right

$$2\alpha + 2\beta \rightleftharpoons 2\alpha\beta \rightleftharpoons \alpha_2\beta_2 \tag{7.4}$$

In addition, mutations involving the change of one amino acid may occur in any of the chains; over a hundred such mutants are now known[94,170]. We can also ring the changes artificially (for examples see reference 8) by (i) forming hybrids with the use of α and β chains from different species, (ii) proteolytic digestion, for example of the C-terminal residues by carboxypeptidase, and (iii) chemical modification, for example of the thiol group of cysteine. On top of this we can, of course, vary the valency of the Fe and the nature of the sixth ligand, and even prepare Hbs in which the state of the Fe is different in the two chains, for example by mixing solutions of $\alpha^{III}CN^-$ and of $\beta^{II}O_2$. The Fe porphyrin can also be removed from many Hbs and Mbs without denaturing the protein, which can then be recombined with Fe complexes of other porphyrins and even with porphyrin complexes of other metals (see sections 7.1 and 7.4). The study of mutants and chemically modified Hbs has greatly extended our knowledge of the mode of action of Hb and reference to some of these will be given in the sections below.

An understanding of the way in which the protein affects the co-ordination of O_2 in Hbs and Mbs requires some discussion of the primary (linear sequence of amino acids in the polypeptide), secondary (localised coiling of segments of the polypeptide into α-helix, etc.), tertiary (folding of the polypeptide chain into a three-dimensional structure) and quaternary (the arrangement of the polypeptide chains into higher aggregates; in this case, the arrangement of the two α and two β chains in the tetrameric Hb) structures. We are interested mainly in the last two.

All the polypeptide chains of vertebrate Hbs and Mbs have a M.W. of 16 000 to 17 000 and contain 140–160 amino-acid residues, of which six appear to be invariant. Mammalian Mbs have 153 residues, Hb α chains 141, and the β, γ and δ chains 146. For references to known amino-acid sequences see reference 118.

The first tertiary structure to be determined was that of sperm-whale Mb at 600 pm resolution, reported by Kendrew and colleagues in 1959[26]. The protein molecules are roughly spherical, the polypeptide chains contain eight helical segments and the Fe porphyrin is held in a hydrophobic region near the surface. Since then virtually the same structure, now termed the 'myoglobin fold', has been reported for: Mbs from the seal[202] and yellow-fin tuna[142]; the α and β chains of both horse and human Hb[27,152,172]; and the monomeric Hbs from the sea lamprey[144], the insect *Chironomus*[110] and the annelid worm *Glycera*[170]. [Note: the structure originally referred to as horse deoxyhaemoglobin[172] was subsequently found to have decomposed to the $Fe^{III}OH_2$ form by the time the X-ray data were collected[169].] The structure of the β chain of Hb is shown in figure 1 (page 22). The eight helical segments are denoted by the letters A to H starting at the N-terminal end of the chain, the corners or internal nonhelical segments by AB, BC, etc., and the two nonhelical segments at the N- and C-terminal ends by NA and HC respectively[172]. Individual amino acids can be designated either by their position in the whole polypeptide chain or by their position in the particular segment of the chain, for example 87 in the α chain and 92 in the β chain can both be called F8. The latter method is obviously more convenient when comparing different polypeptides, and is used here. Figure 1 also indicates the position of three residues of particular interest, namely histidine E7, histidine F8 and tyrosine HC2.

It is very striking that the same myoglobin fold is found in three different phyla (vertebrates, insects and annelids). This is perhaps even more surprising when it is realised that only 20 per cent of the residues in the insect Hb are the same as those in mammalian Mb, though the disposition of polar and nonpolar residues is similar[110].

Major changes in the tertiary structure, that is the reversible or irreversible denaturation of the protein, can be observed under rather drastic conditions (see section 7.4). But so far no major changes in tertiary structure have been detected during the equilibria or reactions to be discussed here, and no major differences observed in the structures of any of the Mbs and normal, mutant or modified Hbs. We are, however, very interested in the small changes in tertiary structure by which the effects of a change at one point in the protein are transmitted to another (section 7.6).

Work on horse Hb and a variety of human Hbs has revealed that in approximately neutral solution they can adopt two different quaternary forms. The T (tense or 'deoxy') form is adopted only by the deoxy or Fe^{II} complex in normal Hbs, the R (relaxed, 'oxy' or 'liganded') form by all others[33]. They differ in the way the four sub-units are arranged together. Crystals of horse $HbFe^{II}O_2$ undergo a drastic change in lattice at pH 5.9 and a further change below pH 5.4, which may represent two additional quaternary forms[169].

The model of horse metHb (that is $Fe^{III}OH_2$) with its R structure is shown in figure 2 (page 23). The four sub-units are arranged tetrahedrally. The haems (Fe porphyrins) are 2.5–3.7 nm apart. The centre of the protein is occupied by a cavity filled with water. The molecule has a dyad axis. There are extensive contacts between dissimilar chains. The $\alpha_1\beta_1$ (and $\alpha_2\beta_2$) contacts involve 34 residues and there are about 110 atoms with contacts of less than 400 pm; the interactions include 4–5 hydrogen bonds. The $\alpha_1\beta_2$ (and $\alpha_2\beta_1$) contacts involve 19 residues and ≈80 atoms with contacts of less than 400 pm; the interactions are all nonpolar, except for one or possibly two hydrogen bonds[172,173]. There are no $\alpha_1\alpha_2$ or $\beta_1\beta_2$ contacts in the R form[27].

The transition from the R form of $Fe^{III}OH_2$ to the T form of Fe^{II} involves a large movement in the $\alpha_1\beta_2$ contacts. The β_2 sub-unit rotates by 13.5° relative to the α_1 sub-unit about a screw axis and moves by 190 pm along it, which leads to relative displacements of atoms at the contact by up to 570 pm. By contrast, the relative displacement of atoms along the $\alpha_1\beta_1$ contact is only about 100 pm. Salt bridges (coulombic interactions between positively and negatively charged sites) are formed at the $\alpha_1\alpha_2$ and $\beta_1\beta_2$ junctions[27,172]. The net result of these changes is that the number of van der Waals interactions decreases, the number of hydrogen bonds remains unchanged, while eight new salt bridges appear. Two salt bridges contribute to each $\alpha_1\alpha_2$, one to each $\alpha_1\beta_2$ and one to each $\beta_1\beta_1$ contact. As will be seen in section 7.6, it is the making and breaking of these salt bridges that links the change in quaternary structure to the small changes in tertiary structure and to the changes at the Fe atom.

7.3 THE PROBLEM OF THE REVERSIBLE CO-ORDINATION OF O_2

The reversible uptake of O_2 by Hbs and Mbs involves the reaction of O_2 with a five-co-ordinate Fe(II) porphyrin complex, in which the axial ligand is the imidazole of histidine F7, to form a six-co-ordinate complex[109,162,169]. The O_2 therefore enters a vacant site in the co-ordination sphere of the Fe. The five-co-ordinate Fe(II) is high spin (ionic) and possesses four unpaired electrons, while the O_2 adduct is low spin (covalent) and is diamagnetic. Since the O_2 also possesses two unpaired electrons, the reaction involves the overall pairing of six spins; this must surely be the greatest change in spin known for any reaction. Other ligands that can react similarly include CO, isocyanides, imidazole, CN^- etc. (see reference 8).

The stereochemistry of the co-ordinated O_2 has not yet been established with certainty, but present evidence (see section 7.4) supports a bent configuration. This is also the structure recently established by X-ray analysis for the complex [Co(bzacen)py.O_2], where bzacen is NN'-ethylene-bis(benzoylacetoniminide) or bis(benzoylacetone)ethylenediamine. The relevant structural details are given below (I); the O—O bond length is close to that of the free superoxide ion (128 pm)[186]. Since all the Co-O_2 complexes so far studied, including the Co-containing analogues of Mb and Hb[106], show similar and characteristic e.s.r. spectra, it can be assumed that the structure of the co-ordinated O_2 remains basically the same. These e.s.r. spectra show g-values similar to those of free O_2^- and ^{59}Co h.f. coupling constants that are much smaller than those of the parent Co(II) complexes, indicating that the unpaired electron must spend most of its time on the O_2^- [134]. The X-ray and e.s.r. spectra, therefore, both show that the Co-O_2 complexes are better represented as Co(III) complexes with a co-ordinated superoxide anion O_2^- than as Co(II) with O_2*. The direct but slight evidence on Hb and Mb (see section 7.4) and the parallel between them and their Co-containing analogues in chemical properties[108,207] both support an analogous structure for the Fe-O_2 unit; the diamagnetism of a $Fe^{III}O_2^-$ complex could be ascribed to antiferromagnetic coupling between the spins of the metal and the ligand. But in order to avoid any confusion we shall here continue to write the complex as $Fe^{II}O_2$.

(I)

The uptake of O_2 is not completely reversible, since the $Fe^{II}O_2$ complexes are slowly converted into $Fe^{III}OH_2$. But red cells contain enzymes for the reduction

---

* *Note added in proof.* The formation of a Co-O_2 complex by the reaction of free O_2^- with a Co^{III} complex has recently been reported by Ellis, J., Pratt, J. M., and Green, M., *Chem. Commun.*, (1973), 781.

of Fe(III) back to Fe(II). It appears that the steady-state concentration of Fe(III) in oxygenated blood is 1–2 per cent[8].

The role of the protein in controlling these and other equilibria and reactions relating to the co-ordination of O_2 can best be put into perspective by looking at the equilibria and reactions of simple protein-free Fe(II) porphyrins.

The study of equilibria involving the binding of axial ligands by Fe(II) porphyrins (in particular Fe(II) protoporphyrin) in aqueous solution is hindered by their low solubility, their tendency to dimerise or form even larger aggregates and their ready autoxidation. The base most frequently studied is pyridine. The study of the imidazole complexes is further hindered by the insolubility of the bisimidazole complex[63]. But a few relevant features have been established. The so-called 'diaquo' complex is high spin[77], though it is not known whether it is really five- or six-co-ordinate. The six-co-ordinate bispyridine complex is low spin[77]. We have already seen that the five-co-ordinate monohistidine complexes of Hb and Mb are high spin, and the same would doubtless be true for the monopyridine complex. Let us now take a solution of the 'diaquo' complex (probably present mainly as the dimer) and slowly increase the concentration of pyridine, in the complete absence of air. Since successive formation constants usually show a fairly regular decrease in magnitude (that is, $K_1 > K_2$ in this case) we should expect to form first the mono- and then the bispyridine complex. In fact, we go straight from the diaquo to the bispyridine complex, and the monopyridine complex is never seen[77,155]; that is $K_1 < K_2$. Such a reversal of the relative order of magnitude of equilibrium constants is usually found where there is a change in spin state or in stereochemistry[193]. A good example of the former is provided by the co-ordination of the three successive molecules of the bidentate ligand dipyridyl by Fe(II), where $K_1 \geqslant K_2 \ll K_3$ and K_3 is associated with a change from high to low spin[20]. The change of spin state associated with K_2 is obviously the major factor that produces the order $K_1 < K_2$ in the Fe(II) porphyrins, though additional complications could be caused by the change in the number of axial ligands. We are unable, therefore, to prepare the five-co-ordinate Fe(II) porphyrin with one base in solution and to study its reaction with O_2.

Solutions of Fe(II) porphyrins in pure dry pyridine do not form any adduct with O_2 and are not oxidised. The addition of small amounts of water has no effect, but larger amounts cause a rapid autoxidation to Fe(III)[62]. Two different groups of workers have studied the kinetics of autoxidation in various solvent mixtures and have both concluded that the initial step involves the substitution of one pyridine in the bispyridine complex to form the O_2 adduct, though they disagree over the nature of the subsequent steps[58,119] (see also section 7.7). It appears, then, that the py–Fe(II)–O_2 complex can be formed in solution, but that under the experimental conditions its concentration in equilibrium with the bispyridine complex is too small to be observable and/or its rate of formation is less than its rate of further reaction*.

---

* *Note added in proof.* An $Fe^{II}O_2$ complex containing a porphyrin-type ligand has recently been prepared in solution and its structure determined in the solid state; Collman, J. P., *et al.*, *J. Am. Chem. Soc.*, 95 (1973), 7868; Collman, J. P., *et al.*, *Proc. natn. Acad. Sci. U.S.A.*, 71 (1974), 1326.

There is, however, evidence that O_2 adducts can be formed in the solid state. Crystalline samples of the bisimidazole Fe(II) complexes of proto- and meso-porphyrin show an increase in weight when exposed to dry air at room temperature for 8 hours, corresponding to 1 mole of O_2 per Fe, which can be reversed by heating the sample in nitrogen for 6 hours at 60–65° [61,63]. Wang has shown that Fe(II) protoporphyrin diethyl ester embedded in a 1 : 3 matrix of polystyrene and 1-(2-phenylethyl)-imidazole can undergo a reversible spectroscopic change on reaction with oxygen or air at room temperature. As prepared, the complex contains CO and one imidazole as the axial ligands; the CO is then removed to give a product with broad absorption bands, which suggests the weak co-ordination of a second imidazole; this is the complex that reacts with O_2. Heating the sample to 80° gives a product, which cannot react with O_2 and shows sharp absorption bands, suggesting that the second imidazole is now firmly bound. The ability to react with O_2 can be restored via treatment with CO [227,229]. Finally, the solid bis-pyridine complexes of several Fe(II) porphyrins will lose one or both molecules of pyridine on heating *in vacuo* to form complexes that react slowly with air; the analytical data agree with the dimeric structure py-Fe-O_2-Fe-py for one of the products and a monomeric structure with O_2 as the only axial ligand for the others [4]; but the reversibility of the reaction with O_2 was not reported. Further reactions leading to irreversible autoxidation are obviously inhibited in the solid state.

The use of a polymeric matrix is particularly relevant because it avoids the relatively large changes in lattice energy that might be associated with equilibria involving crystalline samples; the results are, therefore, more likely to approximate to those that might in principle be observed in solution. In addition, the hydrophobic matrix, possibly with one imidazole near the co-ordinated O_2, will provide an environment similar to that of the amino-acid groups around the O_2 in Hb and Mb (see section 7.4). The results show that O_2 will displace the weakly bound, but not the strongly bound, second imidazole. If we assume that over 90 per cent of the Fe is co-ordinated to O_2 in air and over 90 per cent to the weakly held imidazole in nitrogen, then we can obtain a minimum value for the equilibrium constant for the binding of O_2 by the five-co-ordinate Fe(II) at room temperature; this corresponds to $P_{1/2}$ (the partial pressure required for 50 per cent formation of the O_2-adduct) $\leqslant 2$ or $\log_{10} P_{1/2} \leqslant 0.3$ mm Hg. As will be shown in section 7.5, this is close to the value for many Hbs and Mbs.

We can now see that nature has to overcome several problems in developing O_2-carrying proteins. The main requirements are that the protein must:

(i) Stabilise the five-co-ordinate form of Fe(II); that is, provide one and only one axial ligand. An O_2-carrying protein, which contained six-co-ordinate Fe(II) in the resting state (that is, in the absence of O_2) would be quite conceivable, but the equilibrium constant and rate constant for the binding of O_2 would be very much lower.

(ii) Adjust the equilibrium constant for the binding of O_2 to the needs of the organism.

(iii) Prevent irreversible reactions of $Fe^{II}O_2$, including autoxidation to Fe(III).

(iv) Adjust the $Fe^{II/III}$ couple so as to stabilise Fe(II) in the resting state (that is, in the absence of O_2 and in the presence of reducing power roughly equivalent to that of the H_2 electrode).

A less important requirement is to:

(v) Prevent access to any other extraneous ligands (amino acids, glutathione, phosphates, etc.), which could reversibly or irreversibly 'poison' the protein by co-ordination to the Fe(II) or Fe(III) states.

The protein must also:

(vi) Allow the uptake and release of O_2 to occur at a sufficiently rapid rate.

In the next section we shall examine how the protein controls the basic co-ordination chemistry (requirements (i) and (v)). Sections 7.5 and 7.6 deal with the mechanism for controlling the equilibrium involving O_2 (requirement (ii)). Autoxidation (requirement (iii)) is discussed in section 7.7, and control of the redox potential (requirement (iv)) is mentioned briefly in the summary (section 7.8).

7.4 THE MECHANISM OF CONTROL OF THE BASIC CO-ORDINATION CHEMISTRY

Several Hbs and Mbs have been reversibly dissociated into the apoprotein (globin) and the prosthetic group (Fe porphyrin), by, for example, treating with cold, acidified acetone[6,192]. The reconstituted proteins have properties indistinguishable from those of the original proteins. Transfer or exchange of Fe porphyrins can also be observed in neutral solution at room temperature, for example from Hb to apo-Mb and vice versa[15,191,215], between various free and Hb-bound Fe porphyrins[90] and between Hb F and Hb A containing labelled Fe porphyrin[43]. Lability varies with the state of the protein-bound Fe in the general order $Fe^{III}OH_2 > Fe^{II}CO, Fe^{II}CN^- \gg Fe^{II}$ [43,90,113]; the inertness of the last is hardly surprising since breaking the Fe–histidine bond would in this case produce a very unstable four-co-ordinate intermediate.

We can therefore write the simplified equilibrium

$$\text{Hb or Mb} \rightleftharpoons \text{globin} + \text{Fe porphyrin} \tag{7.5}$$
$$\text{(protein)} \quad \text{(apoprotein)} \quad \text{(prosthetic group)}$$

In practice the equilibria are far more complex, because in aqueous solution Fe(II) and Fe(III) porphyrins without additional axial ligands such as pyridine, CN^- or CO tend to form dimers and higher aggregates. Study of the competition between apo-Mb(G) and the ligands histidylhistidine and pilocarpine (which do not co-ordinate to the Fe in Mb) for the dimeric form of aquo-Fe(III) mesoporphyrin (here written simply as Fe) gave a value of $K = 10^{15}$ M^{-1} [14] at pH 7.5 for the equilibrium

$$Fe_2 + 2G \rightleftharpoons 2Fe \cdot G$$
$$K = [Fe \cdot G]^2 / [Fe_2][G]^2 \tag{7.6}$$

Unfortunately, this value cannot be converted into an equilibrium constant for the binding of the monomeric form of Fe, since no dimerisation constants have been reported. However, if we assume from the existence of dimers at the low concentrations used in spectrophotometric work that the dimerisation constant $K_D =$ $[Fe_2]/[Fe]^2 \gg 10^5$ M^{-1}, then we obtain a minimum value of $K' > 10^{10}$ M^{-1} for the binding of globin by the monomer according to the equilibrium

$$Fe + G \rightleftharpoons Fe . G$$
$$K' = [Fe . G]/[Fe][G] = \sqrt{(KK_D)} \qquad (7.7)$$

A second very approximate value of $K \approx 10^{12}$ M^{-1} can be obtained from the ratio of the rate of reaction of the $Fe^{II}CO$ complex of protoporphyrin with apo-Hb to the rate of loss of the $Fe^{III}OH_2$ complex from Hb, if it is assumed that the $Fe^{II}CO$ and $Fe^{III}OH_2$ complexes have the same rate constants[89]. Competition between apo-Mb and apo-Hb was used, together with the above value of K for Mb, to derive the following values of K for the binding of the first and fourth molecules of $Fe^{III}OH_2$ protoporphyrin by apo-Hb: K_1, 10^{12-13}; K_4, 10^{15-16}; the values depend on pH[15]. Even though rather approximate, these values do indicate a very high affinity of the protein for the Fe–porphyrin complex. This strong and specific binding is shown only by globins that have not been denatured; after denaturation, the proteins may bind a larger number of Fe porphyrins more loosely[192].

X-ray analysis shows that the Fe porphyrins in Hbs and Mbs lie in essentially nonpolar pockets near the surface of the protein, and that there are about 60 contacts of 400 pm or less between the atoms of the protein and those of the complex. These contacts include the co-ordinate bond between the Fe and histidine F8 and hydrogen bonds involving the carboxyls of one or both of the propionic acid side chains at positions 6 and 7 of the porphyrin. All the other contacts are nonpolar, involving van der Waals forces[172]. There are no covalent bonds. Study of the effects of changing the side chains of the porphyrin, of removing the metal and of varying the amino-acid groups of the protein (in mutant Hbs) indicates the relative importance of these types of interaction.

At neutral pH the apoproteins bind Fe(III)-protoporphyrin only 3.0 times as firmly as the metal-free ligand. Neglecting differences in the degree of dimerisation, etc., this clearly indicates that the Fe plays only a relatively unimportant role in the binding of the complex[215]. The apoproteins can be recombined with Fe complexes of a wide range of different porphyrins to give Hbs and Mbs that can reversibly co-ordinate O_2 (see, for example, reference 192). When the two vinyl groups at positions 2 and 4 in protoporphyrin are varied, the rate of reaction with the apoprotein falls in the order: proto- (vinyl) > meso- (ethyl) > deutero- (H) ≫ haemato- (hydroxyethyl) ≫ dimethyldeuteroporphyrin disulphonate ($-SO_3H$), which showed no reaction. The first can also displace all the others under equilibrium conditions[89]. Adducts capable of combining with O_2 can also be prepared from Fe complexes where the hydrophilic propionic acid side chains at positions 6 and 7 in protoporphyrin have been replaced with hydrophobic substituents, but they are much less stable; methylation of the acid groups reduces the affinity for the apoprotein by about 300[192,215]. Finally, we can look at Hb mutants to see the effects of changes in the nature of the residues in contact with the Fe porphyrin[170]. Some of these mutations lead to a more ready loss of the Fe and denaturation of the globin under certain conditions (see, for example, reference 113). Strictly speaking, this is evidence for an increased kinetic lability, not a decreased thermodynamic stability, but the two are probably related. Hb Santa Ana, however, apparently contains no Fe in the β chains; here leucine F4 has been replaced by proline[170].

It is clear that the equilibrium constant for the binding of the monomeric Fe porphyrin to the protein is very high, and that by far the major contribution to the binding energy comes from the van der Waals interaction between the porphyrin ligand and the nonpolar residues of the protein. The protein is therefore able to control the number and nature of the axial ligands presented to the Fe because the energy of interaction between the porphyrin and the protein is far greater than that between the Fe and any potential axial ligands, so that the stereochemical requirements of the former dominate those of the latter.

Reference to figure 1 (page 22) shows that the structure of the Mb fold allows the hydrophilic imidazole groups of two histidines to protrude from helices E and F into the hydrophobic areas on either side of the porphyrin plane. The *proximal* histidine F8 is co-ordinated to the Fe. But the *distal* histidine E7, though tantalisingly close, cannot become co-ordinated without affecting helix E. A few mutants are known in which histidine F8 has been replaced by tyrosine; this occurs in the α chains of Hb M Iwate and in the β chains of Hb M Hyde Park. The abnormal chains stabilise Fe(III) and their ability to co-ordinate O_2 after reduction does not appear to have been investigated[8]. Although extraneous imidazole can co-ordinate to Hb and Mb in both Fe(II) and Fe(III) forms, neither the imidazole of the distal histidine E7 nor any other protein residue becomes co-ordinated as a second axial ligand in normal Hbs and Mbs, except when the protein is denatured. Examples of the reversible denaturation of the protein accompanied by the co-ordination of some nitrogenous base (as shown by the changes in the absorption spectra) are the drying of Hb Fe^{II} *in vacuo*[161] and the heating of or addition of alcohols to solutions of *Aplysia* Hb[42]. Co-ordination of the distal histidine to Fe(III) occurs in the abnormal α chains of Hb M Iwate[94]. But in all normal Hbs and Mbs so far studied the protein provides one and only one axial ligand; that is, the protein enforces five-co-ordination on the Fe(II) ion and histidine is the preferred, but perhaps not unique, axial ligand.

The residues on both the proximal and the distal side of the porphyrin plane play a role in the co-ordination of O_2. Those on the proximal side are particularly important in transmitting the effects of the co-ordination and release of O_2 through helix F to helix H and will be discussed in section 7.6. Here we shall look more closely at the stereochemistry of small ligands that can be co-ordinated in the second or distal position and of the surrounding amino-acid side chains. Slight differences do exist between proteins (for example, between Mb and the α and β chains of Hb[172], but these are probably not significant for the discussion in this section.

The space occupied by small ligands such as O_2, H_2O, N_3^- and CO is bounded by the Fe–porphyrin and by four amino acids, namely the distal histidine E7, valine E11, phenylalanine CD1 and leucine G8 (or leucine B10 in Mb); see, for example, references 8 and 172. The last three amino-acid side chains are hydrophobic and are in contact with the periphery of the porphyrin ring. The imidazole of the histidine, on the other hand, is hydrophilic and is not in contact with the porphyrin, though it is situated closest to the Fe–X axis. The imidazole hangs almost perpendicularly to the haem plane, and one N is 380 pm above the haem plane and 170 pm from the perpendicular through the Fe. The second N of the imidazole is situated at the surface of the protein. The first N forms hydrogen bonds to co-ordinated H_2O and N_3^- in the Fe(III) forms of both Hb and Mb[171,211,232], as shown in structures (II) and (III).

(II)   (III)

Two basic structures have been considered for co-ordinated O_2, involving either 'sideways' π-bonding as found for ligands such as ethylene (see structure (IV)) or 'end-on', but nonlinear co-ordination (see structure (V)) as originally proposed by Pauling and found for N_3^-. The two atoms of co-ordinated oxygen have not yet been definitely located in the electron-density maps of either Hb or Mb, but the difference Fourier maps between the $Fe^{II}O_2$ and $Fe^{III}OH_2$ forms of Mb suggest structure (V), in which the co-ordinated atom is also hydrogen bonded to the distal histidine as shown[232]. Steric hindrance would also seem to exclude structure (IV)[8]. As already mentioned in section 7.3, the O_2 adducts of Co(II) complexes definitely have a structure analogous to (V) and are best formulated as Co(III) complexes with a superoxide ion (O_2^-).

(IV)   (V)

The distal histidine is present in all normal mammalian Hbs and Mbs, and was previously thought to play an essential role in the reversible co-ordination of O_2, by, for example, the formation of a hydrogen bond. But several mutants are now known in which histidine E7 is replaced by other amino acids, for example tyrosine in the α chains (Hb M Boston) or β chains (Hb M Saskatoon) or arginine (Hb Zürich)[8,170]; the Fe in the abnormal chains of at least the second and third proteins can bind O_2, though the rate of autoxidation is increased[153,237]. The tyrosines might be able to form hydrogen bonds to the co-ordinated O_2, but the arginine is apparently too big to be accommodated within the haem pocket and must protrude onto the surface, leaving the haem pocket empty[170]. In the monomeric Hb from the insect *Chironomus* histidine E7 is replaced by glutamic acid. But X-ray analysis shows that this group also protrudes onto the surface and that its place is approximately occupied by isoleucine E11, which cannot form any hydrogen bonds[110]. It appears, then, that the nature of the 'distal' group, and whether it can form hydrogen bonds or not, need not play any significant part in determining the equilibrium constant for the binding of O_2; a possible role in protecting the co-ordinated O_2 from attack by reducing agents is suggested in section 7.7.

The presence of this distal histidine offers no serious steric hindrance to small ligands such as H_2O, or ligands such as N_3^- and O_2, which are not linearly co-ordinated. The N_3^- projects over one of the methine carbons of the porphyrin ring and, in fact, fits quite snugly between the histidine, phenylalanine and valine groups[211]. But the distal histidine will offer very considerable steric hindrance to ligands such as CO and CN^-, which prefer linear co-ordination. This steric hindrance can be reduced either by deviation of the $Fe-\hat{C}-O$ or $Fe-\hat{C}-N$ bond angle from $180°$ and/or by a readjustment of the protein. X-ray analysis of the CO complex of the monomeric Hb from *Chironomus* has revealed a $Fe-\hat{C}-O$ bond angle of $145 \pm 15°$; the isoleucine E11, which occupies roughly the position of the distal histidine in mammalian Hbs, has also suffered a slight displacement[109]. The observation of anomalously low wavenumbers for the stretching frequency of co-ordinated CO in a variety of different Hbs and Mbs that possess the distal histidine, but not in those in which the histidine has been replaced by arginine or tyrosine, has also been ascribed to some interaction (hydrogen bonding or steric) between the histidine and the co-ordinated CO[48]. A Fourier difference map between the $Fe^{III}OH_2$ and $Fe^{III}CN^-$ complexes of Mb, on the other hand, suggests that the $Fe-C-N$ system remains linear and that helix E has been displaced[8]. X-ray analysis therefore provides direct evidence for considerable steric hindrance together with a certain degree of flexibility in the protein surrounding the distal co-ordination position. The ability to co-ordinate much larger ligands such as heterocyclic bases and isocyanides provides indirect evidence for a much greater degree of flexibility. Interesting steric limitations have been observed in, for example, the nature and position of substituents in pyridine and imidazole, which will allow or prevent co-ordination to Mb[123]; for a rationalisation of these steric effects see reference 8. Even the co-ordination of O_2 may require slight movements of the surrounding groups. The cavity between the Fe and helix E in the deoxy or Fe^{II} form is large enough to accommodate O_2 in the α chains of Hb, but one of the γ methyl groups of valine E11 blocks the ligand site in the β chains and must be moved to allow co-ordination[27].

The attraction and repulsion between the co-ordinated ligand and the surrounding amino acids may affect the magnitude of the equilibrium constant, though the size of this effect will be difficult to estimate. There are, unfortunately, no protein-free five-co-ordinate Fe(II) complexes available to provide a comparison of equilibrium constants (see section 7.3). The equilibrium constants for the binding of CN^- by Hb and Mb Fe^{II} appear to be less than 10 M^{-1} [121], which seems very low (compare some of the data in reference 77), and presumably reflects the above-mentioned steric hindrance and the dislike of a charged anion for the more hydrophobic environment in the protein due to a decrease in solvation. Hb binds CO five times more firmly than does an aqueous solution of Fe(II) protoporphyrin in the presence of 5×10^{-4} M pyridine [155], which would seem to imply stabilisation of the Fe−C bond by the protein. But this ratio should, of course, be divided by the equilibrium constant (which is not known) for the binding of the sixth ligand (H_2O or py) by the five-co-ordinate Fe(II); the resulting ratio would probably indicate a significant destabilisation by the protein. But we are mainly interested here in the co-ordination of O_2. It appears from the X-ray data that the arrangement of the residues around the distal co-ordination site is tailor-made to minimise any repulsions and rearrangements of the protein that might reduce the equilibrium constant for binding O_2 − assuming, of course, that it has structure (V); on the other hand, there is no evidence for any significant enhancement of the equilibrium constant by, for example, hydrogen bonding. It appears that this region, if considered in isolation from the rest of the protein, exerts little or no effect on the binding constant of O_2 (see also section 7.5), but reduces the binding constant for most other potential ligands, especially those that prefer linear co-ordination. Variation in the degree of steric hindrance to CO could well be one of the reasons for the wide variation (from 550 to 0.075) in the relative affinities for CO and O_2 [91,122].

Direct access of the ligand to the Fe is, of course, blocked by the amino acids, in particular the distal histidine. As already mentioned, one N of the histidine imidazole ring points towards the Fe, while the other is actually situated on the surface, so that the heterocycle can act as a sort of trap-door, closing the ligand cavity. The co-ordination of any ligand must therefore be a complex process involving some intermediate change in the conformation of the protein, for example a rotation of histidine E7 around its $C_\alpha - C_\beta$ bond or a breathing of helix E [161]. Yet in spite of this the rates of O_2-uptake are extremely fast. The second-order rate constants at $20°C$ for various Mbs fall in the range $1.0-1.9 \times 10^7$ dm^3 mol^{-1} s^{-1} (the three activation energies so far determined are 23.0, 23.0 and 29.3 kJ mol^{-1} (5.5, 5.5 and 7.0 kcal/mole), respectively) and those for the isolated but slightly modified α and β chains in the range $5-8 \times 10^7$ dm^3 mol^{-1} s^{-1} [8], while that of the monomeric Hb from *Chironomus* has the remarkably high value of 3.0×10^8 mol^{-1} s^{-1} [6]. Because of changes in the quaternary structure the Hbs show complex kinetics; but the rate constants are also in the range 10^6-10^7 dm^3 mol^{-1} s^{-1}. Corresponding values for the loss of O_2 are rate constants of $10-70$ s^{-1} and activation energies of 80-88 kJ mol^{-1} (19-21 kcal/mole) for Mbs and $10-50$ s^{-1} and 67-105 kJ mol^{-1} (16-25 kcal/mole) for most Hbs (values very dependent on pH). For references see reference 8. Even if the histidine does significantly decrease the rate constant that might be observed in the absence of protein, the resultant rate is still fast enough for physiological purposes. Mutant

Hbs, in which histidine has been replaced by arginine or tyrosine, show slightly greater rates, at least in their reactions with CO[8]. A few Hbs with very low rates of dissociation ($\approx 10^{-3}$ s^{-1}), which clearly cannot function in supplying O_2, are known among the nematodes[91].

The above discussion shows how the myoglobin fold exerts an effect on the equilibria involving the axial ligands at two very different levels: firstly, a major effect in stabilising the five-co-ordinate Fe(II); secondly, and more tentatively, a minor effect in preventing the co-ordination of unwanted, extraneous ligands. In section 7.2 we saw that when protein-free Fe(II) porphyrins co-ordinate nitrogenous bases or other strong ligands, changes occur in the spin state (and probably in the number of the axial ligands) such that the equilibrium constant for binding the second molecule is greater than for the first; that is, $K_1 < K_2$, and the five-co-ordinate complex containing only one base cannot be observed. In Mb and Hb, however, the energy of interaction between the porphyrin and the protein, together with the disposition of the amino-acid residues, is such that in effect $K_1 \gg K_2$ and the five-co-ordinate form is stabilised.

7.5 OBSERVED VARIATIONS IN THE EQUILIBRIUM CONSTANT FOR BINDING O_2

The equilibrium between Mbs and O_2 shows a simple (hyperbolic) relationship of the type shown in figure 30, which corresponds to the straightforward equilibrium

$$Fe^{II} + O_2 = Fe^{II} O_2 \tag{7.8}$$

Typical mammalian Hbs, on the other hand, give a sigmoid curve of the type shown in figure 31 (page 115), which clearly cannot be described by a single equilibrium. The equilibrium constant for any one of the four Fe atoms in the tetrameric Hb unit depends on the state of oxygenation of the others; that is, there is some co-operative interaction between the four Fe atoms. Comparison of the equilibrium constants of Hb with those of the isolated α and β chains shows that Hb behaves normally at high P_{O_2} and that the anomaly occurs at low P_{O_2}; that is, the equilibrium constant for the binding of O_2 by one Fe is approximately normal when the other three are oxygenated, but anomalously low when the others are de-oxygenated. The O_2-equilibrium curves of Hbs can be described empirically by Hill's equation

$$Y = \frac{Kp^n}{1 + Kp^n} \quad \text{or} \quad \frac{Y}{1 - Y} = Kp^n \tag{7.9}$$

where Y is the fractional saturation of Hb with oxygen, p is the partial pressure of O_2, and K and n are constants. It can be shown theoretically that n should equal 1 at very low and very high P_{O_2}, but have a value of about 3 in the middle range. Absence of haem–haem interaction leads to $n = 1$ throughout[169]. Values of n at $P_{1/2}$ (see below) are therefore often determined and quoted in the literature as a measure of the degree of haem–haem interaction; but we shall not be concerned here with variations in n. Human Hb has $n \approx 2.9$.

For reasons of convenience only one equilibrium constant is usually determined for the binding of O_2 (and other gases such as CO) by Hbs and Mbs, and this is usually reported in the form of $P_{1/2}$, that is the partial pressure (in mm Hg) required to produce 50 per cent oxygenation ($Y = 0.5$). Note, therefore, that the greater the affinity for O_2, the lower the value of $P_{1/2}$. In the case of those haemoproteins that give a simple hyperbolic O_2-equilibrium curve, $P_{1/2}$ can be converted into a true equilibrium constant (in units of mol^{-1}) by insertion of the solubility of O_2 in H_2O.

The physiological reason for the sigmoid shape of the curve in figure 31 is to help the Hb to unload its O_2 and transfer it to the tissue when and where it is most required, namely where the partial pressure is lowest. Most typical Hbs show other unusual variations of $P_{1/2}$, which also serve a physiological function. Figure 32 shows how $\log_{10} P_{1/2}$ varies with pH. This is the so-called *Bohr effect*. Above pH 6

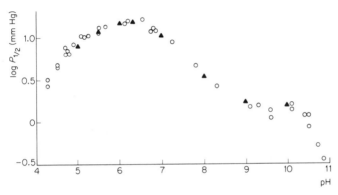

Figure 32 Variation of $P_{1/2}$ with pH (Bohr effect) shown by human Hb in 0.3 M NaCl at $20°C$; data obtained by direct (o) and indirect (▲) methods. From reference 9.

(alkaline Bohr effect) raising the pH increases the affinity of Hb for O_2 and, conversely, the uptake of O_2 involves the liberation of a proton. At physiological pH this corresponds to 0.7 protons per haem[169]. This alkaline Bohr effect has two physiological functions. Firstly, tissues consuming O_2 produce lactic acid, CO_2 and carbonic acid, which help to release the O_2 from Hb; secondly, the local rise in pH due to the release of O_2 will increase the total concentration of dissolved CO_2 (in all forms) by displacing the following equilibrium to the right

$$CO_2 + H_2O \rightleftharpoons H_2CO_3 \rightleftharpoons HCO_3^- + H^+ \qquad (7.10)$$

and hence facilitate its transport from the tissues back to the lungs. The reverse effect of pH, which is observed below pH 6 (acid Bohr effect), may have no physiological function. Other agents that can effect $P_{1/2}$ include organic phosphates such as 2,3-diphosphoglycerate (DPG), which is present in the red cells of humans and many other mammals. It binds to, and therefore stabilises, the deoxy or Fe^{II} form of Hb; and by promoting the release of O_2 allows us to adapt quickly to adverse conditions, for example loss of blood or a change to high altitude. Nature has obviously developed very sophisticated methods for controlling $P_{1/2}$.

Hbs and Mbs can show a wide range of affinity for O_2. The highest recorded affinity is that of Hb from the perienteric fluid of the pig round worm *Ascaris lumbricoides* with $\log P_{1/2}$ at $20°$ of -2.8[91] or -2.5 to -3.0[166]. High affinities are shown by other parasite Hbs[91,122,166] and by leghaemoglobin with $\log P_{1/2} = -1.3$[112]. At the other extreme, many fish Hbs show a large Bohr effect, which can lead to very low affinities. $\log P_{1/2}$ of trout Hb IV, for example, changes from 1.0 at pH 8.5 to 3.1 at pH 6.1, which seems to be the lowest affinity so far recorded[24]. This large Bohr effect is probably related to the mechanism for actively transferring O_2 from the blood into the swim bladder[138,208]. $\log P_{1/2}$ can therefore vary by about 6.

But the vast majority of Hbs and Mbs so far studied have values of $\log P_{1/2}$ within the range covered by human Hb itself (see figure 32), that is from 1.2 to -0.5. Most mammalian Mbs fall within the much more limited range -0.3 to -0.15. Similar values are obtained when the α and β chains of Hb are isolated (-0.3 to -0.4) and from most mutations of Hb (for example, Chesapeake and Rainier, both -0.3), by formation of the complex of Hb with haptoglobin (-0.5), and by many chemical modifications of Hb; however, a few mutations may leave the value unchanged (for example, M Hyde Park and M Saskatoon) or even increase it (for example, Kansas 1.4; compare Hb at pH 11–figure 32). Lamprey Hb, where $\log P_{1/2}$ changes from 0.3 at pH 8.8 to 1.6 at pH 5, is one example that just goes outside the range covered by human Hb. For further data and references see reference 8.

Lamprey Hb is known to possess the same proximal histidine as human Hb[144], and similarities in the absorption spectra of the Fe^{II} and $Fe^{II}CO$ complexes (but not the $Fe^{II}O_2$ complex) strongly suggest that the same is true for the *Ascaris* Hb[238]. We can therefore say that $\log P_{1/2}$ can vary from approximately -2.8 to $+1.6$, that is by ≈ 4.5, without a change in the nature of the ligand. The inclusion of trout Hb IV would extend the range to 6, but unfortunately we do not know whether the axial ligand is the same or not.

Mammalian Mbs show only a very small Bohr effect and until recently it was thought that the occurrence of a Bohr effect was necessarily associated with changes in the quaternary structure of polymeric proteins such as the tetrameric mammalian Hbs. But a well-marked Bohr effect has recently been shown to occur in two of the monomeric Hbs from the fly *Chironomus*; $\log P_{1/2}$ of Hb IV at $25°$ changes from -0.3 at pH 10 to $+0.4$ at pH 5.5[86]. Heterotropic interaction can therefore be associated with changes in tertiary structure alone or in changes in both tertiary and quaternary structure.

If each polypeptide chain contains only one Fe, then homotropic or haem–haem interaction can, of course, only be exhibited by polymeric proteins and is associated with changes in quaternary (as well as tertiary) structure between the so-called R and T forms (see sections 7.2 and 7.6). Equilibrium constants for the binding of O_2 have now been evaluated for both forms of human Hb at pH 7 and of sheep Hb at pH 9; that of the R form is ≈ 250 times greater than that of the T form; that is, $\Delta \log P_{1/2} = 2.4$[204].

The main points to emerge from the above data are:

(i) The protein can alter $\log P_{1/2}$ by ≈ 4.5 (and perhaps even 6) without changing the nature of the axial ligand.

(ii) The protein can couple together the equilibrium involving the co-ordination of O_2 with a second equilibrium; it can change $\log P_{1/2}$ by up to ≈ 2 by heterotropic interaction (Bohr effect) and by up to 2.4 by homotropic (haem–haem) interaction, that is simply by a change in the quaternary structure.

(iii) The coupling of two such equilibria provides our first example of the way in which the protein can drive one thermodynamically unfavourable reaction (in this case the transport of O_2 against the concentration gradient) at the expense of another thermodynamically favourable reaction (indirectly the neutralisation of acid with base).

(iv) We have calculated (section 7.3) that a protein-free Fe(II) porphyrin analogous to that present in Hbs and Mbs would have $\log P_{1/2} \leqslant 0.3$; that is, its affinity for O_2 is comparable at least to that of human Hb and may be greater. We can therefore conclude that, once the protein has stabilised the five-co-ordinate form of Fe(II), there is no need for any further significant (thermodynamic) stabilisation of the Fe–O_2 bond. In other words, the protein uses brute force to effect a major change in the equilibrium between four-, five- and six-co-ordinate Fe(II) porphyrins, but only provides a fine adjustment for the equilibrium involving the co-ordination of O_2.

We can compare these values of $\log P_{1/2}$ for Fe(II) complexes with those recently reported[207] for the Co-containing analogues of two Mbs and for a protein-free Co(II) porphyrin

	Fe(II) 20°C	Co(II) 25°C
Horse and whale Mb	-0.2, -0.3[8]	$+1.8$[207]
Protein-free analogue	$\leqslant +0.3$	$+4.2$[207]

The Co(II) complexes with and without protein all bind O_2 less firmly than their Fe(II) analogues. The protein undoubtedly shows a small but definite stabilisation ($\Delta \log P_{1/2} = 2.4$) of the Co–$O_2$ bond, but the change is within the range that can be considered as 'fine adjustment'. The protein would hardly be expected to have exactly the same effect on Fe(II) and Co(II), because of their differing electronic and steric properties. It is also worth noting the magnitude of the solvent effect that has been observed in the binding of O_2 by a Co(II) porphyrin. 50 per cent oxygenation of the complex of Co(II) protoporphyrin IX dimethyl ester containing 1-methylimidazole as the axial ligand at $-23°C$ requires a partial pressure of 55.6 kN m^{-2} (417 mm Hg) in toluene, but only 1.68 kN m^{-2} (12.6 mm Hg) in dimethylformamide; that is, $P_{1/2}$ decreases by a factor of 33. But allowance for the greater solubility of O_2 in toluene leads to a change in the equilibrium constant by a factor of 60; that is, $\Delta \log K < 2$[212]. This solvent effect, which corresponds to a complete change in the environment of the co-ordinated O_2, is small compared to the effects of the protein we have been discussing, which involve a minimal change in the environment of the O_2.

Many values of ΔH for the co-ordination of O_2 have been determined indirectly from the variation of $P_{1/2}$ (that is ΔG) with temperature, but only two (see below) have been determined directly by calorimetry; all data are taken from reference 8 unless otherwise indicated. Values for myoglobins lie in the range $\Delta H = -54$ to -67 kJ mol^{-1} (-13 to -16 kcal/mole), though that of horse Mb is -86.2 kJ mol^{-1} (-20.6 kcal/mole). The isolated α and β chains of Hb also have values of -56.5 kJ mol^{-1} (-13.5 kcal/mole). The values for haemoglobins show a much greater variation and are usually pH dependent. Calorimetric measurements on ox Hb gave ΔH -39.3 and -56.1 kJ mol^{-1} (-9.4 and -13.4 kcal/mole) at pH 6.8 and 9.5, respectively. Other mammalian Hbs also show ΔH approximately -42 kJ mol^{-1} (-10 kcal/mole) at pH ≈ 7 and -54 to -63 kJ mol^{-1} (-13 to -15 kcal/mole) at pH 9.5. Trout Hb IV shows a much greater change from $\Delta H = -25$ kJ mol^{-1} (-6 kcal/mole) at pH ≈ 7 to -59 kJ mol^{-1} (-14 kcal/mole) at pH ≈ 9[24]. But trout Hb I and tuna Hb both give the very low pH-independent values of -12 and -7.5 kJ mol^{-1} (-3 and -1.8 kcal/mole), respectively[24,190]. All these values relate to the co-ordination of gaseous O_2 and should therefore be corrected for the heat of dissolution of O_2, $\Delta H = -12$ kJ mol^{-1} (-3 kcal/mole)[24]. This correction would lead to $\Delta H = 0$ and $+4$ kJ mol^{-1} (0 and + 1 kcal/mole) for the last two fish haemoglobins. ΔH can therefore vary from $+4$ to -75 kJ mol^{-1} ($+1$ to -18 kcal/mole).

There are several reasons why these values are probably not too meaningful. Firstly, even the calorimetrically determined value of ΔH, which has a definite significance, represents the sum of changes involving the metal and involving the protein, which we cannot separate. Secondly, there are good theoretical reasons for believing that changes in electronic factors (for example, bond energies) may be better reflected in ΔG than in ΔH, when the latter is determined indirectly from the temperature variation of ΔG[23]; this is undoubtedly so in the well-studied case of the dissociation constants of substituted acetic and benzoic acids (see, for example, reference 97). Recorded values of ΔH may also include the effect of the variation of the dissociation of the buffer ions with temperature[242].

Finally, it must be emphasised that O_2 is far from unique in the treatment it receives from Hb and Mb. There are many other equilibria involving the Fe, which also show heterotropic or homotropic interaction, and which clearly involve a very similar mechanism.

The Mbs and monomeric Hbs can only show heterotropic interaction. It is only recently that the existence of a well-defined Bohr effect has been shown in two of the Hbs from *Chironomus*[86] and one from *Glycera*[148]. Mammalian Mbs do not show such pronounced effects, but several examples of weak heterotropic interaction are known, for example the effect of pH (Bohr effect) on the binding of O_2[8] and of CO[228], the effect of the co-ordination of O_2 on the binding of Xe and zinc ions by the protein[129], and the effect of Xe on the co-ordination of CO and CN$^-$ and on the pK of Fe^{III}OH$_2$[56,128]. X-ray analysis shows that bound Xe occupies a single well-defined site next to the co-ordinated imidazole[198], but that the zinc ions are bound at a site far removed from the Fe porphyrin[12].

The usual mammalian Hbs show both homotropic and heterotropic interaction.

As we shall see in the next section, they both involve changes in the salt bridges linking the sub-units and are therefore functionally related. But the two effects can be separated. Certain mutants (such as Hb Rainier) have a diminished or zero haem–haem interaction (n = 1-1.3), but a normal or only slightly reduced Bohr effect, while Hb Hiroshima shows a normal haem–haem interaction, but a diminished Bohr effect[8,170]. The Bohr effect is shown not only in the binding of O_2, but also in the co-ordination of CO and EtNC by Fe^{II}. The value of the mid-point potential for the couple involving Fe^{II} and $Fe^{III}OH_2$ shows a very similar variation with pH; this is called the *oxidation Bohr effect*. Diphosphoglycerate also affects the mid-point potential as well as the affinity for O_2. Homotropic or haem–haem interaction is shown in the co-ordination by Fe^{II} of CO, EtNC and nitrosobenzene as well as O_2, and in the mid-point potential for the $Fe^{II/III}$ couple; and possibly in the reaction of $Fe^{III}OH_2$ with CN^- and N_3^-, but not with F^- or HO^-. For references see reference 8.

These additional equilibria provide further clues and checks for theories of the mechanism of homotropic and heterotropic interaction.

7.6 THE MECHANISM FOR FINE ADJUSTMENT OF THE EQUILIBRIUM CONSTANT FOR BINDING O_2

In this section we want to see firstly how the protein can vary the equilibrium constant for the co-ordination of O_2 without changing the nature of the axial ligand, and secondly how it can couple one equilibrium with another (homo- and hetero-tropic interaction).

The equilibrium constant for the co-ordination of O_2 by Fe^{II} is related to the difference in free energy of formation of the two species involved (Fe^{II} and $Fe^{II}O_2$). These free energies of formation are, in turn, composed of contributions from many sources, the most relevant of which can, for the purposes of discussion, be grouped together under the following three broad but interdependent headings:

(i) The conflicting steric requirements of the metal and the protein will almost inevitably lead to a compromise involving some distortion of the bond lengths and angles of the metal and ligands (porphyrin, O_2, CO, etc.).
(ii) A complementary distortion, or change in the conformation, of the protein.
(iii) The effects of nonbonded interaction between the metal and its ligands on the one hand and the neighbouring amino-acid residues on the other (due to hydrogen bonding, coulombic and van der Waals interactions), which are analogous to the effects of 'solvation' in simpler complexes.

Variations in equilibrium constants between one protein and another are therefore due to variations in the difference between the energies of distortion and solvation of the two species involved in the equilibrium (for example, Fe^{II} and $Fe^{II}O_2$). The same equilibrium constant could, for example, be observed in a protein in which both species are highly distorted and in a protein where they are not; this is bound to complicate any attempt to correlate physical and chemical properties. There are two other factors that limit our ability to sort out these different effects. Firstly, we have so far been unable to prepare and study any simple protein-free and un-distorted Fe porphyrins containing only one co-ordinated imidazole (or histidine), which could serve as a basis for comparison. We can therefore only compare the physical and chemical properties of one protein with another. Secondly, the X-ray analysis of such large molecules can only detect relatively large distortions of the metal and its ligands; smaller but still significant distortions can be detected by spectroscopic techniques, but their correlation with structural details may be difficult. There is sufficient experimental evidence available to show that all these effects can and do contribute, but not to pinpoint the effects or estimate their relative importance.

We have already (section 7.4) met distortion of the ligand in an $Fe^{II}CO$ complex, where the $Fe-\hat{C}-O$ bond angle is $145 \pm 15°$ [109]. Some interesting, and pre-sumably significant, variations have been noted in the position of the Fe atom relative to the porphyrin plane; as will be shown below, displacement of the Fe atom out of the plane is observed in high-spin Fe porphyrins, even in the absence of protein. The following displacements (in pm) of the Fe from the least-squares plane of the four porphyrin nitrogens towards the co-ordinated histidine have been reported (see table 14). We do not, of course, know the distances expected for

TABLE 14   DISPLACEMENT OF Fe FROM LEAST-SQUARES PLANE

	Sperm whale Mb	Horse Hb quaternary structure		Chironomus Hb
		R	T	
Fe^{II}	30		75	30
$Fe^{II}CO$		30		5
$Fe^{III}OH_2$	30	30		30

the undistorted protein-free complexes, but there do appear to be significant difference between the proteins in two of the three sets of analogous complexes. No marked distortion of the co-ordinated histidine has been reported. The porphyrin ring is synclastically curved away from the Fe in Mb $Fe^{III}OH_2$ [232] and probably nonplanar in Hb Fe^{II} [27]; non-planarity is also widely found among simple metal porphyrins [104].

We have also seen (section 7.4) how the anomalously low C−O stretching frequency in the $Fe^{II}CO$ complexes can be ascribed to some strong interaction (hydrogen bonding or steric repulsion) with the protein [48]. The e.s.r. spectra of analogous complexes reveal differences between proteins and their sub-units and provide examples of an unexpectedly low symmetry around the Fe; see, for example, references 8 and 16. The ratio of the high-spin to low-spin forms of the $Fe^{III}OH_2$ and $Fe^{III}OH^-$ complexes may also vary considerably with the nature of the protein; see, for example, references 86 and 103.

The intense u.v.–visible absorption spectra are due mainly to π-π transitions within the porphyrin ligand. Most Hbs and Mbs show slight differences in their spectra which make it possible to identify the species from which they originate. In most cases these differences are greater for the Fe^{II} than for the $Fe^{II}O_2$, $Fe^{II}CO$ and $Fe^{III}OH_2$ complexes. The conversion of the free α and β chains into human Hb A_1 is also accompanied by much larger changes for the Fe^{II} form (where the Hb has the T quaternary structure) than for the $Fe^{II}O_2$ and $Fe^{II}CO$ forms (R quaternary structures); but spectra similar to that of Fe^{II} in the free α and β chains are shown by Hb Fe^{II} in the R quaternary structure, which can be observed as the immediate product of the photolysis of Hb $Fe^{II}CO$, and by various natural and modified haemoglobins that do not show homotropic interaction[40,195]. It appears that the change in quaternary structure, which is the key to homotropic interaction (see below), may alter the equilibrium constant chiefly by affecting the physical properties of Fe^{II}. One notable exception to this generalisation is Hb from *Ascaris* perienteric fluid, where the spectra of the Fe^{II}, $Fe^{II}CO$ and $Fe^{II}CN^-$ complexes are similar to those of mammalian Hb and Mb, while that of $Fe^{II}O_2$ is rather different[238].

There is therefore good experimental evidence for distortion of the Fe and its ligands and for differences in the degree of distortion between one protein and another. The importance of distortion in controlling the chemical properties of haemoproteins was first emphasised in 1959 by Williams, who suggested that variation in the Fe–N (histidine) bond length might be the main mechanism of control by the protein[66, 235]. But distortion of the porphyrin and O_2 or other ligands, and variation in bond angles as well as bond lengths, must be considered also. Vallee and Williams have more recently emphasised the general importance of distortions of all kinds in converting the metal ion into a more reactive (*entatic*) state, and have shown that the metal ions in many metallo-enzymes possess anomalous physical properties, which can be ascribed to distortion[222].

Differences in conformation between one protein and another and changes accompanying the co-ordination of ligands or change in oxidation state of the Fe have been revealed by X-ray analysis. Some examples have already been given in section 7.4 and others will be mentioned below; see also reference 94. Differences around the distal co-ordination site include the presence or absence of hydrogen-bonding groups (section 7.4), that is obvious differences in the 'solvation' of the ligand. Evidence for differences and changes in conformation is also provided by n.m.r, CD and ORD spectra (see, for example, reference 204 and references in reference 8). A particularly interesting result is the finding that the co-ordination of CO or O_2 causes a major change in the CD of Hb, a slight change in the isolated, but chemically modified, β chains, but no change in Mb or the isolated and modified α chains[41]. This strongly suggests that the protein is more flexible in Hb than in Mb, a view supported by a study of models of these two proteins[169]. This difference in flexibility may be connected with the large variation observed in the displacement of the Fe from the porphyrin plane, which is observed in Hb but not in Mb (see table 14).

We conclude that differential distortion of the two metal complexes involved in the equilibrium, complementary differences in the two conformations of the protein and variations in the 'solvation' energy may all play a part in the mechanism by which the protein controls a chemical equilibrium involving the metal. This is equivalent to saying that for each protein we must consider the 'co-operative' changes in configuration, both of the metal with its ligands and of the protein. The same approach will obviously apply to other equilibria and other metallo-proteins.

How are the changes in conformation of the metal + ligands, which accompany some equilibrium at the metal, coupled to the conformational changes of the protein? As an extension to this, how does the protein couple together one equilibrium involving the co-ordination of O_2 at one Fe with a second equilibrium at another site in the protein? The most clear-cut evidence concerns the phenomenon of haem–haem or homotropic interaction in mammalian Hbs, which leads to characteristic sigmoid curves of O_2-uptake of the type shown in figure 31. Since the Fe atoms are 2.5–3.7 nm apart, their interaction must clearly be mediated by the protein. The two questions to be answered are: what is the change in conformation of the protein by which this interaction is relayed and what is the specific change in the conformation of the metal complex, which 'triggers' this change in the protein conformation?

We have already seen (section 7.2) that the four polypeptide sub-units in horse and human Hb can be arranged together in two slightly different ways to give the so-called T or 'deoxy' and R or 'oxy' quaternary structures. The main difference between the R and T forms appears to be the presence of eight salt bridges (involving coulombic interaction) at the $\alpha_1\alpha_2$ and $\beta_1\beta_2$ junctions in the T form. Studies with various natural and modified Hbs have shown that homotropic interaction is observed in the co-ordination of O_2 by Fe^{II} only where there is a change in the quaternary structure during oxygenation[173]. The switch-over from the T to the R form occurs at ≈ 50 per cent oxygenation in human Hb A_1, but in other cases it may occur at much higher or lower values or even fail to be observed; that is, the protein may remain frozen in the R or T form[94,204]. Other equilibria that involve homotropic inter-action (see section 7.5) also involve a change in the quaternary structure[33]. Studies of the difference in spectra (see above) and in $P_{1/2}$ (see section 7.5) between Hb and the isolated α and β chains both show that Hb behaves normally at high loadings of O_2, and that the sigmoid shape of the curve is due to abnormalities at low loadings, that is involving the T form. Equilibrium constants for the binding of O_2 have now been evaluated for both the R and T forms of human Hb at pH 7 and of sheep Hb at pH 9; that of the T form is ≈ 250 times lower than that of the R form[204]. The rate of interconversion of the R and T forms can be measured in the absence of any change involving Fe, because photolysis of Hb containing only $Fe^{II}CO$ (and therefore in the R form) gives Fe^{II} in the R form as the immediate product; this then rapidly undergoes conversion to the normal T form with a half-time of ≈ 2 ms at $3°C$[8,88]. The question now is: how are changes at the Fe converted into changes in the quaternary structure?

In 1970 Perutz[169,173] proposed a mechanism based on a comparison of the structures of the Fe^{II} and $Fe^{III}OH_2$ forms of Hb and of a derivative (BME–Hb), in which the quaternary structure was frozen in the 'oxy' or R form. This enabled ligand-induced changes in the tertiary structure of the sub-units to be isolated from changes in the tertiary structure attendant upon changes in the quaternary structure.

Unfortunately, the reagent (BME) also prevented free movement of the crucial tyrosine HC2 in the β chain and definite conclusions could therefore only be drawn about changes in the β chains.

The key changes involve the C-terminal groups (HC3 arginine in the α chains, HC3 histidine in the β chains), the penultimate HC2 tyrosine (in both chains), helices H and F and, of course, the co-ordinated F8 histidine (see figure 1, page 22). In the deoxy or Fe^{II} form the Fe is displaced 75 pm out of the porphyrin plane towards the co-ordinated nitrogen of histidine F8. Tyrosine HC2 is held in a pocket between helices F and H and is hydrogen bonded to the carbonyl of valine FG5. The terminal HC3 arginine and histidine are doubly anchored by salt bridges (ionic interaction) to neighbouring sub-units as follows:

$$\text{arg HC3 of } \alpha_1 \text{ to val NA1 of } \alpha_2 \text{ and to asp H9 of } \alpha_2$$

$$\text{his HC3 of } \beta_1 \text{ to lys C5 of } \alpha_2 \text{ and to asp FG1 of } \beta_1$$

On conversion to the $Fe^{III}OH_2$ form, the Fe moves closer to the porphyrin plane and thereby moves the histidine closer to the porphyrin plane; that is, the movement of the Fe *vis-à-vis* the porphyrin ring is transformed into a movement of the co-ordinated histidine *vis-à-vis* the other residues, which are pressed hard up against the porphyrin ring and cannot move This somehow induces a movement of helix F towards helix H, which expels tyrosine HC2 from the pocket between the two. The expelled tyrosine then pulls the terminal HC3 groups with it and ruptures the salt bridges, which stabilise the T over the R form.

Subsequent work with another Hb, in which the terminal HC3 arginine has been removed and which is also frozen in the R form, also showed the same movement of the HC2 tyrosine in the α chains. But this movement of HC2 tyrosine does not occur in the β chains of this Hb or of Hb M Iwate[94]. The sequence of ligand-induced changes in the β changes is not known.

The above evidence shows quite dramatically how a minor change in the tertiary structure of the individual sub-unit, involving a limited and linked sequence of changes, can serve to couple together equilibria at two different sites. In this case an equilibrium involving Fe (co-ordination of O_2) and an equilibrium involving the C-terminal residues on the surface does. Since the latter equilibrium also involves the making and breaking of salt bridges to other sub-units: a minor change in the tertiary structure of one sub-unit has been converted into a minor change in the quaternary structure of the whole protein. Further minor changes in the quaternary structure, as other sub-units change, eventually lead to a major change in the quaternary structure between the R and T forms. In other words the equilibrium involving each Fe in its separate sub-unit is eventually coupled to the equilibrium between the R and T forms of the whole protein and hence to the equilibria involving the other Fe atoms.

But the above picture cannot represent the whole story, because we do not know the nature of the ligand-induced changes in the β chains. And since this theory was based on a comparison of the Fe^{II} and $Fe^{III}OH_2$ structures, other ligand-induced changes may be involved in the equilibrium between Fe^{II} and $Fe^{II}O_2$. Nor do we know the order in which the sub-units react with O_2.

The Bohr effect, that is the reciprocal effect of pH on the binding of O_2, is closely related to the above haem–haem or homotropic interaction, though the two effects can be separated in certain Hbs (see section 7.5). It appears to involve several residues, whose pKs are altered by the change in their environment that accompanies the change between R and T forms. These residues include arginine HC3 and valine NA1 in the α chain, and histidine HC3 in the β chain, all of which are directly involved in homotropic interaction, as well as H5 histidine in the α chain, which is only indirectly involved[169,173]. 2,3-Diphosphoglycerate exerts its effect by binding specifically to the T form to give a 1 : 1 complex with the protein[169,173]. Studies of mutant and modified Hbs, involving changes around the C-terminus, provide further support for the important role played by the salt bridges in the mechanism of homotropic and heterotropic interaction in mammalian Hbs[94,130,169,173].

It might be expected that Mb would show a significant Bohr effect via a sequence of changes analogous to those observed in the Hb sub-units. This is, however, precluded by the fact that the C-terminal residues in Mb form part of the H helix and are therefore not free to change their environment, and hence their pK, as they are in Hb[94]. But the monomeric Hb III from *Chironomus* does show a Bohr effect, which obviously cannot involve any change in quaternary structure. Difference maps of the $Fe^{III}OH_2$, Fe^{II} and $Fe^{II}CO$ forms show that the Fe moves into the plane of the porphyrin ring in $Fe^{II}CO$, but does not reveal any significant and extensive changes in the protein of the type found for mammalian Hbs[109]. However, the base responsible for the Bohr effect has been identified as G2 histidine from a study of the effect of pH and oxygenation on the p.m.r. spectra, and a mechanism for the transmission of this Bohr effect from histidine G2 to histidine F9 has been suggested on the basis of the known structure of the protein[205].

Finally, we want to examine more closely the nature of the conformational change in the Fe porphyrin that triggers the conformational change in the protein.

Structures have now been determined for a number of metal porphyrins and related compounds and complexes. Hoard pointed out in 1965[104,105] that the porphyrin ring is often significantly ruffled, folded or domed—that is, it is relatively easily distorted in a direction normal to the plane—but that the size of the central hole can vary only up to a maximum radius of 202 pm and that longer metal–nitrogen bond lengths are found only when the metal is displaced out of the plane of the ring; that is, there is a limit to the distortion of the porphyrin within the plane. This displacement of the metal is well exemplified by two complexes containing five-co-ordinate high-spin Fe(III) ions where the following distances (in pm) have been found for the Fe–N bond length and for the distance of the Fe from the least-squares plane of the four nitrogens.

	Fe–N bond length	distance of Fe from least-squares plane of the four N
Chlorohaemin[133]	206.2	47.5
Methoxy-iron(III)-mesoporphyrin IX dimethyl ester[105]	207.3	45.5

Hoard pointed out that the high-spin Fe(II) would have an even larger radius, but that the radii of low-spin Fe(II) and Fe(III) would be smaller by 10 pm or more. He postulated that the Fe would be significantly displaced (by $\geqslant 35$ pm) from the plane of the nitrogens in all high-spin porphyrin complexes, regardless of oxidation state and co-ordination number, but that the Fe would lie in the plane in low-spin complexes; a subsequent structure determination showed that Fe was indeed coplanar with the porphyrin in the low-spin bis(imidazole)-$\alpha,\beta,\gamma,\delta$-tetra-phenylporphinato-iron(III) chloride[64]. He therefore suggested that the movement into the plane when the high-spin Fe^{II} reacts with O_2 to form the low-spin $Fe^{II}O_2$ complex could provide the trigger for the co-operative movements of the protein. But several facts show that Hoard's hypothesis cannot be the whole story. The experimental data on the displacement of the Fe from the least-squares plane of the porphyrin ring in Hbs and Mbs are summarised in table 15. They suggest that Fe is not coplanar in all low-spin complexes (for example, horse Hb $Fe^{II}CO$ and various Mb complexes) and that a change from high- to low-spin need not be accompanied by a significant change in the displacement of the Fe (compare Mb Fe^{II} and $Fe^{II}O_2$); but it should be noted that the validity of some of these data has been queried[64]. In addition, both homotropic and heterotropic interaction are observed for the redox potentials of Hbs involving Fe^{II} and $Fe^{III}OH_2$ (see section

TABLE 15 STEREOCHEMISTRY OF Fe IN Hbs AND Mbs

Protein	Valency and ligand	$d^\dagger$(pm)	High or low spin	Ref.
sperm-whale Mb	$Fe^{III}OH_2$	30*	HS	232
	$Fe^{III}F^-$	30	HS	231
	$Fe^{III}N_3^-$	?30	LS	211
	$Fe^{III}CN^-$	?<30	LS	231
		?coplanar		33
	$Fe^{III}OH^-$	?<30	HS + LS	231
	Fe^{II}	30	HS	162, 232
	$Fe^{II}O_2$	30	LS	232
	$Fe^{II}CNEt$	30	LS	160, 231
horse Hb	$Fe^{III}OH_2$	30*	HS	27
	Fe^{II}	75*	HS	27
	$Fe^{II}CO$	30	LS	173
Chironomus Hb	$Fe^{III}OH_2$	30*	HS	109
	$Fe^{III}CN^-$	5	LS	109
	Fe^{II}	30	HS	109
	$Fe^{II}CO$	5	LS	109
lamprey Hb	$Fe^{III}CN^-$	<10*	LS	102
	Fe^{II}	?50	HS	144

$\dagger$ d = displacement of the Fe from the least-squares plane of the four porphyrin nitrogen atoms towards the co-ordinated (proximal) histidine. Distances marked * are determined from full 3-d Fourier synthesis, others from difference Fourier synthesis or merely from a comparison of the lattice constants. The minimum difference in d that would be detected is not usually mentioned except in the case of Mb, where differences of 10 pm would have been detected[161].

7.5), both of which are high-spin, and for the uptake of O_2 by the cobalt-containing analogue of Hb, which is low-spin[108].

Bretscher in 1968[33] suggested another very simple hypothesis, which explains all the observations, except for the possible homotropic interaction in the reaction of $Fe^{III}OH_2$ with CN^- and N_3^- (see section 7.5). He pointed out that only the five-co-ordinate Fe^{II} exists in the T form, while all the six-co-ordinate complexes, regardless of valency or spin state, exist in the R form; he suggested that it was simply the presence of the sixth ligand that triggered the conformational change in the protein. This could occur either by the displacement of some group that blocked the co-ordination site or by the formation of a hydrogen bond to the distal histidine.

There is, however, a third possibility for the trigger mechanism, which can include the ideas of both Hoard and Bretscher as special cases. The last two experimental observations (involving $Fe^{II}/Fe^{III}OH_2$ and $Co^{II}O_2$) both suggest that any significant change in the displacement of Fe, whether it is due to a change in spin state or not, could trigger the conformational change of the protein. In the first case, the displacement of the Fe is reduced from 75 pm in Fe^{II} to 30 pm in $Fe^{III}OH_2$ (see table 15). X-ray analysis of protein-free Co complexes shows that a change in the displacement of the metal ion is also possible in the second case. The Co is displaced by 20 pm from the least-squares plane of the four equatorial ligand atoms in the five-co-ordinate, low-spin $[Co^{II}(salen)py]$[45], while the Co is in the plane of the six-co-ordinate $[Co(bzacen)py . O_2]$[186]. A displacement of only 20 pm might be too small to trigger the conformational change of the protein, but it seems reasonable to assume that the protein, whose conformation is tailored to the requirements of Fe, would increase the displacement of the Co towards that found in the native protein.

Arguments based solely on structures observed for protein-free complexes may, in fact, be vitiated by the effect of the protein in distorting the metal–ligand bonds, in particular the bond lengths involving the axial ligands, and, to a lesser extent, by the mutual effect of the axial ligands on each other. Evidence for distortion by the protein has been discussed above. A good example where we can see how the degree of displacement of the metal towards one side of the plane depends on the strength of the binding of the ligand on the other side, is provided by organo-cobalt(III) complexes; here there are no complications due to a change in spin or to steric limitations by the equatorial ligand on the size of the central hole. The Co is coplanar with the equatorial ligand atoms in various six-co-ordinate complexes; it is displaced 6 pm from the plane in the dimeric form of $[Et-Co(salen)]$, where the second axial ligand of one half is a weakly held oxygen atom of the salen ligand of the other half; and it is displaced by 12 pm in the five-co-ordinate $[Me-Co(BAE)]$. For further details and references see reference 182.

The experimental observations can therefore be accommodated by extending Hoard's hypothesis to include any significant change in the displacement of the Fe from the porphyrin plane (whether caused by a change in spin state or co-ordination number or strength of the other Fe–ligand bond), and by allowing for the possible effects of distortion by the protein of the metal-ion stereochemistry. There seems little doubt that the movement of the Fe relative to the plane of the porphyrin is the trigger for the known sequence of changes in the tertiary structure of the α chains; but no conclusions can be drawn about the β chains until the nature of their ligand-induced changes is known.

This section has shown that we now have sufficient experimental evidence to discern the general features, though not the details, of the mechanisms by which the protein can adjust an equilibrium involving Fe and couple it to another equilibrium.

We have already seen (section 7.4) how the protein can exercise coarse control over the number and nature of the axial ligands co-ordinated to Fe (as well as the arrangement of groups around the distal co-ordination site). We now see how distortion of the metal and its ligands, caused by a mismatch in the steric requirements of the protein and of the complex, provides a method for fine control of the metal–ligand interaction. The observed wide variation in the equilibrium constant for the binding of O_2, where there is no change in the nature of the ligand provided by the protein (section 7.5), can therefore be ascribed to variations in the relative distortion of the metal complex and polypeptide and in the 'solvation' energies, as between the two species involved in the equilibrium (Fe^{II} and $Fe^{II}O_2$), or, conversely, to co-operative changes in the metal complex and in the polypeptide during the reaction.

We have also seen (section 7.4) that localised changes in the protein surrounding the distal co-ordination site may result from any change in the steric requirements of the co-ordinated ligand and, of course, from the addition of a ligand to the five-co-ordinate Fe^{II}. The more deep-seated changes in protein conformation observed on the proximal side of the porphyrin in Hb appear to be coupled with (or 'triggered' by) movements of the Fe along a line perpendicular to the porphyrin plane. In the five-co-ordinate Fe^{II}, for example, the Fe is displaced by ≈ 75 pm from the least-squares plane towards the co-ordinated histidine. The degree of displacement will depend on the ionic radius (and hence on the oxidation state and, even more, on the spin state), on the strength of the bond to the other axial ligand and on distortion by the protein.

The mechanism for homotropic and heterotropic interaction in Hb appears to depend on three different kinds of conformation change in the protein: minor changes in the tertiary structure of each polypeptide, and both minor and major changes in the quaternary structure of the four associated polypeptides. In the first, a linked sequence of changes affecting only a limited area of the polypeptide couples together the equilibrium involving Fe with a second equilibrium between free and bound amino-acid residues at a particular spot on the surface of the sub-unit. This type of change alone would be sufficient to explain heterotropic interaction in proteins containing only a single polypeptide chain. But in Hb, the second equilibrium also involves the making and breaking of salt bridges between sub-units, and minor changes in tertiary structure therefore lead automatically to minor changes in quaternary structure. An increasing degree of minor change in quaternary structure eventually leads to a major change in quaternary structure between the R and T forms. In this way

the equilibrium at the Fe in each sub-unit is indirectly coupled to the equilibrium
between the R and T forms of the whole protein.

7.7 AUTOXIDATION

As already mentioned, the $Fe^{II}O_2$ complexes of normal Hbs and Mbs undergo a
slow oxidation to $Fe^{III}OH_2$ (section 7.3), but the protein-free Fe(II) porphyrins
react with O_2 about 10^8 times faster[229]. What reaction paths must the protein
block in order to prevent irreversible loss of O_2, and how is this done? Evidence on
the mechanism of autoxidation of Hb and Mb and of simple Fe(II) porphyrins is
rather limited and can be summarised as follows.

The rate of autoxidation of Hbs and Mbs varies from one particular protein to
another, increases on raising the temperature, the acidity or the concentration of
added salt, but is often decreased by the addition of EDTA, which suggests catalysis
by traces of copper ions, etc. (for references see reference 8). Autoxidation is
reported to be much more rapid in certain mutant Hbs such as Hb M Radom
(= Hb M Saskatoon), where the distal histidine has been replaced by tyrosine in the
β chain[153], and Hb Zürich, where it has been replaced by arginine[237].

Recent work on the autoxidation of Mb $Fe^{II}O_2$ has shown that three-quarters
of the total O_2 is evolved—that is, one mole of O_2 oxidises four Fe^{II} ions—and
therefore casts doubt on some of the earlier kinetics carried out in the presence of
traces of dithionite. The rate depends linearly on the proton concentration over the
range of pH 5-7 and inversely on P_{O_2}. It appears that the rate-determining step may
simply be the loss of O_2, but no explanation was offered for the dependence on pH.
Unfortunately, no study was made of the reaction of $Fe^{II}O_2$ with Fe^{II} [39]. Further
work is obviously needed.

The kinetics of the reactions of various bispyridine-Fe(II)-porphyrin complexes
with O_2 in aqueous and nonaqueous solvents have been studied by two groups of
workers. Kao and Wang[119] found evidence for two paths. The minor path probably
involves electron transfer during collison of an unco-ordinated O_2 with the bispyridine
Fe(II) complex. The major path shows an inverse dependence on pyridine concentration. The authors suggested the following scheme

$$O_2 + py\text{-}Fe^{II}\text{-}py \underset{\text{fast}}{\rightleftharpoons} py\text{-}Fe^{II}\text{-}O_2 + py$$
$$\downarrow \text{slow}$$
$$py\text{-}Fe^{III} + O_2^- \qquad\qquad (7.11)$$

But they also found that as the concentration of dissolved O_2 was increased the
rate at first increased linearly with the concentration, but then tapered off; and yet
the spectra failed to show the build-up of any intermediate py-Fe-O_2. The postulated mechanism cannot therefore be correct. Cohen and Caughey[58] found an even
wider range of kinetic behaviour, including a second-order dependence on Fe(II)

under certain conditions, and suggested the following scheme

$$\text{py-Fe}^{II}\text{-py} \underset{K_2}{\overset{K_1}{\rightleftharpoons}} \text{py-Fe}^{II} + \text{py}$$

$$2\ \text{py-Fe}^{II} + O_2 \xrightarrow{K_3} \text{py-Fe}^{II}\text{-}O_2\text{-Fe}^{II}\text{-py} \tag{7.12}$$

K_3 is unlikely to represent a termolecular reaction and can be separated into two successive bimolecular steps such as

$$\text{py-Fe}^{II} + O_2 \underset{}{\overset{K_{3a}}{\rightleftharpoons}} \text{py-Fe}^{II}\text{-}O_2$$

$$\text{py-Fe}^{II}\text{-}O_2 + \text{Fe}^{II}\text{-py} \xrightarrow{K_{3b}} \text{py-Fe}^{II}\text{-}O_2\text{-Fe}^{II}\text{-py} \tag{7.13}$$

The effect of high concentrations of O_2, observed by Kao and Wang, could then represent the situation where $K_3 \geqslant K_1$. For further qualitative observations on the effect of the nature of the solvent on the rate see reference 47. Here again further work would be welcome.

We shall consider the following three generalised paths for the reaction of co-ordinated O_2 in Hb and Mb, in protein-free Fe(II) porphyrins, and in Co(II) complexes

$$\text{Fe}^{II}O_2 + H_2O \rightleftharpoons \text{Fe}^{III}OH_2 + O_2^- \rightarrow \text{products} \tag{A}$$
$$(\text{or } H_3O^+) \qquad\quad (\text{or } HO_2)$$

$$\text{Fe}^{II}O_2 + \text{Fe}^{II} \rightleftharpoons \text{Fe}^{II}O_2\text{Fe}^{II} \rightarrow \text{products} \tag{B}$$

$$\text{Fe}^{II}O_2 + H\ (\text{and/or} \ominus) \rightarrow \text{Fe}^{III}O_2H, \text{etc.} \rightarrow \text{products} \tag{C}$$

A-C all involve 'inner sphere' electron-transfer reactions; A corresponds to a one-electron transfer, B and C to two-electron transfers. Other paths of higher energy can be envisaged, for example

$$\text{Fe}^{II}O_2 + H_2O \rightleftharpoons [\text{FeO}]^{2+} + H_2O_2 \tag{D}$$

For the reverse reaction see section 8.3. Perhaps paths that involve an 'outer sphere' electron transfer from the porphyrin ring of one complex to a molecule of free O_2 (as above) or to the co-ordinated O_2 or porphyrin ring of a second complex should also be included. But we shall here consider only A-C.

The following redox potentials for the successive one- and two-electron reductions of O_2 at pH 7, taken from the values compiled by George[85], are relevant to the present discussion

$$O_2 \underset{-0.45}{\rule{1.5cm}{0.4pt}} O_2^- \underset{+0.98}{\rule{1.5cm}{0.4pt}} H_2O_2 \underset{+0.38}{\rule{1.5cm}{0.4pt}} HO \underset{+2.33}{\rule{1.5cm}{0.4pt}} H_2O$$

$$\underset{+0.27}{\rule{3cm}{0.4pt}} \qquad \underset{+1.35}{\rule{3cm}{0.4pt}}$$

They show that, although O_2 is a powerful oxidising agent when it can accept two or more electrons, it is a remarkably poor oxidising agent when only one electron is available. The $\text{Fe}^{II/III}$ couples for Hb and Mb in aqueous solution at pH 7 have been given as +0.2 and +0.1 V, respectively[85].

7.7.1 $Fe^{II}O_2 \rightleftharpoons Fe^{III} + O_2^- $ (A)

Wang has suggested that this type of reaction is an essential step in the autoxidation of Fe(II) porphyrins, both with and without protein, and that the formation of O_2^- or HO_2 is made less favourable by the hydrophobic environment in the protein[227,229,230]. But, as George has pointed out[84], comparison of the relevant redox potentials (see above) shows that the redox reaction involving the simple ions

$$Fe^{II} + O_2 \rightarrow Fe^{III}OH_2 + O_2^- \qquad (7.14)$$

will be endothermic and that the reaction

$$Fe^{II}O_2 \rightarrow Fe^{III}OH_2 + O_2^- \qquad (7.15)$$

will be even more endothermic because of the exothermic formation of $Fe^{II}O_2$ from Fe^{II} and O_2. Mechanism A is therefore inherently unlikely to provide a major pathway for the autoxidation of Hb and Mb at pH 7, even if it were not blocked by the protein. Furthermore, the postulate that the protein prevents formation of the anion O_2^- would logically require that the protein very significantly reduces the rate of dissociation of other small co-ordinated anions such as F^- or N_3^- in analogous reactions, for example

$$Fe^{III}N_3^- + H_2O \rightleftharpoons Fe^{III}OH_2 + N_3^- \qquad (7.16)$$

This is not, in fact, observed. We cannot, however, exclude mechanism A as a minor pathway for autoxidation.

Mechanism A could become more important if the redox potential is made less unfavourable. This could happen in three ways:

(1) Lowering the $Fe^{II/III}$ couple either (a) by replacing the co-ordinated histidine by a ligand such as carboxylate or tyrosine, which will tend to stabilise the Fe(III) state, or (b) by some co-operative effect of the protein (see sections 7.5 and 7.6). Mutant Hbs are known in which the histidine F8 has been replaced by tyrosine in either the α or β chains; but this tends to stabilise the Fe(III) state and there appears to be no information on the rate of autoxidation of the Fe(II) ions[170]. Hbs, Mbs and peroxidases all possess the same axial ligand and therefore provide an excellent demonstration of the way in which the protein can influence the rate of autoxidation and the redox potential. Compare the following qualitative observations on the reactions of Fe(II) with O_2 and the values of E° at pH 7.

Hb	$\approx$ +0.2 V[85]	$Fe^{II}O_2$ very stable
Mb	$\approx$ +0.1 V[85]	$Fe^{II}O_2$ very stable
turnip peroxidase P_7	$-$ 0.12 V	$Fe^{II}O_2$ stable for $\geqslant$ 20 mins[185]
turnip peroxidase P_1	$-$ 0.22 V	$Fe^{II}O_2$ detected, but rapidly oxidised to $Fe^{III}OH_2$[177]
horse radish peroxidase	$\approx$ $-$ 0.25 V[85]	very rapid autoxidation, no $Fe^{II}O_2$ detected[55]

There is obviously a parallel between E° and the rate of autoxidation.

(2) Lowering the pH to remove O_2^- as HO_2. This could explain the pH-dependence of the observed slow autoxidation of Mb.

(3) Decreasing the equilibrium constant for the co-ordination of O_2. There is no obvious example of this effect.

7.7.2 $Fe^{II}O_2 + Fe^{II} \rightleftharpoons Fe^{II}O_2Fe^{II}$ (B)

The kinetic data discussed above suggests that this is the main pathway for the reaction of the protein-free Fe(II) porphyrins. Many or most mononuclear Co(II)–O_2 complexes undergo the analogous reaction with excess Co(II) to form the μ-peroxide; see, for example, reference 234. Such dimerisation is, of course, prevented in Hb and Mb by the protein.

7.7.3 $Fe^{II}O_2 + H$ (and/or $\ominus$) $\rightarrow Fe^{III}O_2H^-$, etc. (C)

Rather surprisingly Mb $Fe^{II}O_2$ can be oxidised to $Fe^{III}OH_2$ by reaction with reducing agents such as ferrocyanide, quinol, catechol, resorcinol and o-phenylenediamine[243]. Another surprising and probably analogous reaction is the marked acceleration of the rate of autoxidation of Co(II) cobalamin by reducing agents such as ferrocyanide, quinol, pyrogallol, phenylenediamines, phenylhydrazine, thiols and ascorbic acid[181]. The organic reagents are those that can readily undergo a one-electron oxidation to form a free radical. It was suggested that the presence of the second reducing agent converted the endothermic one-electron reduction of O_2 by Co(II) into a simultaneous and exothermic two-electron reduction (see the potentials listed above). It has subsequently been shown[183] that the fully formed O_2-adduct $[Co^{II}(3\text{-MeO-salen})py . O_2]$ in pyridine at $-40°C$ will react with quinol (QH_2) to form a diamagnetic cobalt complex (and therefore probably Co(III)) and the semiquinone radical (QH) in agreement with the postulated mechanism, for example

$$Co^{II}O_2 + QH_2 \rightarrow Co^{III}O_2H^- + QH \rightarrow \text{further products} \qquad (7.17)$$

The co-ordinated O_2 must therefore be protected from reaction with reducing agents of the above type, whether present as amino-acid residues in the protein or as extraneous molecules and ions in the cell. Two observations may be relevant. Firstly, there are no residues such as cysteine or tyrosine that could act as reducing agents in the immediate vicinity of the co-ordinated O_2 in normal Hbs and Mbs; in Hb M Saskatoon, however, the histidine has been replaced by tyrosine, and the Fe^{II} undergoes much more rapid autoxidation[153]. Secondly, the above-mentioned reactions of Mb $Fe^{II}O_2$ with ferrocyanide, quinol, etc., occur at pH 4.8, but not at pH 7. This suggests that some group (namely, the distal histidine) is protecting the O_2 at pH 7, but can be prised open by protonation at pH 4.8. The histidine can be regarded as a kind of door, which can swing fairly easily on its hinges until a molecule of O_2 enters the ligand cavity and closes the door behind it by formation of the hydrogen bond between the N of the histidine and the co-ordinated atom of O_2. We had no obvious explanation in section 7.4 for the role of the distal histidine, which is present in almost all Hbs and Mbs. And it is here tentatively suggested that its function may be to protect the co-ordinated O_2 from further reaction.

We conclude that the protein is able to protect the co-ordinated O_2 from further irreversible reaction by keeping the $Fe^{II/III}$ couple high enough to prevent any significant reaction according to A, and by physically preventing reaction with any other reagents according to B and C.

7.8 SUMMARY

The development of proteins to bind O_2 is a good example of a problem in co-ordination chemistry, to which nature has found several solutions. The cofactor or prosthetic group that co-ordinates O_2 may be an Fe porphyrin, nonhaem Fe, Cu or perhaps even V; the nature of the ligands is not known except in the case of the Fe porphyrins (section 7.1). In addition, the Fe porphyrin of Hb and Mb can be replaced by the analogous Co porphyrin to give fully functional analogues of Hb and Mb, which shows that nature has not exhausted all the possibilities within the range of metals, amino acids and ligands at her disposal (section 7.1).

The aim of this chapter has been to examine how nature uses a protein to improve and control the inherent O_2-binding capacity of simple Fe porphyrins in order to produce a 'respiratory pigment' tailored to the requirements of the particular organism. As described in section 7.3, this involves overcoming four major and one minor problem. Only one of these is a kinetic problem, namely the inhibition of further reactions of the co-ordinated O_2. The others involve a diversity of thermodynamic problems: a major change in the normal pattern of equilibrium constants for the binding of axial ligands in order to stabilise the five-co-ordinate Fe(II); fine adjustment of the equilibrium constant for the co-ordination of O_2; a decrease in the equilibrium constants for the binding of extraneous ligands, which could 'poison' the Fe; and control of the $Fe^{II/III}$ redox potential in order to stabilise Fe(II).

7.8.1 Stabilisation of five-co-ordinate Fe(II)

O_2 is bound in Hb and Mb by co-ordination to the high-spin five-co-ordinate Fe(II) porphyrin containing a co-ordinated histidine. Analogous protein-free complexes containing one nitrogenous base cannot be prepared in solution, because the high-spin five-co-ordinate complex is thermodynamically unstable with respect to the high-spin complex with no base (the number of co-ordinated solvent molecules is not known) and the low-spin six-co-ordinate complex with two molecules of base, i.e. $K_2 \gg K_1$, where K_1 and K_2 are the formation constants for binding the first and second molecules of base respectively. This reversal of the usual order ($K_1 > K_2$) is probably connected with the change in spin state. The protein must therefore not only bind the Fe porphyrin with a high formation constant, but also provide one and only one axial ligand. High formation constants ($>10^{10}$ M^{-1}) are observed for the binding of Fe porphyrins by apo-Hb and Mb, but very similar affinities are also shown by the metal-free porphyrins! The major contribution to the binding energy therefore comes from the van der Waals interaction between the porphyrin ligand and the nonpolar residues of the protein. This ensures a higher binding constant than could be attained by co-ordination of one ligand alone. It also enables the protein to control the number and nature of the axial ligands presented to the Fe, because the energy of interaction between the porphyrin and the protein is far greater than that between the Fe and any potential ligands and the stereochemical requirements of the former therefore dominate those of the latter (section 7.4).

The polypeptide provides only one axial ligand, the 'proximal' histidine F8. This histidine can be replaced by tyrosine in half the sub-units of certain mutant Hbs, but it is not known whether these abnormal sub-units can reversibly bind O_2 or not (section 7.4). However, the wide range of axial ligands found in $Co^{II}O_2$ complexes (section 7.1) suggests that histidine need not be the unique ligand for Hbs and Mbs, although it is undoubtedly the preferred ligand. A second 'distal' histidine E7 is usually situated close to the Fe on the other side of the porphyrin plane, but cannot become co-ordinated without affecting helix E (section 7.4). Nature has therefore enforced five-co-ordination on the Fe by making use of the greater energy of interaction between the porphyrin and the protein together with the required disposition of potential ligands so that, in effect, $K_1 \gg K_2$.

It is striking that the polypeptide chains of all Hbs and Mbs so far studied by X-ray analysis have the same tertiary structure (the so-called 'myoglobin fold') in spite of the fact that they originate from three different phyla and that only 20 per cent of the amino-acid residues appear to be invariant (see section 7.2 and figure 1, page 22)

7.8.2 Control of the equilibrium constant for the co-ordination of O_2

O_2 can be reversibly co-ordinated by a protein-free Fe porphyrin complex in a polymer matrix, and rough estimates show that the equilibrium constant for the co-ordination of O_2 by the five-co-ordinate Fe(II) is at least as great as those of normal mammalian Hbs (section 7.3). We have concluded that, once the protein has stabilised the five-co-ordinate Fe(II), no further major stabilisation of the Fe–O_2 bond is required, only a fine adjustment to the needs of the particular organism.

The equilibrium constants of Hbs and Mbs so far studied span a range of $\Delta \log_{10}K$ (or $P_{1/2}$) ≈ 6, of which at least 4.5 can be achieved without any change in the nature of the axial ligand provided by the protein, and at least 2.4 merely by a change in the quaternary structure of the Hb. The enthalpy changes (ΔH) show an even greater variation, from +4 to −75 kJ mol^{-1} (+1 to −18 kcal/mole), even though no values are available for those Hbs with the highest values of K (section 7.5). There is now sufficient evidence available to indicate in a general way the nature of the mechanism(s) for adjusting K, but not to specify them in detail. X-ray and spectroscopic techniques provide evidence for: distortion of the metal and ligands in either or both of the Fe^{II} and $Fe^{II}O_2$ forms, owing to a mismatch in the steric requirements of the polypeptide and of the metal complex;

differences between one protein and another; and changes in the tertiary structure of the polypeptide chains (and in the quaternary structure of the Hbs) when Fe^{II} reacts with O_2. We can therefore ascribe the observed large variation in K, where there is no change in the axial ligand, to variations in the relative distortion of the metal complex and polypeptide chain and in the 'solvation' energies—that is, non-bonded interaction—between the Fe^{II} and $Fe^{II}O_2$ forms; or, conversely, to co-operative changes in the metal complex and in the polypeptide during the reaction.

In most Hbs and some Mbs the equilibrium involving the co-ordination of O_2 at one Fe can be coupled together with a second equilibrium such as the protonation of an amino-acid residue (heterotropic interaction) or the co-ordination of O_2 at another Fe (homotropic interaction). Our knowledge of the mechanism of this interaction is due mainly to the X-ray work of Perutz and coworkers on Hb, but is still incomplete. The basic feature, as shown by the α chains in Hb, is that a linked sequence of changes affecting only a limited part of the polypeptide chain—that is, a minor change in the tertiary structure—couples together the equilibrium involving Fe with a second equilibrium (between free and bound forms of the amino-acid residues) at a particular spot on the surface of the sub-unit. As this second equilibrium also involves the making and breaking of salt bridges between sub-units, the minor changes in tertiary structure lead automatically to minor changes in quaternary structure, and an increasing degree of minor change in quaternary structure eventually leads to a major change in quaternary structure between the so-called R and T forms. These changes in tertiary structure are coupled with (or 'triggered' by) movements of the Fe along a line perpendicular to the porphyrin plane. The degree of displacement of the Fe from the least-squares plane of the four porphyrin nitrogen atoms will depend on the ionic radius of the Fe (and hence on the oxidation state and even more on the spin state), on the relative strength of the bonds to the two axial ligands and on distortion by the protein. In the high-spin five-co-ordinate Fe^{II} in Hb, for example, the Fe is displaced by ≈ 75 pm towards the histidine, but it is probably more nearly coplanar in the low-spin six-co-ordinate $Fe^{II}O_2$. For a fuller description of these changes in the conformation of the protein and how they are coupled to the second equilibrium, see section 7.6.

This coupling together of equilibria not only provides a sophisticated mechanism for controlling the equilibrium constant, especially in response to changing conditions (section 7.5), but also provides a mechanism by which the protein can drive the thermodynamically unfavourable transport of O_2 against the concentration gradient (for example, in the inflation of the swim bladder of a fish) at the expense of another thermodynamically favourable reaction, which is essentially the neutralisation of acid with base (see introduction to chapter 7 and section 7.5).

7.8.3 Prevention of the co-ordination of extraneous ligands

In addition to preventing the co-ordination of a second axial ligand derived from one of its amino-acid residues, the protein must also try to prevent access to any other extraneous ligands (such as free amino-acids, glutathione, phosphate), which could reversibly or irreversibly 'poison' the protein by co-ordinating to the Fe(II) or Fe(III) states. It appears from the X-ray data that the arrangement of the residues around the distal co-ordination site is tailor-made to minimise any repulsions or rearrangements of the protein that might decrease the equilibrium constant for binding O_2, and to reduce the binding constant for most other potential ligands, especially those that prefer linear co-ordination (section 7.4).

7.8.4 Control of the $Fe^{II/III}$ couple to stabilise Fe(II)

The functioning of Hb and Mb requires that the potential is adjusted to stabilise Fe(II) in the resting state, that is, in the absence of O_2. Peroxidases, which probably also possess histidine as the axial ligand, are confronted with the opposite problem of stabilising Fe(III) (see chapter 8). This is reflected in the redox potentials for the $Fe^{II/III}$ couple. Most Hbs and Mbs appear to have potentials in the range +0.04 to +0.18 V at pH 6[8]; contrast turnip peroxidase $P_7 = -0.12$ V, $P_1 = -0.22$ V[185] and horse radish peroxidase ≈ -0.25 V[85] at pH 7. Unfortunately, we have no structural data on peroxidases that might indicate the origin of this large difference in potential, though the general mechanisms described under section 7.8.2 will apply here also. The value of the redox potential also plays a part in preventing autoxidation (see below).

7.8.5. Prevention of further reaction of $Fe^{II}O_2$

Complete reversibility of the uptake of O_2 by Fe^{II} requires the complete inhibition of all further reactions of the co-ordinated O_2, especially those that lead to autoxidation, that is, the formation of $Fe^{III}OH_2$. Although the presence of the protein severely inhibits autoxidation, suppression is not complete. The balance between the rate of autoxidation and the rate of enzymatic reduction back to Fe(II) leads to a steady-state concentration of 1–2 per cent $Fe^{III}OH_2$ in oxygenated blood (section 7.3)

The following generalised schemes of reaction probably represent the main paths by which the co-ordinated O_2 can undergo further reaction

$$Fe^{II}O_2 + H_2O \rightleftharpoons Fe^{III}OH_2 + O_2^- \rightarrow products \qquad (A)$$

Since O_2 is a very poor one-electron oxidising agent, this reaction can be suppressed by keeping the $Fe^{II/III}$ couple sufficiently high.

$$Fe^{II}O_2 + Fe^{II} \rightleftharpoons Fe^{II}.O_2.Fe^{II} \rightarrow products \qquad (B)$$

Formation of the dinuclear complex is sterically prevented by the presence of the amino-acid residues around the distal co-ordination site

$$Fe^{II}O_2 + H \text{ (and/or} \ominus) \rightarrow Fe^{III}OOH^-, \text{ etc.} \qquad (C)$$

where the reducing agent may be cysteine, tyrosine, ascorbic acid, etc. The protein takes the precaution of excluding all such amino acids from the immediate vicinity of the Fe-O_2. We have also suggested that the function of the distal histidine, which is present in most Hbs and Mbs, may be to protect the co-ordinated O_2 from attack by external reagents by acting as a kind of door, which closes the ligand cavity by the formation of a hydrogen bond to the co-ordinated oxygen atom (section 7.7).

8 CATALASE AND PEROXIDASE

We have seen how a protein can affect equilibrium constants (chapter 7). We now want to look at the more complex question of how the protein can affect the rates of reaction, and we shall examine the reactions of H_2O_2 with catalases and peroxidases. For a general introduction to these enzymes see the reviews by Nicholls and Schonbaum on catalases[159], by Paul on peroxidases[168] and by Brill on both groups[34], and the book by Saunders, Holmes-Siedle and Stark, *Peroxidases*[196]. Where no specific reference is given, the information in this section is taken from one of these sources.

The following abbreviations are used in this section for three particular enzymes

CPO	chloroperoxidase
CCPO	cytochrome c peroxidase
HRPO	horse-radish peroxidase

Both H_2O_2 and the superoxide ion (O_2^-) are reactive and toxic, and may, unfortunately, be produced by the autoxidation of many organic compounds *in vivo*. The evolution of an atmosphere containing oxygen and of organisms with an aerobic metabolism therefore required the development of enzymes to eliminate H_2O_2 and O_2^- as efficiently as possible. The poisoning of many anaerobic bacteria on contact with air is due to the lack of such enzymes[145]. We shall be concerned here only with the enzymes that remove H_2O_2; for references to the enzymes that catalyse the disproportionation of O_2^- to O_2 and H_2O_2 ('superoxide dismutase'), see reference 145.

Two main methods are available for removing H_2O_2, namely disproportionation, that is

$$2H_2O_2 \rightarrow 2H_2O + O_2 \tag{8.1}$$

and reduction, that is

$$H_2O_2 + 2H \rightarrow 2H_2O \tag{8.2}$$

where the reducing agent may be a purely organic compound, a thiol, iodide, etc. Nature has developed certain groups of enzymes that preferentially catalyse either disproportionation (*catalases*) or reduction (*peroxidases*), though there are others that combine both functions (for example chloroperoxidase; see below); these enzymes are referred to collectively as *hydroperoxidases*. The name 'catalase' may, however, be misleading since these particular enzymes also catalyse the reduction of

H_2O_2 by certain reagents such as alcohols, which prefer to undergo a single two-equivalent oxidation, while the true peroxidases catalyse the reduction of H_2O_2 by reagents such as ascorbic acid and aromatic amines and phenols, which prefer to undergo two consecutive one-equivalent oxidations.

After a general description of the distribution and function of these enzymes and the nature of the coenzymes employed, we concentrate exclusively on those catalases and peroxidases that possess Fe protoporphyrin as the cofactor. These are the only metallo-enzymes where the intermediates can be detected and where the component steps of the catalytic reaction have been identified and studied separately. We shall examine their reactions in some detail: firstly, because their reactions have been extensively studied and there is a large amount of experimental information available; secondly, because significant new data have been published since the last review in 1966[34]; thirdly, because this review differs from previous ones in its emphasis on comparing the reactivity in the presence and absence of protein and trying to establish how the protein enhances the reactivity of the simple, protein-free Fe porphyrin.

Our object in this review is to try to find out how nature uses a protein to increase the catalase and peroxidase activity inherent in a simple Fe porphyrin to a level appropriate to the demands of the metabolism of the organism. After introducing the enzymes in section 8.1, we discuss the reversible dissociation of the enzymes into apoenzyme and coenzyme, and the application of this dissociation to the preparation of synthetic enzymes and to the study of the role of the metal and the porphyrin side chains in the binding to the protein and in the enzymatic reactions (section 8.2), the reaction paths and the nature of the intermediates (section 8.3) and the intriguing question of why the hydroperoxidases apparently always bind a proton whenever they bind a monobasic anion (section 8.4). We then study the effect of changing the protein (and the axial ligand) on the overall catalytic activity (section 8.5), and finally attempt to pinpoint the way in which the protein can increase or modify catalytic activity by comparing the individual rate constants of analogous reactions in the presence and absence of protein (section 8.6). The main points and conclusions are summarised in section 8.7.

8.1 THE DISTRIBUTION, FUNCTION AND NATURE OF THE ENZYMES

Both catalases and peroxidases are widely distributed among animals, plants and aerobic micro-organisms; and at least one virus contains a peroxidase. Catalase comprises nearly 1 per cent of the dry weight of the bacterium *Micrococcus lysodeiktus*. The exact physiological function of these enzymes is not clear. They undoubtedly serve to remove H_2O_2, but probably fulfil some additional role as well. Peroxidases, for example, are particularly abundant in higher plants and have been implicated in various processes, including the control of plant growth (see reference 147 and references therein). Certain peroxidases will catalyse the formation of a carbon–halogen bond according to the overall reaction

$$C-H + X^- + H^+ + H_2O_2 \rightarrow C-X + 2H_2O \qquad (8.3)$$

where X may be Cl, Br or I and C–H may be an aromatic phenol or amine, β-diketone, etc. (see, for example, references 96, 168, 196); for a general review of microbial halometabolites see reference 174. A particularly important example is the iodination of tyrosine residues to form the hormone thyroxine in the mammalian thyroid[65,107]. Some of these halogenations appear to proceed via the intermediate formation of the halogen, others do not. Halogens are probably even more toxic than H_2O_2 itself and are included in the armory of chemical warefare used by the polymorpho-nuclear leukocytes, which are present in the blood and various tissues of mammals to combat bacterial infection. These cells ingest the bacteria and then secrete a peroxidase into the vacuole to catalyse the formation of halogens from H_2O_2 (produced either by the bacteria or the leukocytes) and any halide ion present (mainly Cl^-, but also I^-); the presence of iodide has been shown to lead to iodina-tion of the proteins of the bacterial membrane[157]. But the bombardier beetle can probably claim to make the most ingenious use of these enzymes. They secrete solutions containing hydroquinones and up to 28 per cent H_2O_2. When the beetle is threatened, these secretions are rapidly mixed with catalases and peroxidases in an 'explosion chamber', which results in the ejection with an audible report of a hot toxic gas containing oxygen and quinones[197].

Both the disproportionation of H_2O_2 and its reduction by organic and inorganic compounds can be catalysed by a wide range of simple ions and compounds, both metallic and nonmetallic, homogeneous and heterogeneous (see, for example, references 19 and 60). Yet all the hydroperoxidases so far discovered depend on either a flavin or one of several iron porphyrins as the cofactor, while all the catalases contain specifically Fe protoporhyrin. Nature has, rather surprisingly perhaps, not evolved any such enzymes based on copper, though some of the O_2-carrying proteins based on Cu (see section 7.1) also possess this ability[87]. Further discussion will be restricted to those enzymes that have been shown to possess Fe protoporphyrin as the common denominator and have been subjected to a certain degree of mechanistic study, namely the catalases, which catalyse the decomposi-tion of H_2O_2 to H_2O and O_2 and its reduction by alcohols, formate, etc.; horse-radish peroxidase (HRPO), a typical peroxidase, which catalyses the reduction of H_2O_2 by ascorbic acid, aromatic phenols and amines, ferrocyanide, etc.; cytochrome c peroxidase (CCPO) from yeast, which resembles HRPO in its general reactivity towards donors, but in addition shows an unusually high and specific activity towards the redox protein cytochrome c; and chloroperoxidase (CPO) from the mould *Caldariomyces fumago*, which is unusual in showing reactions typical of both catalase and peroxidase and in addition catalysing the formation of C–Cl bonds according to the equation in the previous section. Nothing is known about the mechanism of action of peroxidases based on flavins or other Fe porphyrins.

Horse-radish peroxidase is not, in fact, a single enzyme. It appears that many, perhaps all, higher plants possess multiple forms of peroxidase (sometimes referred to as isoenzymes), which can be separated by electrophoresis and/or chromatographic techniques; compare the heterogeneity found in Hbs (sections 7.1 and 7.2). Seven forms have been detected in turnip roots[151], and about ten in horse-radishes[131]. Their relative concentration may vary with the season[151] or from one part of the plant to another[147]; this variation is presumably related to the different and changing functions of the peroxidases in the metabolism of the plant. The four isoenzymes of HRPO that have been purified all contain carbohydrate and have essentially the same amino-acid composition[131]; the basis of the heterogeneity is not known. Fortunately, from the point of view of comparing the kinetic data of different workers using different samples, the isoenzymes of HRPO show identical spectra and virtually identical enzymatic activities[131], though the isoenzymes from turnip show differences in spectra[151], while those from the Alaska pea show some differences in kinetics[147]. In this review we shall treat HRPO as a single enzyme.

HRPO has a M.W. of $\approx$40 000, of which 16 per cent represents carbohydrate, and contains one atom of Fe; for the sugar and amino-acid analysis see reference 176. CCPO has a M.W. of 34 000 and contains one Fe, but no carbohydrate. It is interesting that this enzyme is produced by yeast when grown aerobically but, when grown anaerobically, only the apoenzyme is synthesised[245]. CPO has a M.W. of $\approx$42 000, of which 25 per cent represents carbohydrate, and contains one Fe^{217}. The catalases have a more complex structure and are all very similar regardless of source (animals, plants, micro-organisms). They have a M.W. of $\approx$250 000 and consist of four subunits, each of which contains one Fe. Each species probably synthesises a different enzyme, but enzymes from different parts of the same animal (for example, liver and bone-marrow) are serologically indistinguishable[159,196]. The dissociated subunits show peroxidase, but not catalase, activity[117]. Mechanistic studies have been carried out with both bacterial and mammalian catalases; they all show fairly similar activity.

These enzymes form ideal subjects for study. They are readily soluble in water and stable over a wide range of pH without irreversible denaturation of the protein, for example approximately pH 2–11 for catalase[51] and HRPO[53]. They include some of the most thermostable enzymes known; certain peroxidases retain considerable activity even at 90°C[197]. The intense u.v.–visible spectrum of the iron porphyrin (haem) provides a very convenient method of following reactions, and the presence of only one active site per molecule of protein (except in the case of catalase) simplifies the interpretation of the results. They provide a series of enzymes with different but overlapping patterns of reactivity, yet possessing the same cofactor or metal complex. The reactions of these metal–protein complexes can be compared: firstly, with those of the protein-free complexes, which, as we shall see below, show the same basic catalytic properties but to a much lesser degree; secondly, with those of other proteins such as Mb and Hb, which contain the same cofactor but show very little catalytic activity; and, thirdly, with synthetic enzymes prepared by replacing the Fe complex of protoporphyrin with Fe complexes of other porphyrins or with porphyrin complexes of other metals. Finally, the hydroperoxidases are almost unique among enzymes in that intermediates can be detected and studied spectroscopically with fast-reaction techniques, as first shown in the pioneering work of Chance; see, for example, his review in 1951[50]. This has enabled the component steps of the catalase and peroxidase reactions to be identified and studied separately.

By contrast, the protein-free Fe–porphyrin complexes are not so easy to study. They are rather insoluble in water, except in acid and alkaline solution, and tend to form dimers and perhaps higher aggregates. Consequently there is relatively little reliable information available on formation constants for the binding of axial ligands. Solutions of Fe protoporphyrin are sensitive to air, owing probably to autoxidation of a vinyl side chain, and are fairly rapidly destroyed by H_2O_2; data quoted for the protoporphyrin complex may often relate to the product of autoxidation. For references see reference 38. We therefore often lack the necessary thermodynamic and kinetic data on the protein-free complexes that would provide a direct quantitative comparison with the properties of the enzymes.

8.2 THE BINDING OF COENZYME TO APOENZYME AND THE PREPARATION OF SYNTHETIC ENZYMES

The Fe porphyrin can be reversibly dissociated from both HRPO and CCPO without denaturation of the protein and the native enzyme reconstituted without loss of activity. Removal of the Fe porphyrin from catalase, however, has so far proved to be irreversible. The substitution of the Fe protoporphyrin by other Fe porphyrin complexes and by porphyrins containing other metals or no metal allows the study of some of the factors that affect the binding of the coenzyme (or cofactor) to apoenzyme and the enzymatic activity of the enzyme (or holoenzyme).

As in the case of Mb and Hb (see section 7.4) we can write the simplified equilibrium (neglecting the axial ligands on Fe)

$$\text{apoenzyme} + \text{Fe porphyrin} \rightleftharpoons \text{enzyme} \qquad (8.4)$$

but no equilibrium constants have yet been determined for the formation of the enzyme. The apoenzyme from CCPO will firmly bind the Fe complexes of haemato-, meso- and deutero- as well as the native protoporphyrin, which shows that the vinyl side chains at positions 2 and 4 in protoporphyrin IX are not essential for binding; but the Fe complexes of aetioporphyrin and the esters of protoporphyrin, on the other hand, are held very weakly, which indicates the importance of the carboxylic acids at positions 6 and 7 in protoporphyrin[11]. The apoenzyme will also bind metal-free porphyrins[10], though a qualitative comparison of the binding of the iron complexes and metal-free forms of protoporphyrin[10,244] and of aetioporphyrin[11] shows that the iron complex is bound rather more firmly in both cases. The evidence shows that the key factor that controls the binding of the Fe porphyrin to the apoenzyme is the interaction of the protein on the one hand with the porphyrin ring and its side chains on the other, though the co-ordination of an amino-acid side chain by the Fe can make an additional contribution. More quantitative work with Hb and Mb led to the same conclusion (section 7.4).

A preliminary sequencing of the 505 amino acids in the single-chain sub-unit of bovine-liver catalase has been reported[201], but no X-ray analysis has yet been carried out on any hydroperoxidase. Similarities in the spectrum of the adduct between metal-free protoporphyrin and the apoenzyme of CCPO with that of the porphyrin in organic solvents led to the conclusion that the porphyrin is held in a hydrophobic environment[10], as is known from X-ray data to be the case for Mb and Hb (sections 7.2 and 7.4).

Adducts have also been made between the apoenzyme of HRPO and porphyrins containing Mn, Co, Ni and Cu[92,216,248] and between the apoenzyme of CCPO and porphyrins containing V, Mn, Co, Zn and Ag[246,247,248]. Of these only the Mn analogues showed catalytic activity. In fact, although the Mn analogue of CCPO showed lower activity than the native enzyme towards ferrocyanide and Fe(II) cytochrome c, it showed greater activity towards ascorbic acid[246]. This strikingly shows that there is nothing unique about Fe and that 'synthetic' enzymes can be prepared with a different metal (compare the analogues of Mb and Hb in which Fe can be replaced by Co without loss of activity—see section 7.1). Because the co-factor cannot be removed from catalase without denaturing the protein the effect of changing the metal on catalase activity cannot be examined.

The effect of different porphyrin side chains on the catalytic activity has been studied with synthetic derivatives of both HRPO and CCPO. Variation of the Fe porphyrin in HRPO can either increase or decrease its activity in catalysing the reduction of H_2O_2 by mesidine (compare the relative activities with protoporphyrin IX 100, haemato- and mesoporphyrin IX: ≈ 150, and coproporphyrins I and III: ≈ 1). High activity appears to be associated with the presence of carboxylic-acid side chains at positions 6 and $7^{[167]}$. Variation of the Fe–porphyrin complex in CCPO apparently has a much smaller effect on the activity of this enzyme in catalysing the reduction of H_2O_2 by ferrocyanide, except in the case of esters with long alkyl groups (compare the relative activity with proto- 100, meso- 100, deutero- 97, haemato- 88 and aetioporphyrin 53, and the dimethyl and dipentyl esters of the first 40 and 2, respectively). As might be expected, catalysis of the reduction of H_2O_2 by the protein ferrocytochrome c is more sensitive to changes in the porphyrin side chains[11]. The adducts of the apoenzyme of CCPO with metal-free porphyrins are inert[10].

8.3 INTERMEDIATES IN THE ENZYMATIC REACTIONS

The hydroperoxidases all contain Fe(III) in the resting state, that is in the presence of air and water and the absence of H_2O_2. Analysis of the u.v.–visible spectra suggests that the ligand provided by the protein is the imidazole of a histidine in the peroxidases (as in Mb and Hb) and a carboxylate in the catalases, while the sixth co-ordination position is 'free', that is probably occupied by H_2O[206]. The co-ordinated H_2O ionises with a pK of 9.5–11 in HRPO and the P_1 isoenzyme of turnip peroxidase, though the pK of the P_7 isoenzyme is as low as 8.4; no pK can be detected in catalase before the onset of denaturation at pH ≈ 11 [34,185]. The Fe(III) ion is high spin in the aquo- complexes of catalase and HRPO and low spin in the hydroxo-complex of HRPO[34]. The spectra suggest that the P_1 isoenzyme of turnip peroxidase is also high spin, while P_7 is a thermal mixture of high- and low-spin forms[185]. CCPO at room temperature and in neutral solution is claimed to consist of an equilibrium mixture of the aquo- and hydroxo- complexes, both present in high- and low-spin forms[245]. There is obviously little difference in energy between the high- and low-spin forms.

The Fe(III) is reduced to Fe(II) only with difficulty; since the enzymatic reactions are inhibited by ligands such as F^- and CN^-, which bind to Fe(III), but not by CO, which only binds to Fe(II), we can neglect the Fe(II) state in discussing the enzymatic reactions. These involve two further types of complex, commonly referred to as compounds I and II, which have been shown to contain two and one additional oxidising equivalents, respectively; we shall refer to them as Fe^V and Fe^{IV} complexes. Their structures have not yet been conclusively established, though recent evidence has narrowed the range of possibilities. And, although the electronic spectra suggest similar structures for the analogous complexes from different enzymes, this may not always be the case; and a given complex may even be present as a mixture. The Fe^{IV} compounds can also react with excess H_2O_2 to form yet another type of complex (Fe^{VI}, compound III); but these are not intermediates in the enzymatic reactions.

Similar or identical Fe^V complexes are formed by the reaction of HRPO with alkyl and acyl peroxides as well as H_2O_2[199], and by other strong oxidising agents as diverse as HOCl, $S_2O_8^{2-}$ + Ag^+ and $IrCl_6^{2-}$ [168]. By studying the transfer of labelled oxygen in the reaction of CPO with *m*-chloroperbenzoic acid to give Fe^V and the reaction of the latter with further per-acid to liberate O_2, it was shown that the Fe^V complex contains one O atom derived from the per-acid[95], while the stoichiometries of some of these reactions have been established as follows

$$1Fe^{III} + 1H_2O_2 \rightarrow 1Fe^V$$
$$1Fe^{III} + 1EtOOH \rightarrow 1Fe^V + 1EtOH \qquad (8.5)$$
$$1Fe^{III} + 1ArCO_3H \rightarrow 1Fe^V + 1ArCO_2^- + 1H^+$$

where $ArCO_3H$ is *m*-nitroperbenzoic acid; no loss or gain of a proton was observed except where indicated[199]. Magnetic susceptibility studies indicate the presence of three unpaired electrons. The most likely structures are then either (a) a d^3 Fe(V) complex with three unpaired electrons, probably containing co-ordinated oxide in the form of a ferryl ion $[FeO]^{3+}$, or (b) a d^4 Fe(IV) ion in the form of $[FeO]^{2+}$, together with an organic radical, for example a porphyrin radical formed by a one-electron oxidation. The Fe^V complex of CCPO does, in fact, show an additional e.s.r. signal with $g \approx 2$, which can be attributed to an organic radical[245]; this suggests structure (b). But no such signal has yet been detected in catalase or HRPO. For further references to physical measurements and to a discussion of other possible structures see reference 199. The Fe^{IV} complexes possess two unpaired electrons and are generally considered, but not proved, to contain the ferryl ion $[FeO]^{2+}$, while the diamagnetic Fe^{VI} complexes may be $Fe^{II}O_2$ complexes analogous to those of Hb and Mb. The only example of the direct spectrophotometric observation of the formation of such porphyrin complexes in the absence of protein appears to be the reaction of Fe(III) deuteroporphyrin with H_2O_2 in aqueous solution to give a complex, which is assumed to be Fe^V [178].

The individual steps in these reactions are shown in the following scheme, those involved in the catalase and peroxidase reactions being marked — and - - -, respectively

$$Fe^{III} \underset{k_{53}}{\overset{k_{35}}{\longleftrightarrow}} Fe^V \underset{k_{43}}{\overset{k_{54}}{\longleftrightarrow}} Fe^{IV} \tag{8.6}$$

Fe^{III} reacts with H_2O_2 to give Fe^V; this can then react with a second molecule of H_2O_2 (catalase reaction) or with donors such as ethanol or formic acid (nonclassical peroxidase reaction) to regenerate Fe^{III} in a single step, or with donors such as pyrogallol (classical peroxidase reaction) to form first Fe^{IV} and then Fe^{III}. Labelling experiments have shown that both atoms of O_2 are derived from the same molecule in the decomposition of H_2O_2 catalysed by catalase[95,114], but from different molecules in the decomposition of m-chloroperbenzoic acid catalysed by CPO[95]. It is perhaps worth noting that the spontaneous decomposition of organic and inorganic per-acids can also proceed by mechanisms in which the two oxygen atoms originate either in the same or in different molecules[38]. There is indirect evidence that catalase evolves O_2 in the singlet state[7].

The essential correctness of the above scheme has been shown by a quantitative comparison of the rates of the individual steps with those of the overall catalytic reaction and with the steady-state concentrations of the intermediates; slight anomalies observed in k_{35} are discussed below. Further support for the scheme is the fact that the substrates for the classical peroxidase reaction are known to undergo one-electron oxidations readily and that radicals derived from quinol, ascorbic acid and dihydroxyfumaric acid have been detected during the enzymatic reaction[177]. Formate and ethanol, on the other hand, prefer to undergo a single two-equivalent oxidation by the loss of a hydride ion[71], and no radicals have yet been reported during the reactions of these substrates. It is probably relevant that ethanol will rapidly reduce the Fe^V complex of CPO to Fe^{III}, but will not touch Fe^{IV} [218]. However, strictly speaking we cannot exclude the occurrence of a two-step reaction such as

$$Fe^V + CH_3CH_2O^- \rightarrow Fe^{IV} + CH_3CH_2O\cdot$$
$$Fe^{IV} + CH_3CH_2O\cdot \rightarrow Fe^{III} + CH_3CHO + H^+ \tag{8.7}$$

in which Fe^{IV} is rapidly reduced by the radical before it can diffuse away; but this would be kinetically indistinguishable from a single two-equivalent reaction. In principle H_2O_2 may be oxidised either directly to O_2 in a single step or indirectly via the superoxide anion O_2^- in two steps. The potentials (in volts) at pH 7 favour the single two-equivalent reaction; O_2/O_2^-: -0.45; O_2^-/H_2O_2: $+0.98$; O_2/H_2O_2: $+0.27$[85]. It seems reasonable to conclude that the catalase and non-classical peroxidase reactions involve only a single intermediate (Fe^V), which is reduced back to Fe^{III} in a single two-equivalent step, while the classical peroxidase reactions involve two intermediates (Fe^V, Fe^{IV}) and two consecutive one-equivalent reductions.

There is no kinetic evidence to suggest any interaction between the four Fe atoms in the catalase tetramer.

Several side reactions should be mentioned. Firstly, Fe^{IV} compounds react reversibly with high concentrations of H_2O_2 to give Fe^{VI}; these complexes are not involved in the catalytic cycle, though their formation can affect the kinetics. Secondly, the Fe^V complexes are 'spontaneously' reduced by amino-acid residues and/or the carbohydrate attached to the protein to Fe^{IV} and then to Fe^{III}. Since some part of the protein must have been altered in the process, this does not regenerate the truly 'native' enzyme. These reactions occur slowly with catalase and the peroxidases; but when the Fe^{III} complexes of Hb and Mb are treated with H_2O_2 the first detectable product is Fe^{IV}, due probably to a very rapid reduction of Fe^V (see references in reference 249).

It should not be assumed that all reagents must react directly with the Fe; the donors in the peroxidase reactions, for example, might well transfer their electrons via the porphyrin ring. It will, however, be tacitly assumed in the subsequent discussion that the bond between the Fe and the amino-acid side chain of the protein (histidine or carboxylate) remains intact throughout these reactions, simply because there is no evidence to the contrary.

Finally, it is worth mentioning the redox potentials. Although the peroxidases apparently have the same axial ligands (histidine and H_2O) as Mb and Hb, they are much more difficult to reduce. This can be understood in terms of their function. Mb and Hb act as carriers of O_2 by virtue of their ability to co-ordinate O_2 to the Fe^{II} form, and no other oxidation state is required. The catalytic activity of the peroxidases and catalases, on the other hand, involves Fe^{III}, Fe^{IV} and Fe^V; and the formation of Fe^{II} would be tantamount to a poisoning of the catalyst. Certain peroxidases, can however, also catalyse the autoxidation of particular substrates ('oxidase' activity), possibly via the formation of the $Fe^{II}O_2$ complex; a good example is the P_7 isoenzyme of turnip peroxidase, where the $Fe^{II}O_2$ is relatively stable[185]. This change in function is paralleled by the change in redox potential of the Fe^{III}/Fe^{II} couple (in volts) at pH 7: Hb $\approx +0.2$; Mb $\approx +0.1$[85], P_7 -0.12; P_1 -0.22[185]. HRPO ≈ -0.25[85]. Possible methods by which the protein can alter the redox potential (and equilibrium constants) without changing the ligands have been discussed in section 7.6. Catalase, which probably possesses a carboxylate instead of histidine as the axial ligand, is even more difficult to reduce, and the potential has not been determined. The Fe^V/Fe^{IV} and Fe^{IV}/Fe^{III} couples of HRPO both have potentials of approximately +1 volt[85].

8.4 THE BINDING OF PROTONS AND ANIONS

The hydroperoxidases share one intriguing property (apart from their catalytic activity), which distinguishes them from Mb and Hb. The latter reversibly bind anions (X^-) according to simple equilibria such as

$$Mb + X^- \rightleftharpoons [Mb.X^-] \qquad (8.8)$$

either by co-ordination to the metal or at some other site in the protein. The binding of monobasic anions to the hydroperoxidases, on the other hand, shows a dependence on pH, which indicates either the simultaneous uptake of a proton, for example

$$Cat + H^+ + X^- \rightleftharpoons [Cat.H^+.X^-] \qquad (8.9)$$

or the equivalent loss of hydroxide, for example

$$[Cat.OH^-] + X^- \rightleftharpoons Cat.X^- + OH^- \qquad (8.10)$$

Such pH effects are shown by catalase, HRPO and CPO, both in the equilibrium constants for the binding of anions, and in the rate constants for redox reactions involving anions. Justifiably or not, it is tempting to think that nature must have developed this property for some purpose and that it contributes in some way to the enzymatic activity. So it is worth looking at these effects more closely.

   A proton is taken up (or hydroxide lost) simultaneously with the binding of F^-, N_3^-, CN^-, formate and acetate by catalase Fe^{III} [52,159], of Cl^-, Br^- and I^- by CPO Fe^{III} and all four halides by HRPO Fe^{III} (see reference 218 and references therein) and of CN^- by HRPO Fe^{II} [175]. Changes in magnetic properties show that CN^- is co-ordinated by the metal; and spectroscopic changes suggest, but do not prove, co-ordination in most of the other cases. The fact that the ligand displaced by the anion is H_2O and not HO^- tends to suggest that the equilibrium involves the uptake of H^+ rather than the loss of HO^-. But there are other examples where the anion appears not to be co-ordinated and must be held by the protein alone. The rates of the spontaneous reduction of the Fe^V and Fe^{IV} complexes of catalase, for example, are increased by the presence of anions (F^-, CN^-, etc.), and evaluation of the change in rate with the change in anion concentration yields equilibrium constants that also show the uptake of one proton; no spectroscopic changes could be detected, at least in the case of Fe^{IV} [159]. In addition, it seems unlikely that the anion would displace either the oxide ligand, which is probably attached to the metal in the Fe^{IV} and Fe^V complexes, or the carboxylate of the protein, since that would require a considerable change in the interaction between the protein and the porphyrin.

   The rates of all redox reactions so far studied between anionic reducing agents and the Fe^{IV} and Fe^V complexes of hydroperoxidases appear to increase linearly with the proton concentration, though additional inflections in the pH profile due to other protonations are also observed. By contrast, the rates of reaction with neutral reducing agents (for example, phenols, ascorbic acid) are generally pH independent around pH 7; compare the pH profiles in reference 68. Examples of such proton-dependent reductions are: formate with catalase Fe^V [159]; ferrocyanide with catalase Fe^V [159] and HRPO Fe^V and Fe^{IV}, though in the last case the rate is pH independent above pH 8 [100]; nitrite with catalase and HRPO Fe^V and Fe^{IV} and, apparently in this one case, Mb Fe^{IV} [159]; iodide with HRPO Fe^V and Fe^{IV} [187,188]. The CPO-catalysed chlorination of monochlorodimedone also requires acid, and there is excellent agreement between the pH dependence of the binding of Cl^- by CPO Fe^{III} and the rate of chlorination [218], which serves to link together the equilibrium and kinetic effects of the proton. But we do not know whether all the quoted examples involve the binding of a proton at the same site or not.

These examples show: that the binding of an anion together with the uptake of a proton (or just conceivably the loss of hydroxide) is a common feature of the hydroperoxidases; that the effect of the proton can be seen in both equilibrium and kinetic properties; that, at least in the case of CPO, there is a correlation between the equilibrium and kinetic effects; and that the anion need not be co-ordinated to the metal. The case of catalase and formate is particularly significant, since formate is a substrate for the nonclassical peroxidase reaction of catalase and its mechanism of reaction with Fe^V is probably analogous to that of H_2O_2. We have seen that a proton is required when formate is co-ordinated to Fe^{III}, when it is bound (though apparently not co-ordinated) to Fe^{IV} and when it reduces Fe^V.

Are the proton and the anion bound together in the form of the undissociated acid or are they bound separately? The binding of the undissociated acid would seem energetically reasonable for weak acids such as HCN, HCOOH and HF, and there are examples outside the field of porphyrin complexes of the co-ordination of undissociated acids such as HF, HN_3 and HCN or HNC. But it seems intuitively unlikely that HCl, HBr and HI, whose pKs are -7, -9, and -10, respectively [23] should be bound as such. Kinetic measurements have, in fact, shown that in the reaction of I^- with HRPO Fe^V and Fe^{IV} [187,188] and in the overall reduction of H_2O_2 by I^- [25] the reacting species cannot be HI, because this would require a rate constant greater than the diffusion-controlled limit by factors of up to 10^{10}; that is, H^+ and I^- must react independently with the enzyme. The kinetics of the reversible binding of the weak anion F^- to HRPO Fe^{III} were also found to be compatible only with the independent binding of H^+ and F^- at separate sites [75]. We do not, of course, know whether this conclusion applies to all weak anions, but there is no need to invoke a second mechanism for other anions until there is evidence for it.

If the proton and anion are bound at separate sites, there must obviously be some co-operative interaction between these sites. This is similar to the effects we have already noted with Mb and Hb (section 7.5), whereby the binding of one reagent (for example H^+) at one site can influence the equilibrium constant for the binding of a second reagent (for example O_2) at another site. But the effects appear to be quantitatively greater with the hydroperoxidases. The binding of each reagent can be studied separately in the case of Mb and Hb, and the binding of one reagent changes the equilibrium constant for the second by up to $10^{2.5}$. The binding of H^+ and F^- or Cl^- by HRPO Fe^{III} obeys the relationship

$$K = [\text{HRPO} . \text{HX}] / [\text{HRPO}] [H^+] [X^-]$$

within experimental error over the range of pH 2.8–5 and no change in spectrum due to protonation alone could be seen using solutions of H_2SO_4 down to pH 2.8[53]; in this case the binding of F^- must affect the basic dissociation constant by more than 10^2 and vice versa. But the kinetics suggest that the protonation of HRPO Fe^{III} increases the rate of binding of F^- by about 10^4 without significantly affecting the rate of dissociation[75].

We conclude that the hydroperoxidases have developed the basic mechanism of co-operative interaction between two sites observed in Mb and Hb in a particular direction, which enables them to bind a proton together with a monobasic anion; the greater degree of interaction in the former could be due to the fact that both reagents are charged ions. This could provide a mechanism for facilitating the binding at the active site of the anions of strong inorganic acids such as Cl^- and I^- (for example, in enzymatic halogenation), which do not readily co-ordinate to metals or form ion pairs, according to the equilibrium

$$\text{Enz} + H^+ + X^- \rightleftharpoons [H^+ . \text{Enz} . X^-] \qquad (8.11)$$

But if we now consider a very weak acid whose pK is much greater than the physiological pH of ≈ 7, then we can write

$$\text{Enz} + HX \rightleftharpoons [H^+ . \text{Enz} . X^-] \qquad (8.12)$$

This shows that the co-operative interaction could also provide a method for increasing the effective dissociation of HX and hence increasing the rate at physiological pH of those reactions requiring anions that are normally present in sufficient quantity only at a very much higher pH. We shall discuss the probable use of this device to activate H_2O_2 in section 8.6.

8.5 THE EFFECT OF THE PROTEIN ON THE OVERALL CATALYTIC ACTIVITY

Keilin and Hartree[120] made an interesting comparison of the catalytic activity in aqueous solution at pH 5.9 of horse-liver catalase, HRPO, horse-heart myoglobin (Mb) and, in one reaction, iron protoporphyrin (haematin). They expressed the catalytic activities in the form of pC, the negative logarithm of the concentration of Fe porphyrin (molarity) required to cause 50 per cent of the H_2O_2 generated under their standard conditions to be used in the oxidation of various hydrogen donors, which were present at a concentration of $\approx 1.7 \times 10^{-2}$ M (except, of course, in the case of H_2O_2 itself, that is for the catalase reaction). The pC value therefore probably increases roughly with the logarithm of the catalytic activity. Their data are given in table 16 together with some additional results on the protein-free complex from other workers.

TABLE 16 RELATIVE CATALYTIC ACTIVITIES OF Fe PROTOPORPHYRIN IN THE REACTION OF H_2O_2 WITH VARIOUS DONORS IN THE PRESENCE AND ABSENCE OF DIFFERENT PROTEINS

Substrate/Donor	Reactivity (pC)			
	Fe protoporphyrin (haematin)	Myoglobin	Peroxidase (HRPO)	Catalase
H_2O_2 (i.e. $\rightarrow O_2 + 2H_2O$)	≈ 5	≈ 4	5.9	9.6
nitrite		4.5	6.2	5.1
pyrogallol	+a	4.8	9.3	5.5
p-cresol		4.5	6.5	4.5
ascorbic acid	+a, b	*	7.7	–
$[Fe^{II}(CN)_6]^{4-}$		5.5	8.2	≈ 0
guiacol		5.0	8.3	≈ 0
adrenaline		5.3	7.9	≈ 0
ethanol	–c	–	–	5.5
formate	–c	≈ 0	–	5.1

Notes: Data from reference 120 except where indicated: a = reference 196; b = reference 136; c = reference 115. For definition of pC see text.
– no activity observed
≈ 0 slight activity observed
+ activity observed by other workers (therefore cannot be given a pC value)
* reduction to Fe(II), which is catalytically inactive.

Comparison of the data in table 16 for the enzymatically active catalase and HRPO with those for the relatively inert Fe protoporphyrin and Mb shows how the presence of the protein in the former can enhance or suppress (or even leave unchanged) the rates of particular reactions. Comparison of Mb with Fe protoporphyrin shows that the presence of protein does not automatically lead to enhanced activity. Changes in rate by factors of 10^5 or more are observed both in enhancement and suppression.

The catalytic activity of CPO[217] provides an interesting comparison. It resembles other peroxidases in using substrates such as guiacol, pyrogallol and ascorbic acid, though the reactivity towards the first is about one tenth that of HRPO. It also resembles catalase in being able to catalyse the liberation of O_2 from H_2O_2, though less efficiently than catalase (≈ 2 per cent of the rate at pH 4.5), and the oxidation by H_2O_2 of ethanol to acetaldehyde and probably also of formic acid; it can even catalyse the liberation of O_2 from ethyl hydroperoxide and from *m*-chloroperbenzoic acid, which catalase is unable to do. The catalytic properties of CPO are therefore intermediate between those of catalase and the typical peroxidases; it appears to be slightly less efficient than HRPO in catalysing the oxidation of a substrate such as guiacol, and slightly less efficient than catalase in decomposing H_2O_2. In addition, it catalyses chlorination reactions, which HRPO and catalase do not; but we shall be concerned here only with the typical peroxidase and catalase reactions. The most important point is probably that CPO is a monomer, while catalase is a tetramer. This shows that the tetrameric structure is not essential to the basic mechanism of the catalase reaction, though it may modify the reaction in some way.

The above data also suggest that such changes in overall activity by a factor of 10^5 cannot be due to changes in the nature of axial ligand (X), which might be the first thought of the co-ordination chemist, but must be ascribed to more subtle effects on the protein. Table 16 shows that Mb (X = histidine) is very different from HRPO (X also histidine) but similar to Fe protoporphyrin (X = H_2O or HO^-). Catalase activity is exhibited both by the tetrameric catalase itself (X = carboxylate) and by CPO (X = histidine). The monomeric sub-unit of catalase, on the other hand, exhibits classical peroxidase, but no catalase, activity; it would be interesting to know whether X has remained carboxylate as in the tetramer or has changed to histidine as in the peroxidases, but unfortunately no spectroscopic data have yet been reported[117]. Nor have kinetic data been reported for well-defined Fe porphyrin complexes containing one nitrogenous base, but qualitative results for the change in catalytic activity observed on adding excess base suggest that the addition of various nitrogenous bases does not increase either the catalase or the peroxidase activity by more than about an order of magnitude[13,140,209]; however, the simultaneous addition of two different bases (for example, histidine and guanidine) seems to have a considerably greater effect, at least on peroxidase activity[154].

We see then that a protein can, independently of any change in the axial ligand, significantly increase or decrease the rates of the catalysed reactions of H_2O_2 with many different substrates. Does each substrate, where there is a significant change in rate, require a different structural feature or 'device' of the protein? Or, conversely, are there any common denominators in the structural features of the protein required to cause these effects as between the different substrates or between the different enzymes? Since a decrease in rate may well be due simply to such factors as steric or coulombic repulsion, we shall focus attention mainly on the mechanism of increasing the rate. The fact that an enhancement of catalase activity is apparently accompanied by an enhancement of nonclassical peroxidase activity towards ethanol and formic acid strongly suggests the existence of some common denominator in these two types of reaction. The data of table 16 show that the features required to accelerate the catalase and nonclassical peroxidase reactions are not the same as those required to accelerate the classical peroxidase reaction with donors such as pyrogallol. We therefore need to invoke (at least) two features or devices in order to explain the effect of the proteins in enhancing the catalytic activity of the simple Fe porphyrin complexes. But the above data on the overall reactions cannot tell us whether the classical catalase and peroxidase reactions each require a separate device, or whether one of the reactions involves two devices, one being shared with the other reaction. This can be established by looking at the individual steps in the reaction.

8.6 THE EFFECT OF THE PROTEIN ON THE INDIVIDUAL RATE CONSTANTS

Let us now look more closely at the effect of protein on the individual rate constants (see scheme on p. 161): firstly, the reaction of Fe^{III} with H_2O_2 to give Fe^V, which is common to both catalase and peroxidase reactions (k_{35}); secondly, the one-equivalent reductions of Fe^V by donors in the classical peroxidase reaction (k_{54} and k_{43}); thirdly, the two-equivalent reduction of Fe^V by H_2O_2 in the catalase reaction and by donors such as alcohols in the nonclassical peroxidase reaction (k_{53}).

It should be noted that the rate constants to be compared do not always relate to the same conditions of temperature, pH and ionic strength, and reference must be made to the original papers for full experimental details. But the changes in rate due to a change in the experimental conditions are probably insignificant compared to the changes due to the protein and will therefore be neglected. Reference will be made to the reactions of Fe deuteroporphyrin; since the substitution of Fe proto-porphyrin in CCPO by Fe deuteroporphyrin gives a synthetic enzyme with 97 per cent of the activity of the native enzyme (see section 8.2), we shall assume that the two protein-free complexes also have very similar activities.

8.6.1 $Fe^{III} + H_2O_2 \xrightarrow{k_{35}} Fe^V$

In the light of the discussion in section 8.3 this reaction can probably be written

$$[Fe.OH_2]^{3+} + H_2O_2 \rightarrow [FeO]^{3+} + 2H_2O \tag{8.13}$$

where the porphyrin and second axial ligand are omitted. However, the kinetics of the reaction of Fe^{III} deuteroporphyrin indicate that the rate-determining step is preceded by a rapid pre-equilibrium[178], and the break in the Arrhenius plot of the temperature variation of the reaction rate of bacterial catalase[210] shows that the reaction consists of more than one step. The rate constant also appears to decrease at very high concentration (≈ 4 M) of H_2O_2[163]. This reaction is therefore not a simple one. But second-order kinetics are nevertheless observed under most conditions; these rate constants (k_{35}) are listed in table 17. The reactions all appear to be virtually independent of pH, at least around pH 7, in agreement with the above formulation of the reaction. The rate of the analogous reaction of catalase with peroxyacetic acid, which has a pK of 8.2, does vary with pH; quantitative evaluation of the data shows that the undissociated acid (CH_3CO_3H) reacts far faster than the anion[116], as might be expected if the reactions really were analogous. Values of k_{35} have also been determined indirectly for Fe protoporphyrin from the kinetics of the overall reaction at $0°C$[83]; however, evaluation of the data required making certain assumptions, which led to two very different values of k_{35} (2.7×10^4 and 5.8 M^{-1} s^{-1}), and it is difficult to know what weight to give these results.

The rates in table 17 show that the values of k_{35} are all significantly higher for the enzymatically active proteins than for the denatured enzymes, the inactive Mb and the protein-free complex. The presence of the protein in the enzymes accelerates k_{35} by a factor of 10^3–10^5 by comparison with Fe deuteroporphyrin. If we assume that the anomaly, which is observed in the kinetics of catalase only above 0.4 M H_2O_2, is due to the same type of pre-equilibrium that has an effect on the kinetics of the much slower reaction of Fe deuteroporphyrin at $\approx 10^{-3}$ M H_2O_2, then we can conclude that the protein must also increase the rate of pre-equilibration (k_a) by a factor of $\approx 10^7$ and that the intermediate involves the iron porphyrin and H_2O_2, but not the protein. A reasonable scheme to explain these observations would be

$$Fe^{III} + H_2O_2 \underset{k_{-a}}{\overset{k_a}{\rightleftharpoons}} \begin{array}{c} [Fe.H_2O_2]^{3+} \\ (\text{or } [Fe.HO_2]^{2+} + H^+) \end{array} \xrightarrow{k_{35}} [FeO]^{3+} + H_2O \quad (8.14)$$

There is, unfortunately, no evidence to suggest whether k_a is pH dependent or not, and to distinguish between the two possible formulations for the intermediate; but the lower structure might be preferred on the grounds that HO_2^- would be a better ligand than H_2O_2.

TABLE 17 SECOND-ORDER RATE CONSTANTS (k_{35}) FOR THE REACTION:
$$Fe^{III} + H_2O_2 \rightarrow Fe^V$$

Compound	Temp (°C)	k_{35} $(M^{-1}s^{-1})$	Ref.
	Nonenzymes		
Fe deuteroporphyrin (Fe protoporphyrin—see text)	21	$\approx 10^3$	178
myoglobin†	23	1.4×10^2	249
catalase denaturated with 17 M formamide	13–17	30	139
catalase denaturated with 7.2 M guanidine	13–17	2×10^3	139
	Enzymes		
CPO	R.T.	1.5×10^6	218
catalases	25	0.6–1.1×10^7	159
HRPO	25–30	0.9×10^7	50
CCPO	23	1.4×10^8	249

† the first detectable product is Fe^{IV}, due probably to a very rapid spontaneous reduction of the initially formed Fe^V (see section 8.3).

It is clearly impossible yet to specify the effect of the protein any more closely. We could envisage the thermodynamic stabilisation of the intermediate $[Fe.H_2O_2]^{3+}$ or $[Fe.HO_2^-]^{2+}$, the product $[FeO]^{3+}$ or the HO_2^- if required as the reagent (see section 8.4), and the kinetic acceleration of either or both steps (k_a and k_{35}), via the formation of specific hydrogen bonds with favourably positioned amino-acid groups in the protein, perhaps coupled with co-operative movements within the protein of the type discussed in section 7.6.

8.6.2 $Fe^V + 1\ominus \xrightarrow{k_{54}} Fe^{IV}$; $Fe^{IV} + 1\ominus \xrightarrow{k_{43}} Fe^{III}$

The classical peroxidase reaction involves the one-electron reduction of Fe^V to Fe^{IV} and of Fe^{IV} to Fe^{III}. Reducing agents include aromatic amines, phenols, ascorbic acid, ferro-cytochrome c, ferrocyanide, iodide and nitrite; reactions with the last three reagents all require a proton (see section 8.4).

Catalase shows a striking similarity to HRPO in its proton-dependent reduction by nitrite. Compare the following rate constants (in $M^{-1} s^{-1}$) calculated on the basis of undissociated HNO_2 [159]

catalase	$k_{54} = 1.4 \times 10^7$	$k_{43} = 1.5 \times 10^6$
HRPO	2.0×10^7	2.4×10^5
(cf. also Mb		2.5×10^5)

We see that k_{54} is greater than k_{43} in both cases, as expected in view of the greater oxidising power of Fe^V over Fe^{IV}, and that the analogous rate constants of the two enzymes are numerically quite similar.

But differences appear when we look at reactions with reducing agents such as ascorbic acid, aromatic amines and phenols. These reagents react at rates that are essentially pH independent, at least around pH 7. We still find that $k_{54} > k_{43}$ for both catalase and HRPO, but both rate constants are very much smaller for catalase. Both guiacol and p-aminobenzoic acid reduce HRPO Fe^V ($k_{54} = 9 \times 10^6$ and $5 \times 10^4 M^{-1} s^{-1}$, respectively) about 25–30 times faster than Fe^{IV} ($k_{43} = 3 \times 10^5$ and 2×10^3)[34]; no analogous rate constants have been reported for catalase. Moreover pyrogallol reduces catalase Fe^V ($k_{54} = 2.4 \times 10^3 M^{-1} s^{-1}$ [159] about 30 times faster than Fe^{IV} ($k_{43} = 80$)[159]. But this latter rate is several orders of magnitude slower than the analogous rates for HRPO ($k_{43} = 3 \times 10^5$) or for the protein-free Fe protoporphyrin in the presence of excess histidine ($k_{43} = 6 \times 10^4$)[219]. Compare also the rate of reaction of ascorbic acid at pH 7 with catalase Fe^V ($k_{54} = 300 M^{-1} s^{-1}$)[159], with HRPO Fe^V ($k_{54} = 1.8 \times 10^5$)[49] and with the protein-free Fe^V deuteroporphyrin ($k_{54} \approx 10^7$)[178]; the reaction with catalase Fe^{IV} is too slow to be observed. The low activity of catalase in catalysing the reaction of H_2O_2 with pyrogallol and ascorbic acid (see table 16) is therefore due to the unusually low values of k_{54} and k_{43}. It appears that the protein in catalase actually inhibits the reduction of both Fe^V and Fe^{IV} by large molecules such as pyrogallol and ascorbic acid (by a factor of $\geqslant 10^3$), but not by a small ion such as nitrite. Reference to table 16 indicates that the inhibition is even greater (by a factor of $\approx 10^8$) for reagents such as guiacol and adrenaline. That this inhibition can probably be ascribed to steric hindrance is suggested by a comparison of the rates of oxidation of Fe^{III} to Fe^V with the peroxides ROOH, where R = H, Me and Et. The rate falls by a factor of 3×10^2 for mammalian catalase and 4×10^3 for bacterial catalase as R is changed from H to Et[159], while those for HRPO and CCPO fall by factors of only 3 and 6, respectively, and that for Mb actually rises by about 10^2 [245,249]. Compare also the effect of increasing the size of the alkyl group on the rate (k_{53} in $M^{-1} s^{-1}$) of the reaction of catalase Fe^V with the following alcohols: MeOH and EtOH both ≈ 1000; $Pr^n OH$, 17; $Bu^n OH$, 2; and isoamyl alcohol 0.1 [49]. There is obviously a very considerable steric hindrance to the approach of large substrate molecules, which is absent in the other haemoproteins. It seems that the low peroxidase activity of Mb (see table 16) is caused mainly by the low value of k_{35} (see table 17).

Some data are available which allow HRPO to be compared with protein-free Fe porphyrins. Second-order rate constants have been reported for the reduction of Fe^V deuteroporphyrin to Fe^{III} by various reducing agents at pH 7.4[178]. The kinetic data showed no deviation from that expected for a simple one-step reaction. Assuming that the initial complex was correctly identified as Fe^V (and not Fe^{IV}), this means either that the reaction proceeds in a single two-electron step or that $k_{54} < k_{43}$ and that the observed rate constant is k_{54} and not k_{43}. But, since k_{54} and k_{43} do not appear to differ by more than two orders of magnitude in HRPO and we are looking for changes in rates that are considerably greater, we are probably justified in neglecting this uncertainty. The dependence of the rate on pH was not investigated. Rate constants for the reactions of HRPO and Fe deuteroporphyrin that involve the same reducing agent under comparable conditions are listed in table 18.

TABLE 18 SECOND–ORDER RATE CONSTANTS (k_{54} AND k_{43}) FOR THE ONE-ELECTRON REDUCTION OF Fe^V AND Fe^{IV}

	Rate constants (M^{-1} s^{-1})		
complex	HRPO		Fe deuteroporphyrin
pH	7.0		7.4
temperature	25–30°C		25°C
reference	50		178
Reducing agent	k_{54}	k_{43}	k_{54}
p-hydroxydiphenyl		8×10^7	$\geqslant 10^7$
hydroquinone		3×10^6	$\geqslant 10^7$
catechol		2×10^6	1.4×10^5
resorcinol		3×10^5	0.6×10^5
aniline		7×10^4	1.4×10^5
p-aminobenzoic acid	$2 \times 10^{4\dagger}$	1×10^3	5.6×10^3
ascorbic acid	1.8×10^5	$1.3 \times 10^{3\ddagger}$	$\geqslant 10^7$

† reference 54
‡ reference 49

As the authors point out[178], the protein appears to have little effect on the rate constants k_{54} and/or k_{43} compared to the large effect on k_{35} and k_{53}; that is, it neither increases all rates generally nor influences the substrate specificity by changing the relative rates towards different reducing agents. Rate constants have also been determined for Fe protoporphyrin in the presence of 0.3 M histidine at pH 6.3–6.5 with the reducing agents leucomalachite green, guiacol and pyrogallol[219]. These rate constants were evaluated from the overall catalytic activity and it was assumed, by analogy with HRPO, that the rate-limiting step corresponded to k_{43} (k_4 in their nomenclature); but the possibility cannot be excluded that the rate constants represent k_{54}, as in the case of Fe deuteroporphyrin. Their results showed that the values of k_{43} were about 3, 5 and 10 times greater for HRPO than for the protein-free complex for the three reagents mentioned. Here again the presence of the protein seems to have relatively little effect on the rate.

We therefore conclude that the values of k_{54} and k_{43} are not significantly changed by the presence of the protein in HRPO (also presumably in other peroxidases, and probably in other haemoproteins such as Mb), but that the protein in catalase may, because of steric hindrance, very significantly decrease both k_{54} and k_{43} (by factors of 10^3–10^8) when the reducing agent is a large molecule.

8.6.3 $Fe^V + H_2O_2$ (RCH_2OH, $HCOOH$, etc.) $\xrightarrow{k_{53}}$ $Fe^{III} + O_2$ ($RCHO$, CO_2, etc.)

Rate constants for the reaction with H_2O_2 have been obtained only for certain catalases; they all fall in the range 2–4 x 10^7 M^{-1} s^{-1} and are pH independent over the range 3–9[159]. The reactions with alcohols are also pH independent, but much slower; compare the values (in M^{-1} s^{-1}) of k_{53} for mammalian catalase with MeOH $\approx$EtOH ($\approx 10^3$) > PrOH (17). The reaction with formate increases with acidity and involves HCOOH with the much higher rate constant of 0.9 x 10^6 M^{-1} s^{-1} [159]. No anomalies have been observed for the reaction with H_2O_2, either in the rate constants or in their temperature variation[210], to indicate that this particular reaction is not a simple one. But the rate constants for the reactions of Fe^V with MeOH and EtOH, which probably occur by an analogous mechanism, fall off at high alcohol concentrations[49], suggesting the occurrence of some pre-equilibrium; and experiments with (−)-ethanol-1-d apparently indicate a stereospecific oxidation of the alcohol, presumably involving the formation of some complex between the donor and the protein[159]. Taken together, these results suggest a rapid pre-equilibrium, involving the binding of the alcohol to a part of the protein close to the FeO^{3+} ion, followed by an oxidation–reduction reaction. It is reasonable to assume that a similar two-step mechanism occurs with H_2O_2, even though the first step is too fast to be detected.

We can obtain additional information relating to k_{53} by examining the overall rates of the catalytic reaction, in particular the effect of pH, since we know the values of k_{53} for the enzyme and that they are essentially pH independent. Figure 33 shows the results reported for the influence of pH on the catalytic activity of two catalase enzymes and the protein-free Fe protoporphyrin (as well as the simple aquated Fe(III) ion)[35]. The difference is striking. The rates of both enzymes are pH independent, while that of the protein-free complexes depends linearly on the concentration of hydroxide ion (compare also reference 83). Since k_{35} is pH independent both in the presence and absence of protein, this difference in pH dependence must be associated with the second step of the reaction (that is with k_{53}). As Jones and coworkers point out, extrapolation of the two graphs in figure 33 would lead to intersection at $\approx$pH 13, which is approximately the pK of H_2O_2 ($\approx$12), and 'the key feature of catalase action lies not so much in the absolute value of its specific catalytic activity but in its ability to use molecular H_2O_2[35]. It should be mentioned that the catalase activity of CPO does show an insignificant pH effect, which probably reflects protonation of certain parts of the protein; the rate rises by a factor of 3 as the pH is increased from 2.5 to 4.5 and then falls by the same amount as the pH is increased to 6^{217}.

The intriguing question now is; has the enzyme adopted and adapted the basic mechanism of the simple Fe porphyrin involving dissociated HO_2^-, or has it evolved a totally different mechanism based on undissociated H_2O_2? The reasonable agreement between the pK of H_2O_2 and the pH of intersection of the extrapolated graphs could be purely coincidental; but it surely suggests that there is a common denominator in the mechanisms and that the protein may act by providing HO_2^- at the active

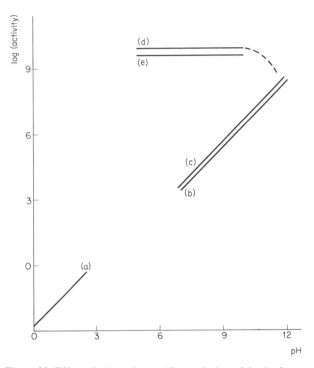

Figure 33 Effect of pH on the specific catalytic activity in the decomposition of H_2O_2 at $0°C$ of (a) the simple aquated Fe(III) ion, (b) and (c) monomeric Fe protoporphyrin, determined by two different methods, (d) bacterial catalase and (e) ox-liver catalase. From reference 35.

site even in neutral solution; that is, that it binds H_2O_2 in the form of the separate ions H^+ and HO_2^-. Such an interpretation presupposes:

(i) That k_{53} is fairly similar in the presence and absence of the protein. This cannot be checked independently, but the relatively small effect of the protein on several other rate constants discussed above (for example the overall catalase activity of Mb and HRPO, k_{54} and k_{43} of HRPO) shows that this is not improbable.

(ii) That the overall rate in the protein-free complex is determined by k_{53}, and not by k_{35}, at least up to pH 12. This is probably the case since the enzyme contains 30 per cent Fe^V and 70 per cent Fe^{III} in the steady state[137]; that is, k_{35} and k_{53} lead to comparable rates only at the higher rates, which would correspond to the reaction of the protein-free complex at pH 13.

If we accept that H_2O_2, alcohols and formate probably react with catalase Fe^V by analogous mechanisms, we can draw together several different lines of evidence, which all point in the same direction, and draw some possible conclusions.

(i) The above-mentioned coincidence of the extrapolated graphs of the pH variation of the overall rates in the presence and absence of protein (see figure 33) suggests that the protein supplies H_2O_2 in the form of HO_2^- at the active site. No such comparisons are available for the reactions of alcohols and formate in the presence and absence of protein.

(ii) Alcohols and formate are known to act as better reducing agents in their anionic than neutral forms by the transfer of a hydride ion[71]. Nothing is known about the relative reducing power and rates of reaction of H_2O_2 and HO_2^-.

(iii) We have already seen that catalases and peroxidases all have the ability simultaneously but separately to bind a proton and a monobasic anion (section 8.4). More significantly, we have seen that formate is co-ordinated to catalase Fe^{III} with the uptake of a proton; that it is also bound to Fe^{IV} with the uptake of a proton, but apparently not co-ordinated to the metal; and that its reduction of Fe^V also requires a proton. We concluded that the formate and proton were probably bound separately. There is no reason for supposing that H_2O_2 and alcohols (ROH) are not bound analogously in their reaction with Fe^V, that is, as H^+ with HO_2^- or RO^-.

The simplest conclusions to draw from these three lines of argument are, firstly, that with all these substrates the protein binds the anion (HO_2^-, $HCOO^-$, RO^-) at one site close to the FeO and the proton at a separate site and, secondly, that the redox reaction probably involves the transfer of a hydride ion. H/D isotope effects have been observed[159], but the results can, of course, be interpreted on the basis of several different mechanisms. Examination of the structure of those compounds that can act as substrates to catalase (and CPO) but not to HRPO, namely H_2O_2, primary and secondary alcohols, formaldehyde (probably as the hydrate $H_2C(OH)_2$) and formic acid, shows that they are not as diverse as might appear at first sight, but share a common stereochemistry, shown here for the anionic forms

(VI)

It seems that the proteins in catalase (and probably CPO) possess the correct steric requirements for binding molecules with this stereochemistry in their anionic form such that the hydrogen atom to be transferred as the hydride is held close to the oxygen atom of the ferryl ion.

We therefore conclude that the protein-free Fe porphyrin and catalase both react by analogous mechanisms, but that the protein dramatically enhances the rate in neutral solution by binding and stabilising the substrate in the form of HO_2^-, that is not by a strictly kinetic effect, but by affecting the equilibrium involving the substrate.

8.7 SUMMARY

The catalysis of the reduction of H_2O_2 by various donors provides another example (compare the co-ordination of O_2; section 7.1) of a problem, to which nature has found more than one solution; the cofactors of peroxidases can be either flavins or Fe porphyrins (section 8.1). Furthermore, fully active synthetic enzymes can be prepared in which the Fe has been replaced by Mn (section 8.2), which shows that the Fe is not unique. By contrast, nature has developed only one type of enzyme designed specifically for the decomposition of H_2O_2 into H_2O and O_2; they are all based on Fe protoporphyrin, are all tetramers and have very similar M.W.s, regardless of the source (section 8.1).

Both the catalases and, even more so, the peroxidases, provide good examples of enzymes that are remarkably unspecific in their selection of substrate—in contrast to many other enzymes. They are also among the most active enzymes known and their bimolecular rate constants (up to 1.4×10^8 M^{-1} s^{-1}; see table 17) approach the diffusion-controlled limit.

As stated in the introduction, the object of this section is to try to find out how nature uses a protein to increase the catalase and peroxidase activity inherent in the simple Fe porphyrin. Let us start with the Fe porphyrin in solution and see what nature has to do and how she sets about doing it.

(1) The Fe porphyrin must be attached to the protein with a high formation constant, but at least one co-ordination site must be left 'free' for reaction with the substrate (H_2O_2). This is essentially the same problem as in the case of Mb and Hb and is solved in a similar way. The apoenzyme of cytochrome c peroxidase (CCPO) can bind certain metal-free porphyrins almost as strongly as the Fe complexes themselves. This shows that the energy of interaction between the porphyrin + side chains and the protein is greater than that between the metal and the ligand (histidine) provided by the protein, which ensures a higher binding constant than could be attained by co-ordination of the ligand alone. It also means that the protein can control the nature of the axial ligands by suitable location of potential ligands relative to the porphyrin ring, in this case by providing only one histidine in a position suitable for co-ordination to the Fe. Similar considerations presumably apply to the binding and co-ordination of the Fe porphyrin in other peroxidases and catalases. As pointed out in section 8.5, it appears that the exact nature of the axial ligand (histidine or carboxylate) provided by the protein is not crucial; catalase activity is exhibited by

chloroperoxidase (with histidine) as well as by the true catalases in their tetrameric form (with carboxylate), while peroxidase activity is shown by the dissociated monomeric form of catalase (presumably still with carboxylate) as well as by the true peroxidases (with histidine).

(2) The magnitude of the Fe^{II}/Fe^{III} redox potential is much more important than the nature of the axial ligands. The functioning of the hydroperoxidases involves the reaction of H_2O_2 with Fe(III), while that of the O_2-carriers Mb and Hb (and probably of enzymes with oxidase activity) involves the co-ordination of O_2 to Fe(II). Nature can therefore switch from reactions involving Fe(II) and O_2 to those involving Fe(III) and H_2O_2 by stabilising either Fe(II) or Fe(III) in the resting state within the cell. This is reflected in the redox potentials, which are much more negative for the peroxidases—that is, the Fe(III) state is much more stable—than for Mb and Hb, even though they all possess the same ligands (section 8.3). Nature is clearly able to alter the redox potential without having to change the axial ligand. We cannot pinpoint the mechanism, but possibilities include variation in the Fe-histidine bond lengths and angles, changes in the groups surrounding the co-ordinated water, co-operative changes in the protein, etc. (see section 7.6). Unfortunately, we do not have the redox potential of the protein-free monohistidine complex as a basis for comparison. The catalases, which possess a carboxylate instead of histidine, are even more difficult to reduce.

(3) Nature must next increase the rates of reaction. The reaction paths and formal valencies of the Fe are shown as follows

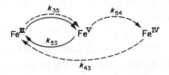

where Fe^V and Fe^{IV} may be the ferryl complexes $[FeO]^{3+}$ and $[FeO]^{2+}$, respectively, and the catalase and peroxidase reactions are denoted by —— and - - - -, respectively. All steps are essentially pH independent both in the presence and absence of protein *except* for k_{53}, which increases with pH in the absence of protein. There is evidence for some pre-equilibrium in the reactions corresponding to k_{35} and k_{53}; that is, they are, in fact, composite steps.

We can identify the steps that are affected by the protein by comparing the rates in the presence and absence of protein (section 8.6), but only in one case can we suggest how the rate is affected.

(i) The protein increases k_{35} in both catalase and peroxidase by a factor of 10^3-10^5 and the rate of attainment of the pre-equilibrium by a factor of up to 10^7. This reaction can probably be written

$$[H_2O \rightarrow Fe^{III}] + H_2O_2 \rightleftharpoons [H_2O_2 \rightarrow Fe^{III}] + H_2O \xrightarrow{k_{35}} [FeO]^{3+} + 2H_2O \quad (8.15)$$

$$(\text{or } [HO_2^- \rightarrow Fe^{III}] + H^+ + H_2O)$$

The protein could enhance the rate by thermodynamically stabilising $HO_2{}^-$, the intermediate or the product, or by kinetically accelerating either or both of the steps by the formation of suitable hydrogen bonds, perhaps coupled with co-operative movements of the protein; but since neither the nature of the intermediate nor the pH dependence of the two steps are known, we cannot be any more specific.

(ii) The protein has no significant effect on k_{54} and k_{43} in peroxidase. But in catalase the rates may be decreased by factors of 10^3-10^8 due to steric hindrance to the approach of the large substrate molecules.

(iii) The protein increases k_{53} in catalase by a factor of 10^5 at pH 7; more significantly, the protein makes k_{53} pH independent and comparable to the rate that would be expected (by extrapolation) for the protein-free complex at pH ≈ 12, that is equal to the pK of H_2O_2. For reasons given in section 8.6 we have concluded that the protein does indeed deliver the anion $HO_2{}^-$ to the active site, and that it is able to do this by making use of the property, which is apparently common to all hydroperoxidases, of simultaneously binding a proton and a mono-basic anion at two separate sites, which show co-operative interaction of the type already seen in Mb and Hb (sections 7.5 and 7.6); that is, there is a pH-independent equilibrium of the type

$$\text{Catalase} + \text{HX} \rightleftharpoons [\text{H}^+ . \text{Catalase} . \text{X}^-] \qquad (8.15)$$

where HX may be H_2O_2, ROH, etc. The protein therefore enhances the rate by a thermodynamic rather than kinetic mechanism, that is by influencing the dissociation of H_2O_2 and other substrates.

Reference to the above scheme of reaction paths would suggest that the evolution of peroxidases occurred first and involved the elaboration of some device for increasing k_{35} (nature has apparently not bothered about increasing k_{54} and k_{43}), and that the subsequent evolution of catalases depended either on the elaboration of some additional device to increase k_{53} or some further subtle modification of the first device. Both devices presumably involve a specially tailored portion of the protein close to the active site that will form hydrogen bonds with the substrate (H_2O_2, $HO_2{}^-$, RO^-, etc.) and possibly the active site (in the form of $Fe . OH_2$ and FeO). The exact steric requirements will obviously be different in the two cases, though there may be considerable overlap. We have already suggested that the second device, which serves to hold the substrate in the form of its anion ($HO_2{}^-$, RO^-, etc.,) near the active site, must show co-operative interaction with another site in the protein, which binds the proton. The pronounced steric hindrance found in catalase, which decreases k_{54} and k_{43} (see section 8.6), may be the natural consequence of the greater degree of tailoring of the protein required around the active site in catalase.

Hydroperoxidases have also evolved in a third direction, leading to catalysis of the oxidation of halides by H_2O_2 and the halogenation of substrates by halides and H_2O_2; these reactions have only been briefly mentioned (sections 8.1 and 8.5). This has presumably required the elaboration of some third device, which is neither identical with the first two nor incompatible with them; chloroperoxidase, for example, shows peroxidase and catalase as well as halogenation activity. But this third device resembles the device for increasing k_{53} in that it appears to depend on the simultaneous binding of the halide near the active site and a proton at some other site (section 7.4). It may even be wondered whether this unusual property of simultaneously binding a proton and a monobasic anion, which is common to all hydroperoxidases, is not also involved in increasing the rate of the reaction, common to all hydroperoxidases, namely k_{35}. For example, if the intermediate involved in the pre-equilibrium before k_{35} were the hydroperoxide complex $[HO_2^- \rightarrow Fe^{III}]$, then it is very likely that its rate of formation would be enhanced by any device that stabilised the anion HO_2^- either before or during co-ordination.

This interpretation would suggest that the devices for increasing the rates of k_{35}, k_{53} and the oxidation of halides by H_2O_2 are all different manifestations of the same basic properties of hydroperoxidases, namely that of being able to hold mono-basic anions (including those of very weak acids such as H_2O_2 and alcohols) near the active site, and that the conversion of one device into another or the modification of one to include another may only require minor changes around the active site. The apparently reversible conversion of the tetrameric units of the typical catalases into monomers with peroxidase but no catalase activity[117] supports the idea that a device that increases k_{35} can be modified into one that increases k_{53} as well by a slight change in the configuration of the protein. But this is mere speculation; we do not have sufficient data even to check the possible pH dependence of the component steps of k_{35}.

We can, however, state that the presence of a protein can increase k_{35} and k_{53}, but not k_{54} or k_{43}; and we can with some degree of assurance conclude that the protein provides one or more supplementary sites close to the real 'active' site (Fe) that play an essential role in increasing the rate, and that the protein increases k_{53} specifically by stabilising the substrate in the form of its anion (HO_2^-, etc.), that is by a thermodynamic rather than a truly kinetic effect.

(4) Nature must also prevent the destruction of the porphyrin ring and its side chains. Solutions of Fe protoporphyrin are attacked slowly by O_2 and rapidly by H_2O_2, the reaction probably involving oxidation of the vinyl side chains; H_2O_2 will also slowly attack the porphyrin ring itself. Catalase and peroxidase are also attacked, but much more slowly. It has been suggested that the proteins of catalase and peroxidase, like that of Mb, serve to provide a protective, hydrophobic environment for sensitive groups such as the vinyl side chains[36,37].

The above evidence and arguments suggest that the environment of the Fe porphyrin in catalase and peroxidase is very similar to that in Hb and Mb. The porphyrin ring and vinyl side chains are probably buried in (and held by) hydrophobic amino-acid residues. But the hydrophobic region is penetrated, on one side, by the histidine or carboxylate that is co-ordinated to the Fe (compare the proximal histidine in Hb and Mb) and, on the other side, probably by the hydrogen-bonding groups needed to handle H_2O_2, etc. (compare the distal histidine in Hb and Mb).

High rates of reaction are observed (up to 1.4×10^8 M^{-1} s^{-1}; see table 17), in spite of the postulated presence of these groups and the evidence for steric hindrance around the active site; compare the high rates observed for the uptake of O_2 by Hbs—up to 3×10^8 M^{-1} s^{-1}—in spite of the fact that the co-ordination site is completely surrounded by amino-acid residues (section 7.4).

Co-ordination chemists will probably be surprised that a change in the axial ligand has only a small effect on the reactivity of the Fe porphyrin (section 8.5). This point is emphasised even more dramatically by comparing the catalase activity of the Fe protoporphyrin with that of the simple aquated Fe(III) ion. The latter can only be studied in acid media, which prevent the formation and precipitation of polymeric hydroxo complexes; but similar kinetics suggest a similar mechanism in the two cases[35,135], and reference to figure 33 shows that the two would have similar activity under comparable conditions (but see reference 226). As Jones and coworkers have emphasised[35], the co-ordination of the porphyrin ligand does not essentially alter the activity of the Fe, but merely allows the reaction to proceed at higher pH. This apparent lack of effect of the ligands on reactivity can probably be ascribed to the fact that the catalase and peroxidase reactions are both very exothermic and involve redox couples with large potentials (≈ 1 volt), which would tend to swamp the smaller changes in potential due to any change in the axial ligand.

The catalase reaction was quoted in chapter 6 as a good example of the different rates to be observed in the presence and absence of a protein; the gross catalytic activity for the decomposition of H_2O_2 at pH 7 by equal molar quantities of the simple aquated Fe(III) ion, simple Fe(III) porphyrins and the enzyme catalase, respectively, varies in the ratio $1 : 10^3 : 10^8$ [35]. We can now see that the first jump in catalytic activity is due to the effect of the porphyrin ligand in keeping Fe(III) in solution, and the second is probably due, at least in part, to the ability of the protein to stabilise $HO_2{}^-$; that is, the porphyrin and protein combined can bring together at pH 7 forms of the catalyst and substrate, which would otherwise only be found at the opposite extremes of the pH scale.

9 NITROGEN FIXATION

The primeval atmosphere probably contained an abundance of fixed nitrogen in the form of NH_3. But as the atmosphere became more oxidising, NH_3 was replaced by N_2, and the supply of fixed nitrogen was limited to nitrates, etc., produced by the action of lightning; and the need arose to develop enzymes capable of 'fixing' N_2 itself. Micro-organisms, mainly those living symbiotically in the root nodules of legumes, now fix about 10^8 tons of nitrogen per year[73]; just about half this amount was fixed by industrial processes in 1971[57].

The biochemistry of nitrogen fixation has developed rapidly since 1960, when consistent fixation of N_2 by cell-free extracts was first reported. For a recent survey of the field see the reviews by Burris[44] and by Hardy *et al.*[98]; except where otherwise indicated, all the biochemical data in this section are taken from these sources. The so-called 'nitrogenase' enzymes catalyse the reductive fixation of N_2 to give NH_3; no oxidative fixation (to give NO_3^-) has been reported. We still know relatively little about the mechanism of the reaction or the nature of the active sites, but certain unusual features have emerged, and interest centres on the mechanistic pathway that nature has devised for the reaction.

The following abbreviations are used in this section:

ATP   adenosine triphosphate
ADP   adenosine diphosphate
P_i inorganic phosphate

The free energy of hydrolysis of ATP $\rightarrow$ ADP + P_i provides the driving force for many thermodynamically unfavourable reactions, for example the condensation of amino acids to form peptides.

9.1 THE PROBLEM OF FIXING N_2

The reduction of N_2 to $2 NH_3$ requires the addition of six electrons. Since there are no established examples among simpler ions, atoms or molecules of the simultaneous transfer of more than two reducing equivalents, it is reasonable to assume that this reaction must consist of several steps, each probably involving the transfer of no more than two equivalents. The simplest mechanism would require three successive additions of two reducing equivalents to give first di-imine (HN=NH), then hydrazine (H_2N-NH_2) and finally ammonia. These products could be present during the enzymatic reaction in neutral, protonated or anionic form, free or co-ordinated to one or two metal ions. Ammonia and hydrazine are well-known, stable compounds. Di-imine, on the other hand, is a very unstable compound (see the review by Hünig *et al.*[111]), although solid complexes have recently been reported containing di-imine in both neutral[164] and ionised forms[74]; it occurs as both *cis* and *trans* isomers. The heats and free energies of formation in kJ mol^{-1} (kcal per mole) or, in the case of NH_3, per 2 moles, of these three products of reduction are given in table 19

TABLE 19 HEATS AND FREE ENERGIES OF FORMATION OF N_2H_2, N_2H_4 AND NH_3†

Molecule	State	ΔH_f^0		ΔG_f^0		Ref.
N_2H_2	gaseous	204 ± 21	(48.7 ± 5)			80
N_2H_4	liquid	50.44	(12.05)			141
	aqueous	34.2	(8.16)	127.9	(30.56)	141
2 NH_3	gaseous	-92.43	(-22.08)	-33.3	(-7.95)	141
	aqueous	-161.7	(-38.64)	-53.25	(-12.72)	141

† Values are given in kJ mol^{-1} with the corresponding figure in kcal/mole in parentheses

They show that, although the overall reduction of N_2 to NH_3 by H_2 (and the reducing power of cellular material is close to that of the standard hydrogen potential) is thermodynamically favourable, the formation of free hydrazine is un-favourable, and that of di-imine extremely unfavourable. As already pointed out by the author[70,179], the real problem in the reductive fixation of N_2 by micro-organisms (and by co-ordination chemists!) is to circumvent the highly endo-thermic addition of the first two electrons. When the demand for a nitrogenase arose, nature was faced with a choice of either (1) using a mechanistically simple pathway, which would involve di-imine but have very unfavourable thermodynamics and therefore require the elaboration of methods for changing the thermodynamics, or (2) devising a more complicated pathway, which would avoid the intermediate formation of any compound resembling di-imine in electronic structure and thermo-dynamic properties and, with luck, offer favourable thermodynamics at all stages. The thermodynamics of the first step, namely

$$N_2 + 2H \rightarrow N_2H_2 \qquad (9.1)$$

can be made more favourable by using a more powerful reducing agent than H_2 and/or by stabilising the diimine.

Recent work on the co-ordination chemistry of N_2 complexes and N_2 fixation offers few clues, except that the reduction of N_2 by mild reagents under mild conditions is very difficult to achieve. Complexes are known in which N_2 is linearly co-ordinated (that is σ-bonded $N \equiv N \rightarrow M$) to one metal as in $[Ru^{II}(NH_3)_5N_2]^{2+}$[5,30], two similar metals as in $[(NH_3)_5Ru^{II}N_2Ru^{II}(NH_3)_5]^{4+}$[99,220] or two dissimilar metals as in $[(NH_3)_5Os^{II}N_2Ag]^{3+}$[70]. But co-ordination does not appreciably decrease the magnitude of the negative potential required to reduce N_2 (see, for example, references 67 and 70). The only known catalysts for the reduction of N_2 by H_2 are the industrially important heterogeneous catalysts of the Haber type, which operate at moderately high temperatures and pressures (see, for example, reference 29), and mixtures of sodium with certain metal phthalocyanines, which can react slowly even at room temperature[213]. N_2 can, however, be reduced at room temperature by much more powerful reducing agents. The best studied reaction is that with solutions of $Ti(C_5H_5)_2Cl_2$ reduced with Grignard reagents, sodium metal, etc.; it appears that N_2 reacts via the formation of an unstable dark blue dinuclear complex $[(C_5H_5)_2Ti \cdot N_2 \cdot Ti(C_5H_5)_2]$ to give nitride complexes, which only liberate NH_3 on decomposition by aqueous acid[21,149,150,158,203,223]. Co-ordination chemists have so far failed to reduce N_2 with a reagent as mild as H_2 at room temperature and pressure at any significant rate.

9.2 THE DISTRIBUTION AND NATURE OF THE ENZYMES

The ability to fix N_2 enzymatically is confined to certain bacteria and blue-green algae. In every case so far studied the nitrogenase system consists of two proteins, which are termed protein 1, fraction 1, molybdoferredoxin or the MoFe-protein, and protein 2, fraction 2, azoferredoxin or the Fe-protein, respectively. The MoFe-proteins have a M.W. of 100 000–300 000 and contain 8–40 Fe and 1–2 Mo atoms; some can be dissociated into sub-units. The Fe proteins characterised so far have a M.W. of $\approx 50\,000$ and contain 1–4 Fe, but no Mo atoms. Both proteins are brown and contain nonhaem iron and labile sulphide, that is, they are ferredoxins; but otherwise little is known about the co-ordination of the metal ions. The Mössbauer spectra of the ^{57}Fe nuclei show the presence of several different types of site[126]. The reduction of N_2 to NH_3 requires the presence of (1) both proteins (in one case maximum activity was obtained with a ratio of 2Fe-protein to 1MoFe-protein), (2) a reducing agent ($Na_2S_2O_4$ is commonly employed, but H_2 can be used if a hydrogenase is added, showing that a very negative potential is not essential), (3) ATP and (4) magnesium ions. As already mentioned, the reaction involves both the reduction of N_2 and the simultaneous hydrolysis of ATP to ADP; that is

$$N_2 + 6\ominus + 6H^+ + n\text{ATP} \rightarrow 2NH_3 + n\text{ADP} + nP_i \qquad (9.2)$$

where n is 12–15.

Tests have shown that fixation can be achieved by combining the two proteins not only from the same organism but also from two different organisms. It would seem as if nature has been able to devise only one mechanism for nitrogen fixation, but that considerable variation on this theme can be tolerated. It is interesting that the nitrate reductase ($NO_3^- \rightarrow NO_2^-$) of *Neurospora crassa* also consists of two proteins, one of which contains Mo, and that an active nitrate reductase can be reconstituted on replacing this particular Mo-protein by Mo-containing proteins (usually after treating with acid) from several completely different enzymes, including xanthine, aldehyde and sulphite oxidases and the MoFe-protein of nitrogenase from two different bacteria[156]. The reverse reconstitution of nitrogenase from the normal Fe-protein and an unrelated Mo-protein has not yet been reported. Nason *et al.*[156] have suggested the possible existence of a Mo-containing sub-unit that can be considered as a cofactor, shared in common by all Mo-containing enzymes from animals, plants and micro-organisms, and functioning both as an electron carrier and as a link to bind together the sub-units of the enzyme. If this is the case, it would appear that the basic electron-transfer reaction is carried out by the Mo-protein, while the second protein modifies the reaction, tailoring it to the particular substrate concerned. Thus, the key event in the evolution of the nitrogenase enzymes was the development of the Fe protein to modify and adapt the reactions of the existing group of Mo-containing enzymes. It is perhaps surprising then to find that Mo is not in fact essential; V can satisfy the requirement for Mo in some N_2-fixing bacteria, and cell-free extracts containing proteins with V instead of Mo are able to fix N_2, though at a lower rate[146].

9.3 EVIDENCE FOR THE MECHANISM OF ENZYMATIC FIXATION

The nitrogenase enzymes can catalyse other, at first sight unrelated, reactions and reduce several other substrates. If we start with the full complement required for nitrogenase activity (N_2, reducing agent, ATP, Mg ions, two proteins) and then exclude N_2, we observe an ATP-dependent evolution of H_2; that is, H_2 is evolved and ATP is hydrolysed to ADP and P_i. No reaction is observed if ATP or the reducing agent or one of the proteins is omitted. But if H_2 is used as the reducing agent together with a hydrogenase, then we merely observe the catalysed hydrolysis of ATP, since H_2 is evolved by the nitrogenase and taken up again by the hydrogenase. This ATP-dependent evolution of H_2 by nitrogenase is not inhibited by CO, as is the activity of ordinary hydrogenases, which catalyse the reversible liberation and uptake of H_2. Both N_2 fixation and H_2 evolution show the same variation of activity with pH and the same ratio of ATP per pair of electrons. These two reactions are obviously intimately related and it must be concluded that the evolution of H_2, which might appear to be a completely unnecessary and wasteful side reaction, plays an essential role in the mechanism of fixation. The irreversible evolution of H_2 shows that it is generated at a more negative potential than that of a reversible hydrogenase, and the hydrolysis of ATP is obviously being used to drive the reaction. The evidence would suggest that one of the devices used by nature to overcome the very unfavourable first step in the reduction of nitrogen is to use a more powerful reducing agent than H_2, and supports the idea, proposed by the author in 1968[180], that the role of ATP in nitrogen fixation is to generate this reducing agent. There is no evidence to indicate whether this highly reducing site of the enzyme contains a metal or not.

Other substrates that can be reduced by nitrogenase include N_2O ($\rightarrow N_2$ before further reduction), N_3^- ($\rightarrow N_2 + NH_3$), CN^- ($\rightarrow CH_4$, NH_3 and some CH_3NH_2), CH_3NC ($\rightarrow CH_4$, C_2H_6, C_2H_4 and CH_3NH_2), C_2H_2 ($\rightarrow C_2H_4$) and higher homologues of the last two. Of these only N_2O, which is a fairly common substrate and product of bacterial metabolism in the soil, would be a potential substrate *in vivo*; it is worth pointing out that the number of known reactions with simple metal complexes is even less for N_2O than for N_2 (see, for example, reference 17). It appears that ATP is required for all these reductions. H_2 and CO act as inhibitors and, of course, the reduction of one substrate will inhibit the reduction of another. NO destroys the enzyme irreversibly[126]. C_2H_4 is not, however, an inhibitor[72]. The nature of these substrates and inhibitors provides good, though circumstantial, evidence that N_2 co-ordinates to a transition-metal ion. But co-ordination to Fe, at least, has not been observed under equilibrium conditions; the complex Mössbauer spectra shown by the ^{57}Fe nuclei in the two proteins are effected by the presence of substrates and inhibitors (H_2, N_2, acetylene, CN^-, CO), but no new signals were detected corresponding to the co-ordination of a new ligand[126]. Inhibition between the various substrates and inhibitors is sometimes competitive, sometimes non-competitive, and it is not clear how many active sites are involved; but, considering the large number of Fe and Mo atoms present in the enzyme, it is not surprising that some differences in mutual inhibition are observed. The sites that bind N_2 are obviously distinct from those that evolve H_2 and are insensitive to CO. There is no evidence yet to suggest how many and which of the metal atoms play an essential role or even which protein contains which of the two main active sites.

The reduction of acetylene is of particular interest. Firstly, it is isoelectronic with N_2 and may therefore show some analogy to N_2 in its reactions. Secondly, it is reduced only as far as ethylene, which allows attention to be focused on the addition of the first two electrons, which is probably the crucial step in the fixation of N_2. Finally, in contrast to the reaction of N_2 with H_2 to give di-imine, which is thermodynamically very unfavourable ($\Delta H^\ominus \approx 210$ kJ mol^{-1} = +50 kcal/mole), the reaction of acetylene with H_2 to give ethylene is very favourable ($\Delta H^\ominus$ −180 kJ mol^{-1} = −42 kcal/mole; $\Delta G^\ominus$ −140 kJ mol^{-1} = −34 kcal/mole)[141]; yet the reduction of acetylene also requires the presence of ATP. There is obviously a very close analogy between the reduction of acetylene and of nitrogen, which, as pointed out by Dilworth[72], provides evidence that the reduction of N_2 proceeds by an initial two-electron reduction to di-imine. Studies in D_2O have shown that the reduction of acetylene to ethylene gives entirely (*Azotobacter vinelandii*) or mainly (*A. chrooco cum*) *cis*-$C_2H_2D_2$—that is, the reaction involves the completely or partially stereo-specific *cis* addition of two deuterium atoms; and that no tri-deuterated ethylene is formed—that is, the reaction does not appear to involve a co-ordinated acetylide ion[72,125]. The incorporation of D also shows that protons from the solvent can reach the active site. The simplest explanation is that acetylene forms a π-complex with a metal ion at one active site (M) and receives two deuterium atoms (or electrons) from the second active site (E); that is

$$(9.3)$$

The analogous reaction with N_2 would proceed via the formation of a π-complex to give the *cis*-di-imine; the di-imine might remain co-ordinated to the metal through either a σ- or π-bond while being reduced further, that is

$$(9.4)$$

It is worth emphasising that all the simple complexes of N_2 so far studied by co-ordination chemists possess a linear or σ-bonded ($N\equiv N \rightarrow M$) group and have formation constants high enough to enable them to be handled and isolated; by contrast, nature appears to make use of a π-bonded complex with a very low formation constant.

Little is known about the further steps in the fixation of nitrogen. No intermediates have been detected before the appearance of free NH_3, and the reaction of possible intermediates such as N_2H_4 with purified enzymes does not appear to have been studied.

9.4 SUMMARY

The reduction of N_2 to NH_3 provides a good example of a reaction that nature is unable to 'catalyse' in the true sense of the word. As pointed out at the beginning of this chapter, the overall reduction to NH_3 (using reducing agents with a potential similar to that of H_2) is thermodynamically favourable, but the addition of the first two electrons to give di-imine is very unfavourable. Only one method of fixing N_2 has been evolved and the above analysis suggests that this depends on making the thermodynamics of the first step favourable by using a more powerful reducing agent. The irreversible evolution of H_2 shows that a reducing agent that is more powerful than H_2 itself (but of unknown potential) is being generated by the use of ATP, while the analogy with acetylene provides circumstantial evidence that the first product is *cis*-di-imine. The enzyme therefore catalyses two distinct reactions, namely (1) the conversion of reducing equivalents (2H) supplied at a potential equivalent to that of H_2 into a more powerful reducing agent (2H*) at the expense of the hydrolysis of ATP; that is

$$2H + n\text{ATP} \rightarrow 2H^* + n\text{ADP} + nP_i \qquad (9.5)$$

where n can vary up to 5, and (2) the actual reduction of N_2; that is

$$N_2 + 2H^* \rightarrow N_2H_2 \qquad (9.6)$$

followed by its further reduction to NH_3. The potential of H* may be sufficient in itself to drive reaction (2) to the right; but the enzyme could in addition serve to stabilise di-imine if the formation constant for co-ordinating N_2H_2 is greater than for N_2 and/or if the surrounding functional groups form hydrogen bonds preferentially with N_2H_2, in both cases with or without any co-operative movements of the protein (section 7.6). Co-ordination to the metal could also increase the rate of reaction by holding the N_2 close to the reducing site.

The enzyme requires (at least) two active sites for the first step in the reduction of N_2. One site complexes N_2 and probably contains a metal. The other site provides the low-potential reducing equivalents; there is no evidence as to whether this site contains a metal or not. The enzyme system does, in fact, consist of two proteins, but there is no evidence to indicate which protein contains which active site(s).

If we focus attention on the first step in the reaction, then nitrogen fixation appears to provide an example in which the main role of the protein is not simply to increase the rate of reaction, but actually to allow the reaction to proceed by making the thermodynamics favourable. It apparently does this by catalysing and coupling together two separate reactions by the provision of two active sites close together.

10 CONCLUSIONS

In part 2 we have tried to pinpoint some of the ways in which the presence of a protein may affect the reactivity of transition-metal complexes. We have examined in some detail three case studies chosen to illustrate how the protein can vary (1) the thermodynamics of an individual step (for example, the equilibrium constant for the co-ordination of O_2 by Fe(II) in Hb and Mb—chapter 7), (2) the kinetics of individual steps (for example, the reactions of the Fe in peroxidase and catalase with H_2O_2 and other substrates—chapter 8) and (3) the thermodynamics of the overall reaction (for example driving the endothermic reduction of N_2 to N_2H_2 with a second thermodynamically favourable reaction—chapter 9). In each case we have also considered what the simple protein-free complex or ion can and cannot do, and hence what problems nature has to overcome and how she has solved them. This has thrown up additional examples of the ways in which the protein can affect thermodynamic and kinetic properties.

We now summarise the above experimental evidence (and throw in a few additional examples and theoretical possibilities) for the mechanisms by which a protein can affect the thermodynamic and kinetic properties of transition-metal complexes (section 10.1) and then look at the isomerase reactions catalysed by cobalt corrinoids in the light of these ideas (section 10.2).

10.1 GENERAL PRINCIPLES

10.1.1 Thermodynamic effects
(1) The protein can determine the number and nature of the ligands co-ordinated to the metal ion, when the energy of interaction between the amino-acid residues in the tertiary structures of the polypeptide (and, if necessary, between the polypeptide and ligands such as porphyrin) is greater than that between the polypeptide and the metal ion. This is one of the major variables that determines the function of the Fe porphyrin in a protein, for example electron transfer, O_2 transport, oxidase, oxygenase, catalase or peroxidase activity; see, for example, reference 206. Stabilisation of the five-co-ordinate Fe(II) porphyrin with only one co-ordinated histidine provides an excellent example (section 7.4). Analogous five-co-ordinate Fe^{II} porphyrins are apparently also present in certain mono- and di-oxygenase enzymes[101,124]. Even greater opportunity to ring the changes with the ligands is provided by metal ions such as Cu and Mo, where there is no porphyrin to keep four positions unchanged.

(2) The protein can vary the equilibrium constant K for the co-ordination of a particular ligand. We can distinguish three main methods:
(a) K will obviously be affected by changes of the type listed under (1), namely by changes in the number, nature and stereochemistry of the other ligands. There are

no clear-cut examples of the influence of the *trans* ligand on equilibria in the Fe porphyrins, but the importance of the *trans* effect is well documented in the Co corrinoids[181].

(b) K can also be reduced (compared to the value in the protein-free complex) when the amino-acid residues around the co-ordination site offer steric hindrance (as in the co-ordination of CO to Fe^{II} in Hbs; see section 7.4) or an unfavourable environment (for example the hydrophobic environment probably contributes to the low K observed for the co-ordination of the charged CN^- by Fe^{II} in Hb and Mb; see section 7.4). K might also be increased by favourable solvation, hydrogen bonding, etc., but this effect alone would probably be rather small.

(c) $Log_{10}K$ for the co-ordination of O_2 by Fe^{II} in Hb and Mb can be varied by 4.5 or more by changes in the relative distortions of the Fe^{II} and $Fe^{II}O_2$ forms and complementary (or co-operative) changes in the configuration of the polypeptide, and by 2.5 by changes in the quaternary structure alone (sections 7.5 and 7.6). This provides a useful mechanism for either increasing or decreasing K as required.

(3) The protein can vary the redox potential, for example of the $Fe^{II/III}$ couple in Fe porphyrins. As in (2), the value of E^0 will depend on the nature of the ligands, the relative distortions of the two metal complexes and complementary changes in the polypeptide. X-ray analysis of the Fe^{II} and Fe^{III} forms of cytochrome c has established that the redox reaction is accompanied by a considerable change in the conformation of the protein[214]. E^0 may vary by over 0.4 V even where there is apparently no change in the axial ligand: compare Hb $\approx$ +0.2 V with horse-radish peroxidase $\approx$ −2.5 V at pH 7 (sections 7.8 and 8.3). Here again there is even greater scope for the protein to vary the potential when there is no porphyrin to keep four positions unchanged.

(4) Other types of equilibria involving the metal and its ligands can be envisaged, which could be affected by the protein, for example

$$Co-R \rightleftharpoons Co^{II} + R\cdot \tag{10.2}$$

where Co−R represents the B_{12} coenzyme (5-deoxyadenosylcobalamin); this possibility is discussed in section 10.2.

(5) The protein can couple one equilibrium involving the metal with another at a second site, via a linked sequence of changes in the intervening amino-acid residues, as shown by the so-called homotropic and heterotropic interactions observed in Hb (sections 7.5 and 7.6).

(6) A subtle variation of (5) is to couple together changes in the thermodynamic properties of the metal ion with the binding of the substrate at the active site. Good examples are provided by two enzymes based on Fe porphyrins. The binding of the substrate by the enzyme tryptophane-2,3-dioxygenase increases the formation con-

stant for the binding of CN^- by Fe^{III} by about 100 and of CO by Fe^{II} by about 50, while the $Fe^{II}O_2$ complex can only be observed in the presence of substrate[101]. The binding of camphor by the Fe^{III} form of camphor-5-mono-oxygenase is accompanied by a change in the number of unpaired electrons from one to five (together with changes in the u.v.–visible and e.s.r. spectra) and in the redox potential from -0.38 to -0.17 V[221]. This could provide a sophisticated mechanism for ensuring that an essential, but very reactive, intermediate is formed only in the presence of the substrate. The isomerase reactions may provide another example (see section 10.2).

(7) Similar principles could obviously be used to control the binding of a substrate at a nonmetal site and to affect the equilibria involving that substrate. We have suggested that the polypeptide chain in catalase is able to deliver HO_2^- to the active site at pH 7 by co-operative interaction between two sites, which bind H^+ and HO_2^-, respectively (section 8.6).

10.1.2 Kinetic effects

We have met examples where the rates of reaction are markedly increased (for example catalase and peroxidase Fe^{III} with H_2O_2 by a factor of 10^3–10^5, and catalase Fe^V with H_2O_2 at pH 7 by a factor of 10^5; see section 8.6), decreased (for example catalase Fe^V with guiacol by a factor of $\approx 10^8$; see section 8.6, and further reactions of Hb $Fe^{II}O_2$ leading to autoxidation by a factor of $\approx 10^8$, see section 7.7) or left more or less unchanged (for example peroxidase Fe^V with reducing agents; see section 8.6) compared to the rates found for the protein-free complexes. Some reactions may be fairly specific (for example the reaction of catalase Fe^V with reducing agents; see section 8.6); others are virtually nonspecific (for example peroxidase Fe^V and Fe^{IV} with reducing agents; see section 8.6).

It may not always be possible to pinpoint the mechanism by which the protein affects the rate, but certain conclusions and generalisations can be made.

(1) Many changes in rate probably result from a change in the thermodynamics of some step involving either the metal or the substrate, and may therefore be brought about by the mechanisms discussed above. For example, the autoxidation of Hb and Mb via the initial step

$$Fe^{II}O_2 + H_2O \rightarrow Fe^{III}OH_2 + O_2^- \tag{10.2}$$

is probably inhibited by making the redox potential more positive (section 7.7). We have suggested that the increase in the rate of reaction of catalase Fe^V with H_2O_2 at pH 7 is due to the ability of the protein to deliver the substrate in the form of HO_2^-, that is to influence the equilibrium $H_2O_2 = H^+ + HO_2^-$ (section 8.6).

Reactions involving the co-ordination of O_2 to Fe porphyrins (as in mono- and di-oxygenases and probably also cytochrome oxidase) will obviously be enhanced if the protein stabilises the reactive five-co-ordinate Fe^{II} ion as in Hb and Mb.

There are no clear-cut examples of the effect of changing the axial ligand on the rates of ligand substitution in Fe porphyrins, either with or without a protein, but this *trans* effect is well exemplified in the Co corrinoids, both for normal ligand-substitution reactions and for fission of the Co—C bond[181]. Nitrogen fixation represents an extreme case of the effect of thermodynamics on rate, where one of the major roles of the protein appears to be to allow the reaction to proceed by making the reduction of N_2 to N_2H_2 thermodynamically favourable by the production of a powerful reducing agent (section 9.4).

(2) The protein could also affect the transition state, as distinct from the ground states of the initial and final complexes, and the result would be reflected in the kinetics, as distinct from the thermodynamics, of the reaction. As Vallee and Williams have pointed out, distortion by the protein could lower the activation energy by forcing a compromise structure, closer to that of the transition state, on either or both the initial and final complexes. This could be particularly important in, for example, electron-transfer reactions where the two oxidation states prefer different symmetries, for example the normally tetrahedral Cu(I) and tetragonal Cu(II)[222]. Distortion could also be used to increase the activation energy and hence decrease the rate.

(3) Tight coupling of the vibrations of the metal complex and the polypeptide could further modify the kinetics by two different effects. Firstly, complementary movements of the protein could either increase or decrease the activation energy required to reach the transition state. Secondly, any increase in the number of oscillators that can transfer energy into the critical oscillator (that involved in the making or breaking of the new bond) should increase the pre-exponential (Arrhenius) factor in the rate equation, and hence increase the rate.

(4) The amino-acid residues around the co-ordination site can play a major part in controlling rates. The protein can increase the rate by binding the substrate near the metal in a pre-equilibrium, which would increase the contact time, and perhaps hold it in an orientation more favourable for reaction with the metal. This is, of course, part of the mechanism for delivering HO_2^- to the active site (see above) and could provide a mechanism for the increased rate of reaction of catalase and peroxidase Fe^{III} with H_2O_2 (section 8.6). The porphyrin ligand itself could play a part in the binding of hydrophobic substrates. Such binding of substrates by the protein and ligands will probably be weak and may involve hydrogen bonds, coulombic or van der Waals interaction, depending on the nature of the substrate. An obvious extension in reactions requiring several compatible active sites is to hold them in close proximity; this appears to be the case in nitrogen fixation, which requires one site (probably a transition metal) to hold the N_2 and a second site to produce the powerful reducing agent (section 9.3). The protein may also decrease the rate (and thereby protect a particular complex, ligand or group) by sterically hindering the approach of potential reagents. This is probably the reason for the severe inhibition of the reaction of catalase

Fe^V with many reducing agents (section 8.6) and one of the reasons for the stability of $Fe^{II}O_2$ in Hb and Mb (section 7.7). The porphyrin ligand is protected from attack by O_2 and H_2O_2 by being buried in the hydrophobic amino-acid residues (section 8.7). Substrate specificity may, of course, be induced either by sterically hindering the approach of unwanted substrates or by selective binding of the desired substrate as above.

(5) It is also worth noting that the Fe porphyrins, Co corrinoids and ferredoxins are probably all unusually reactive complexes, even in the absence of protein, because they contain very polarisable ligands (porphyrin, corrin, cysteine and sulphide), which exert a strong labilising effect. The corrin ligand appears to accelerate ligand-substitution reactions at the Co(III) ion by a factor of $\approx 10^7$ compared to the *cis* ligands $(DMG)_2$, $(CN)_4$ and $(NH_3)_4$ [181].

The above examples show some of the ways in which the protein can affect the thermodynamic and kinetic properties of transition-metal complexes and indicate how nature can ring the changes in reactivity and specificity with a fairly limited selection of metal ions and complexes.

Two points stand out and are worth re-emphasising

(i) Much of the enhanced catalytic activity of metallo-enzymes can probably be ascribed to the effect of the protein on the equilibrium constants rather than the rate constants, that is on the thermodynamic rather than the kinetic properties, of the metal complex and/or substrate.

(ii) Much of the effect of the protein on these equilibrium constants can be ascribed to the occurrence of associated changes in the conformation of the protein.

The first point does not appear to have been adequately emphasised previously.

10.2 POSTSCRIPT: THE ISOMERASE REACTIONS OF VITAMIN B_{12}

Finally, let us take a brief look at the enzymatic reactions involving cobalt corri-noids (derivatives of vitamin B_{12}), where the mechanisms have not yet been established, and see whether the general principles outlined in section 10.1 can provide any further ideas or insight. The structure of the best known group of corrinoids, the cobalamins, is shown in figure 34. For background information on the bio-chemistry and co-ordination chemistry of these complexes see the review by Barker[18] and the author's book *Inorganic Chemistry of Vitamin B$_{12}$* [181], respectively. Except where otherwise indicated, all the information in this section is taken from these two sources.

Enzymes containing cobalt corrinoids catalyse three types of reaction:

(i) The transfer of methyl groups, for example in the formation of methionine and methyl-mercury compounds.

(ii) The so-called 'isomerase' reactions (see table 20), which involve the 1,2-shift of a C, N or O atom.

(iii) The reduction of the $-CHOH-$ group of ribonucleotide triphosphates to $-CH_2-$.

Reactions of group (i) probably involve the intermediate formation of a $Co-CH_3$ complex; those of groups (ii) and (iii) both require the complex to be present in the

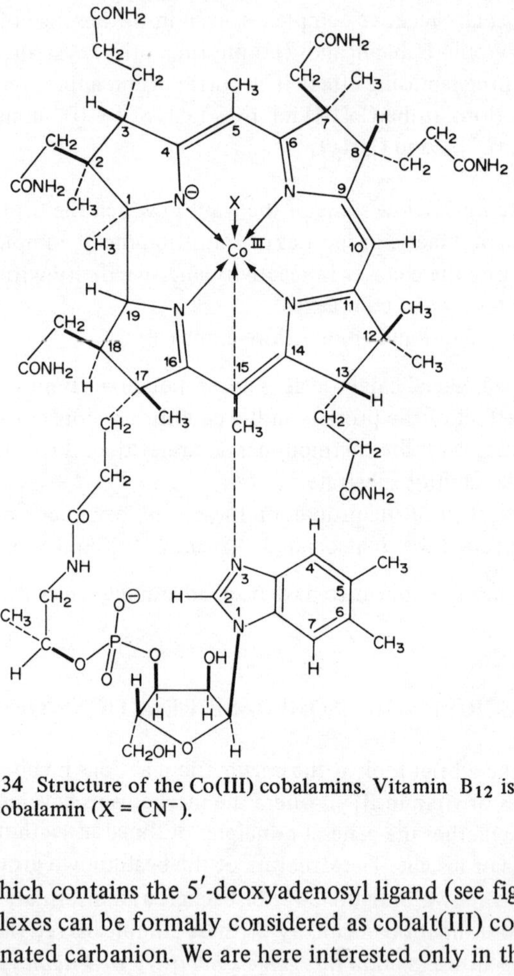

Figure 34 Structure of the Co(III) cobalamins. Vitamin B_{12} is cyanocobalamin ($X = CN^-$).

'coenzyme' form, which contains the $5'$-deoxyadenosyl ligand (see figure 35). These organo-cobalt complexes can be formally considered as cobalt(III) complexes containing a co-ordinated carbanion. We are here interested only in the isomerase reactions.

The Fe porphyrins and Co corrinoids provide an interesting contrast. We have a reasonable idea of the nature of certain of the steps and intermediates in the enzymatic reactions of Fe porphyrins, but we are unable to prepare and study the true 'coenzyme' form (for example the high-spin, five-co-ordinate Fe^{II}) in the absence of protein. With the Co corrinoids, on the other hand, we can obtain and

TABLE 20 'ISOMERASE' REACTIONS REQUIRING 5-DEOXYADENOSYLCORRINOIDS

Substrate	R_1	R_2	R_3	Product
			irreversible reactions	
ethane-1,2-diol	H	OH	OH	acetaldehyde
propane-1,2-diol	CH_3	OH	OH	propionaldehyde
glycerol	$HOCH_2$	OH	OH	β-hydroxypropionaldehyde
ethanolamine	H	NH_2	OH	acetaldehyde + NH_3
			reversible reactions	
L-glutamate	H	$-CHNH_2 . COOH$	COOH	threo-β-methylaspartate
succinyl-coenzyme A	H	$-CO . SR$	COOH	L-methylmalonylcoenzyme A
α-methyleneglutarate	H	$-\underset{\underset{CH_2}{\parallel}}{C} . COOH$	COOH	β-methylitaconate
ornithine	H	NH_2	$-CH_2 . CHNH_2 . COOH$	2,4-diaminovalerate
L-β-lysine	H	NH_2	$-CH_2 . CHNH_2 . CH_2 . COOH$	3,5-diaminohexanoate
D-α-lysine	H	NH_2	$-CH_2 . CH_2 . CHNH_2 . COOH$	2,5-diaminohexanoate

$$R_1 - \underset{H}{\overset{(R_2)(H)}{C}} - \underset{H}{\overset{}{C}} - R_3 \quad \rightleftharpoons \quad R_1 - \underset{H}{\overset{}{C}} - \underset{H}{\overset{(H)(R_2)}{C}} - R_3$$

study the whole coenzyme (with no change in any of the ligands) in the absence of the protein, but we have only a few clues as to the nature of the individual steps in the enzymatic reactions. This situation has, of course, provided the ideal breeding ground for speculation on the mechanism, and allows us to add our own speculations here.

However, certain features of the isomerase reactions have now been established and in some cases shown to be shared by several reactions, which suggests some underlying similarities in mechanism:

(i) There is no exchange of hydrogen atoms between the substrate and solvent; that is, no ionisable protons are involved.

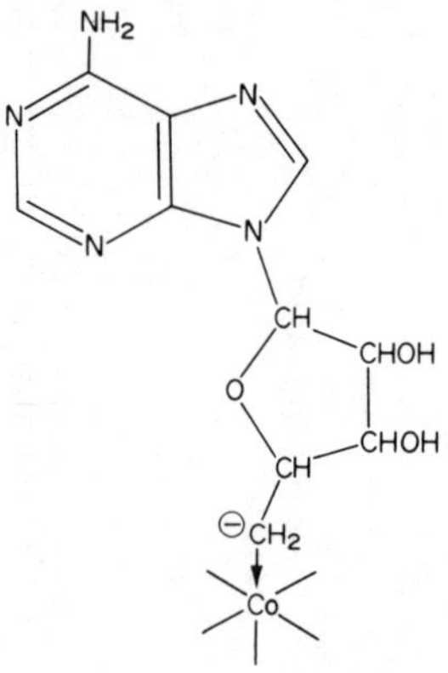

Figure 35 The 5′-deoxyadenosyl ligand present in the corrinoid 'coenzymes'.

(ii) The hydrogen atoms of the substrate do, however, exchange with the hydrogen atoms on the co-ordinated ($C_5{}'$) atom of the organo-ligand; that is, a hydrogen atom or hydride ion is transferred between the substrate and $C_5{}'$.

(iii) Although the chemical and physical properties of the coenzyme are not affected by binding to the apoenzyme (but see below), it has been shown that with certain enzymes the addition of substrate leads to the appearance of the Co^{II} complex (detected by its e.s.r. and u.v.–visible spectra), which disappears again when the substrate has been consumed; that is, the Co—C bond is broken. The observation of Co^{II} suggests, but does not prove, that the bond undergoes homolytic fission.

(iv) Substitution of —O— by —CH_2— in the furanose ring of the coenzyme does not inhibit the propanediol reaction; that is, this O atom does not play any role in the cleavage of the Co—C bond.

(v) Small amounts of 5′-deoxyadenosine (derived from the ligand by the addition of a third hydrogen atom to $C_5{}'$) can be isolated on stopping the enzymatic reaction and degrading the enzyme, thus providing further evidence for cleavage of the Co—C bond.

We now wish to follow these clues in the light of the known chemistry of the protein-free corrinoids, the known chemistry of possible intermediates such as organic free radicals and the general ideas on metallo-enzymes outlined in section 10.1.

It is known that the Co—C bond in organo-corrinoids and in many other organo-cobalt(III) complexes can be made or broken by reactions involving Co^I, Co^{II}, Co^{III} or Co—H complexes[182]. Most of the speculation about the mechanism of the isomerase reactions revolves around the type of cleavage of the Co—C bond (to give Co^I, Co^{II} or Co^{III}) and the form of the substrate that actually undergoes rearrangement (free radical or ligand on the cobalt). As already pointed out by other authors (see, for example, reference 184), the reactions shown on the middle line of the following scheme provide the simplest, but not necessarily the correct, explanation of the experimental observations. The Co—C bond in the coenzyme (Co—R) undergoes homolytic fission to give Co^{II} and the free radical R·, which abstracts a hydrogen atom from the substrate SH to give the new radical S·. Rearrangement could occur either at the free-radical stage or after reaction with Co^{II} to form Co—S. Analogous reactions then lead to the re-formation of Co—R and the rearranged product. Does the known chemistry of the protein-free organo-corrinoids and free radicals provide any support for such a scheme?

$$
\begin{array}{c}
Co^{III} + R^- \\
+ \ SH \\
\updownarrow \\
Co\text{—}R \xrightleftharpoons{(A)} \ \underset{+ \ SH}{Co^{II} + R\cdot} \xrightleftharpoons{(B)} \ \underset{+ \ S\cdot}{Co^{II} + RH} \xrightleftharpoons{} \ \underset{+ \ RH}{Co\text{—}S} \quad (10.3) \\
+ \ SH \qquad\qquad\qquad\qquad (C) \\
\updownarrow \qquad\qquad \updownarrow \\
Co^I + R^+ \ \rightleftharpoons \ Co^I + RH \\
+ \ SH \qquad + \ S^+
\end{array}
$$

Can the CO—C bond in the coenzyme or other organo-corrinoids undergo reversible homolytic fission to give Co^{II} and the free radical according to equation (A)? The occurrence of certain thermal and photochemical reactions shows that this equilibrium does theoretically exist. Effects that can be ascribed to the reversible, and purely thermal, homolytic fission can be observed at or above $90°C$—for example, the interconversion of the isomers of methylcorrinoids, which differ according to whether the methyl group is bound to the cobalt above or below the asymmetric corrin ring. But the stability of most organo-corrinoids, including the coenzyme, towards O_2 (which would react rapidly and irreversibly with any free radical formed) shows that the equilibrium is displaced so far to the left at room temperature as to make the free radical kinetically unavailable as a reaction intermediate, unless the protein can significantly alter the position of the equilibrium and/or the rate of equilibration.

We suggest here that the protein and substrate together cause a very significant displacement of this equilibrium in favour of dissociation (compare the effect of the substrate on the thermodynamic properties of Fe porphyrin in mono- and di-oxygenase enzymes; see section 10.1). The protein alone probably causes some change, since the coenzyme does undergo slow decomposition in certain enzymes (for example ethanolamine ammonia lyase) even in the absence of substrate; but the binding of the substrate causes a further, and perhaps the major, change in the position of this equilibrium. Such a displacement of the equilibrium could be achieved either by stabilising the products of dissociation (for example by improved hydrogen bonding) or by destabilising the coenzyme. The former mechanism alone seems unlikely to provide a large enough change in free energy. We therefore suggest that the protein can destabilise the coenzyme by distorting the co-ordination sphere of the cobalt (compare the distortion of the Fe—N(histidine) bond length in Hb; see section 7.6). The simplest mechanism would be distortion of the Co—$\hat{C}$—C bond angle, but other methods can be envisaged. We have pointed out elsewhere (page 70 of reference 181) that 'the corrinoids are remarkably flexible (not only the side-chains, but also the corrin ring) and that effects may readily be transmitted from one part of the molecule to another (either by electronic interaction between the axial ligands, the cobalt ion and the conjugated corrin ring, by steric repulsion or by attraction through the formation of hydrogen bonds)'. Thirdly, we suggest that the shift in the equilibrium that apparently occurs on binding the substrate is due to a change in the conformation of the protein, which increases the distortion of the cobalt co-ordination sphere, and which is reversed when the substrate has reacted and diffused away. Such a mechanism, which only permits the formation of the very reactive R while the substrate is bound at the active site, will also tend to suppress the irreversible reactions of the C_5' atom with O_2 and with the adenine ring, which are observed in solution.

Reactions such as (B) involving the abstraction of a hydrogen atom from an organic compound by a free radical are, of course, well known. A good example is provided by the photolysis of ethylcobalamin in the presence of isopropanol (SH), which yields pinacol (S.S) via the following sequence of reactions

$$
\begin{array}{cccc}
\text{Co—Et} \xrightarrow{\text{light}} & \text{Co}^{II} + \text{Et} \longrightarrow & \text{Co}^{II} + \text{EtH} \longrightarrow & \text{Co}^{II} + \text{EtH} \\
+\text{SH} & + \text{SH} & + \text{S·} & + \tfrac{1}{2}\text{S.S}
\end{array}
\qquad (10.4)
$$

where S· is $Me_2\dot{C}OH$. This may provide a model for step (B) in the enzymatic reactions of diols. But the photolysis of ethyl- (and higher alkyl-) cobalamins also produces some Co^I by a subsequent redox reaction between the ethyl radical and the Co^{II} ion. It is not, of course, known whether this occurs via the transfer of an electron or a hydrogen atom, but the occurrence of simple electron transfer has been established[132] in reactions of the type

$$ Cu^{II} + R· \rightarrow Cu^I + R^+ \qquad (10.5) $$

We therefore suggest that the analogous reaction or equilibrium

$$ Co^{II} + R· \rightleftharpoons Co^I + R^+ \qquad (10.6) $$

should be seriously considered when discussing the enzymatic reactions, since it opens up the possibility of carbonium-ion mechanisms, and we have included it as step (C) in the above scheme. Other theoretically possible, but less likely, paths are marked in the scheme with broken lines. It should be noted that since R is a (substituted) primary alkyl group while S is in most cases a (substituted) secondary alkyl group, electronic factors will tend to favour the formation of S^+ rather than R^+, while steric factors will tend to destabilise Co–S relative to Co–R; the formation of Co–S would be still further hindered if the $C_5{}'$ atom of R remained close to the cobalt. We now wish to consider whether rearrangement of the substrate is most likely to occur in the form of the free radical (S·), the carbonium ion (S^+) or when co-ordinated to the cobalt (Co–S).

It is a general observation that organic rearrangements occur far more readily via carbonium ions than via free radicals or carbanions. For example, the simple 1,2 shift of an alkyl group, which would provide a model for the glutamate-β-methyl-aspartate rearrangement, is a well-known example of the involvement of carbonium ions; but the chemical literature is full of vain attempts to observe such a shift in a free radical[236]. Very little work has so far been reported on the possible rearrangements of compounds containing as many functional groups as are present in the enzymatic substrates, and it is therefore not at all certain what influence they might have on possible free-radical reactions. Since carbonium ions react readily with the lone pairs of heteroatoms (to form cyclic ethers, etc.), it might be expected that the presence of such heteroatoms outside the 1,2 positions would competitively inhibit any 1,2 shift proceeding via a carbonium-ion intermediate, unless the protein suppressed such side reactions by, for example, anchoring the heteroatoms by hydrogen bonding. The Co–S group has undoubtedly been the most popular candidate for the entity undergoing isomerisation, due basically to a certain mystique surrounding the catalytic abilities of transition metals. But the only example of such a rearrangement involving the cobalt is the decomposition of β-hydroxyethyl-cobalt(III) complexes by alkali according to the reaction.

$$HO^- + Co-CH_2CH_2OH \rightarrow H_2O + Co^I + CH_3CHO \qquad (10.7)$$

for which three different mechanisms have been proposed (see reference 182). No rearrangements more relevant to the glutamate-β-methylaspartate reaction have yet been reported.

We clearly do not yet know the full range of rearrangements possible for either S radicals or Co–S groups, especially in the presence of several functional groups; and we do not know whether all these 'isomerase' reactions involve the same basic mechanism or not. All we can say is that, according to the present state of knowledge, only a carbonium-ion mechanism will allow a 1,2 shift of an alkyl group similar to that which occurs in the glutamate-β-methylaspartate reaction; and that, if all the isomerase reactions involve the same mechanism, then this is most likely to involve carbonium ions.

In this analysis we have assumed that we need consider only the coenzyme and the substrate—that is, that no functional groups of the protein are involved—and that the initial step involves homolytic fission of the Co–C bond of the coenzyme. Although both assumptions may be invalidated by further work, they have enabled us to make the following tentative suggestions:

(i) that the main effect of the protein (+ substrate) on the coenzyme (cobalt corrinoid) is to displace the following equilibrium to the right

$$Co-R \rightleftharpoons Co^{II} + R\cdot \tag{10.7}$$

that is, to affect the thermodynamic rather than the kinetic properties of the metal complex.

(ii) that the protein affects the position of this equilibrium mainly by distorting the co-ordination sphere of the cobalt (perhaps specifically the $Co-\hat{C}-C$ bond angle), hence destabilising $Co-R$.

(iii) that a substrate-induced change in the conformation of the protein ensures that R is produced only when the substrate is bound at the active site.

We have also suggested, even more tentatively, that rearrangements may involve carbonium-ion intermediates produced by steps (i)–(iii) in the above scheme.

Finally, it is worth pointing out that the magnitude of any such change in an equilibrium constant, which is induced by the binding of the substrate, is not limited by the magnitude of the equilibrium constant for the binding of the substrate by the protein, as can be seen by the following argument. Let us arbitrarily separate the changes involving the active site X (e.g. the metal) and the rest of the protein P. We can then write the following generalised scheme for the binding of the substrate accompanied by a change in the conformation of the protein (from P to P′) and by a change in the properties, including equilibrium properties, of the active site (from X to X′):

$$[P.X] + S = [P'.X'.S] \tag{10.9}$$

This can formally be treated in two separate equilibria, viz:

$$[P.X] = [P'.X'] \tag{10.10}$$

$$[P'.X'] + S = [P'.X'.S] \tag{10.11}$$

even though the change in conformation may, in fact, only occur on binding the substrate S. These equations serve to emphasise that large changes in the free energy of an equilibrium at X can occur even if the overall change in free energy for the binding of S is very low, provided it is accompanied by large compensating changes in free energy due to the change in conformation of P.

REFERENCES

1. Abel, E. W., Pratt, J. M., and Whelan, R., *Chem. Commun.* (1971), 449.
2. Abel, E. W., Pratt, J. M., and Whelan, R., *Inorg. nucl. Chem. Lett.,* 7 (1971), 901.
3. Abel, E. W., Pratt, J. M., and Whelan, R., *Inorg. nucl. Chem. Lett.,* 9 (1973), 151.
4. Alben, J. O., Fuchsman, W. H., Beaudreau, C. A., and Caughey, W. S., *Biochemistry*, 7 (1968), 624.
5. Allen, A. D., Bottomley, F., Harris, R. O., Reinsalu, V. P., and Senoff, C. V., *J. Am. chem. Soc.,* 89 (1967), 5595.
6. Amiconi, G., Antonini, E., Brunori, M., Formanek, H., and Huber, R., *Eur. J. Biochem.,* 31 (1972), 52.
7. Anbar, M., *J. Am. chem. Soc.,* 88 (1966), 5924.
8. Antonini, E., and Brunori, M., *Hemoglobin and Myoglobin in Their Reactions with Ligands,* North-Holland, Amsterdam (1971).
9. Antonini, E., Wyman, J., Brunori, M., Bucci, E., Fronticelli, C., and Rossi Fanelli, A, *J. biol. Chem.,* 238 (1963), 2950.
10. Asakura, T., and Yonetani, T., *J. biol. Chem.,* 244 (1969), 537.
11. Asakura, T., and Yonetani, T., *J. biol. Chem.,* 244 (1969), 4573.
12. Banaszak, L. J., Watson, H. C., and Kendrew, J. C., *J. molec. Biol.,* 12 (1965), 130.
13. Bancroft, G., and Elliott, K. A. C., *Biochem. J.,* 28 (1934), 1911.
14. Bannerjee, R., *Biochim. biophys. Acta,* 64 (1962), 368.
15. Bannerjee, R., *Biochim. biophys. Acta,* 64 (1962), 385.
16. Bannerjee, R., Alpert, Y., Leterrier, F., and Williams, R. J. P., *Biochemistry*, 8 (1969), 2862.
17. Banks, R. G. S., Henderson R. J., and Pratt, J. M., *J. chem. Soc.(A)* (1968), 2886.
18. Barker, H. A., *Ann. Rev. Biochem.,* 41 (1972), 55.
19. Baxendale, J. H., *Adv. Catalysis,* 4 (1952), 31.
20. Baxendale, J. H., and George, P., *Trans. Faraday Soc.,* 46 (1950), 55.
21. Bayer, E., and Schurig, V., *Chem. Ber.,* 102 (1969), 3378.
22. Bayston, J. H., King, N. K., Looney, F. D., and Winfield, M. E., *J. Am. chem. Soc.,* 91 (1969), 2775.
23. Bell, R. P., *The Proton in Chemistry,* Methuen, London (1959).
24. Binnotti, I., Giovenco, S., Giardina, B., Antonini, E., Brunori, M., and Wyman, J., *Archs Biochem. Biophys.,* 142 (1971), 274.
25. Björkstén, F., *Eur. J. Biochem.,* 5 (1968), 133.
26. Bodo, G., Dintzis, H. M., Kendrew, J. C., and Wyckoff, H. W., *Proc. R. Soc.,* A253 (1959), 70.
27. Bolton, W., and Perutz, M., *Nature, Lond.,* 228 (1970), 551.
28. Bonaventura, J., and Riggs, A., *Science, N.Y.,* 158 (1967), 800.
29. Bond, G. C., *Catalysis by Metals,* Academic Press, London (1962).
30. Bottomley, F., and Nyburg, S. C., *Acta. crystallogr.,* B24 (1968), 1289.
31. Braun, V., Crichton, R. C., and Braunitzer, G., *Hoppe-Seyler's Z. physiol. Chem.,* 349 (1968), 197.
32. Bray, R. C., and Swan, J. C., *Struct. Bond.,* 11 (1972), 107.
33. Bretscher, P. A., *Nature, Lond.,* 219 (1968), 606.
34. Brill, A. S., in *Comprehensive Biochemistry,* Vol. 14 (ed. M. Florkin and E. H. Stotz) Elsevier, Amsterdam (1966), p. 447.
35. Brown, S. B., Dean, T. C., and Jones, P., *Biochem. J.,* 117 (1970), 741.
36. Brown, S. B., and Jones, P., *Trans. Faraday Soc.,* 64 (1968), 999.
37. Brown, S. B., Jones, P., and Suggett, A., *Trans. Faraday Soc.,* 64 (1968), 986.
38. Brown, S. B., Jones, P., and Suggett, A., *Prog. inorg. Chem.,* 13 (1970), 159.
39. Brown, W. D., and Mebine, L. B., *J. biol. Chem.,* 244 (1969), 6696.
40. Brunori, M., Antonini, E., Wyman, J., and Anderson, S. R., *J. molec. Biol.,* 34 (1968), 357.

41. Brunori, M., Engel, J., and Schuster, T. M., *J. biol. Chem.*, **242** (1967), 773.
42. Brunori, M., Giacometti, G. M., Antonini, E., and Wyman, J., *J. molec. Biol.*, **63** (1972), 139.
43. Bunn, H. F., and Jandl, J. H., *Proc. natn. Acad. Sci. U.S.A.*, **56** (1966), 974.
44. Burris, R. H., in *The Chemistry and Biochemistry of Nitrogen Fixation* (ed. J. R. Postgate), Plenum, London (1971), pp. 105–60.
45. Calligaris, M., Minichelli, D., Nardin, G., and Randaccio, L., *J. chem. Soc. (A)* (1970), 2411.
46. Carlisle, D. B., *Proc. R. Soc.*, **B171** (1968), 31.
47. Caughey, W. S., Alben, J. O., and Beaudreatu, C. A., in *Oxidases and Related Redox Systems* (ed. T. E. King, H. S. Mason and M. Morrison), Wiley, New York (1965), p. 97.
48. Caughey, W. S., Alben, J. O., McCoy, S., Boyer, S. H., Charache, S., and Hathaway, P., *Biochemistry*, **8** (1969), 59.
49. Chance, B., *Acta chem. Scand.*, **1** (1947), 236.
50. Chance, B., *Adv. Enzymol.*, **12** (1951), 153.
51. Chance, B., *J. biol. Chem.*, **194** (1952), 471.
52. Chance, B., *J. biol. Chem.*, **194** (1952), 483.
53. Chance, B., *Archs Biochem. Biophys.*, **40** (1952), 153.
54. Chance, B., *Archs Biochem. Biophys.*, **41** (1952), 416.
55. Chance, B., in *Oxidases and Related Redox Systems* (ed. T. E. King, H. S. Mason and M. Morrison), Wiley, New York (1965), p. 504.
56. Chance, B., Mela, L., and Harris, E. J., *Fedn Proc. Fedn Am. Socs exp. Biol.*, **27** (1968), 902.
57. *Chem. Week* (Sept. 27th 1972), 11.
58. Cohen, I. A., and Caughey, W. S., *Biochemistry*, **7** (1968), 636.
59. Coleman, J. P., *Prog. bio-org. Chem.*, **1** (1971), 159.
60. Connor, J. A., and Ebsworth, E. A. V., *Adv. inorg. Chem. Radiochem.*, **6** (1964), 279.
61. Corwin, A. H., and Bruck, S. D., *J. Am. chem. Soc.*, **80** (1958), 4736.
62. Corwin, A. H., and Erdman, J. G., *J. Am. chem. Soc.*, **68** (1946), 2473.
63. Corwin, A. H., and Reyes, Z., *J. Am. chem. Soc.*, **78** (1956), 2437.
64. Countryman, R., Collins, D. M., and Hoard, J. L., *J. Am. chem. Soc.*, **91** (1969), 5166.
65. Coval, M. L., and Taurog, A., *J. biol. Chem.*, **242** (1967), 5510.
66. Cowan, M. J., Drake, J. M. F., and Williams, R. J. P., *Discuss. Faraday Soc.*, **27** (1960), 217.
67. Creutz, C., and Taube, H., *Inorg. Chem.*, **10** (1971), 2664.
68. Critchlow, J. E., and Dunford, H. B., *J. biol. Chem.*, **247** (1972), 3714.
69. Crumbliss, A. L., and Basolo, F., *J. Am. chem. Soc.*, **92** (1970), 55.
70. Das, P. K., Pratt, J. M., Smith, R. G., Swinden, G., and Woolcock, W. J. U., *Chem. Commun.* (1968), 1539.
71. Deno, N. C., Peterson, H. J. and Saines, G. S., *Chem. Rev.*, **60** (1960), 7.
72. Dilworth, M. J., *Biochem. Biophys. Acta*, **127** (1966), 285.
73. Donald, C. M., *J. Aust. Inst. agric. Sci.*, **26** (1960), 319.
74. Dobinson, G. C., Mason, R., Robertson, G. B., Ugo, R., Conti, F., Morelli, D., Cenini, S., and Bonati, F., *Chem. Commun.* (1967), 739.
75. Dunford, H. B., and Alberty, R. A., *Biochemistry*, **6** (1967), 447
76. Er-el, Z., Shaklai, N., and Daniel, E., *J. molec. Biol.*, **64** (1972), 341.
77. Falk, J. E., *Porphyrins and Metalloporphyrins,* Elsevier, Amsterdam (1964).
78. Fernández-Morán, H., van Bruggen, E. F. J., and Ohtsuki, M., *J. molec. Biol.,* **16** (1966), 191.
79. Floriani, C., and Calderazzo, F., *J. chem. Soc(A)* (1969), 946.
80. Foner, S. N., and Hudson, R. L., *J. chem. Phys.*, **28** (1958), 719.
81. Fox, H. M., *Proc. R. Soc.*, **B111** (1932), 358.
82. Fox, H. M., and Vevers, G., *The Nature of Animal Colours*, Sidgwick and Jackson, London (1960).
83. Gatt, R., and Kremer, M. L., *Trans. Faraday Soc.*, **64** (1968), 721.
84. George, P., in *Haematin Enzymes* (ed. J. E. Falk, R. Lemberg and R. K. Morton), Pergamon, Oxford (1961), p. 103.
85. George, P., in *Oxidases and Related Redox Systems* (ed. T. E. King, H. S. Mason and M. Morrison), Wiley, New York (1965), p. 3.

86. Gersonde, K., Sick, H., Wollmer, A., and Buse, G., *Eur. J. Biochem.*, **25** (1972), 181.
87. Ghiretti, F., in *Oxygenases* (ed. O. Hayaishi), Academic Press, New York (1962), p. 517.
88. Gibson, Q. H., *Biochem. J.*, **71** (1959), 293.
89. Gibson, Q. H., and Antonini, E., *J. biol. Chem.*, **238** (1963), 1384.
90. Gibson, Q. H., and Antonini, E., in *Hemes and Hemoproteins* (ed. B. Chance, R. W. Estabrook and T. Yonetani), Academic Press, New York (1966), p. 67.
91. Gibson, Q. H., and Smith, M. H., *Proc. R. Soc.,* **B163** (1965), 206.
92. Gjessing, E. C., and Sumner, J. B., *Arch. Biochem.*, **1** (1942), 1.
93. Gmelin's *Handbuch der anorgische Chemie, Sauerstoff*, Vol. 3, Verlag Chemie (8th edn, 1958), Verlag Chemie, p. 460.
94. Greer, J., *Cold Spring Harb. Symp. quant. Biol.*, **36** (1972), 315.
95. Hager, L. P., Doubek, D. L., Silverstein, R. M., Hargis, J. H., and Martin, J. C., *J. Am. chem. Soc.*, **94** (1972), 4364.
96. Hager, L. P., Morris, D. R., Brown, F. S., and Eberwein, H., *J. biol. Chem.*, **241** (1966), 1769.
97. Hambly, A. N., *Rev. pure appl. Chem.*, **15** (1965), 87.
98. Hardy, R. W. F., Burns, R. C., and Parshall, G. W., *Adv. Chem.*, **100** (1971), 219.
99. Harrison, D. F., Weissberger, E., and Taube, H., *Science, N.Y.*, **159** (1968), 320.
100. Hasinoff, B. B., and Dunford, H. B., *Biochemistry*, **9** (1970), 4930.
101. Hayaishi, O., *Ann. N.Y. Acad. Sci.*, **158** (1969), 318.
102. Hendrickson, W. A., and Love, W. E., *Nature, Lond.*, **232** (1971), 197.
103. Henry, Y., and Bannerjee, B., *J. molec. Biol.*, **73** (1973), 469.
104. Hoard, J. L., in *Hemes and Hemoproteins* (ed. B. Chance, R. W. Estabrook and T. Yonetani), Academic Press, New York (1966), p. 9.
105. Hoard, J. L., Hamor, M. J., Hamor, T. A., and Caughey, W. S., *J. Am. chem. Soc.*, **87** (1965), 2312.
106. Hoffman, B. M., and Petering, D. H., *Proc. natn. Acad. Sci. U.S.A.*, **67** (1970), 637.
107. Hosoya, T., and Morrison, M., *J. biol. Chem.*, **242** (1967), 2828.
108. Hsu, G. C., Spilburg, C. A., Bull, C., and Hoffman, B. M., *Proc. natn. Acad. Sci. U.S.A.*, **69** (1972), 2122.
109. Huber, R., Epp, O., and Formanek, H., *J. molec. Biol.*, **52** (1970), 349.
110. Huber, R., Epp, O., Steigemann, W., and Formanek, H., *Eur. J. Biochem.*, **19** (1971), 42.
111. Hünig, S., Müller, H. R., and Thier, W., *Angew. Chem. int. Edn*, **4** (1965), 271.
112. Imamura, T., Riggs, A., and Gibson, Q. H. *J. biol. Chem.*, **247** (1972), 521.
113. Jacob, H. S., Brain, M. C., Dacie, J. V., Carrell, R. W., and Lehmann, H., *Nature, Lond.*, **218** (1968), 1214.
114. Jarnagin, R. C., and Wang, J. H., *J. Am. chem. Soc.*, **80** (1958), 786.
115. Jones, P., personal communication.
116. Jones, P., and Middlemiss, D. N., *Biochem. J.*, **130** (1972), 411.
117. Jones, P., and Suggett, A., *Biochem. J.*, **108** (1968), 833.
118. Jukes, T. H., and Holmquist, R., *J. molec. Biol.*, **64** (1972), 163.
119. Kao, O. H. W., and Wang, J. H., *Biochemistry*, **4** (1965), 342.
120. Keilin, D., and Hartree, E. F., *Biochem. J.*, **60** (1955), 310.
121. Keilin, D., and Hartree, E. F., *Biochem. J.*, **61** (1955), 153.
122. Keilin, D., and Wang, Y. L., *Biochem. J.*, **40** (1946), 855.
123. Keilin, J., in *Hemes and hemoproteins* (ed. B. Chance, R. W. Estabrook and T. Yonetani), Academic Press, New York (1966), p. 173.
124. Keller, G. M., Wütrich, K., and Debrunner, P. G., *Proc. Natn. Acad. Sci., U.S.A.*, **69** (1972), 2073.
125. Kelly, M., *Biochim. Biophys. Acta*, **191** (1969), 527.
126. Kelly, M., and Lang, G., *Biochim. Biophys. Acta*, **223** (1960), 86.
127. Keresztes-Nagy, S., and Klotz, I. M., *Biochemistry*, **4** (1965), 919.
128. Keyes, M., and Lumry, R., *Fedn Proc. Fedn Am. Socs exp. Biol.*, **27** (1968), 895.
129. Keyes, M., and Lumry, R., quoted in reference 8, p. 146.
130. Kilmartin, J. V., and Hewitt, J. A., *Cold Spring Harb. Symp. quant. Biol.*, **36** (1972), 311.
131. Klapper, M. H., and Hackett, D. P., *Biochim. Biophys. Acta*, **96** (1965), 272.
132. Kochi, J. K., in *Free Radicals*, Vol. 1 (ed. J. K. Kochi), Wiley, New York (1973), p. 591.
133. Koenig, D. F., *Acta crystallogr.*, **18** (1965), 663.

134. Kon, H., and Sharpless, N. E., *Spectrosc. Lett.*, **1** (1968), 49.
135. Kremer, M. L., *Trans. Faraday Soc.*, **61** (1965), 1453.
136. Kremer, M. L., *Trans. Faraday Soc.*, **63** (1967), 1208.
137. Kremer, M. L., *Biochim. Biophys. Acta*, **198** (1970), 199.
138. Kuhn, W., and Kuhn, H. J., *Z. Elektrochem.*, **65** (1961), 427.
139. Kurozumi, T., Inada, Y., and Shibata, K., *Arch. Biochem. Biophys.*, **94** (1961), 464.
140. Langenbeck, W., Hutschenreuter, R., and Rottig, W., *Ber. dt. chem. Ges.*, **65** (1932), 1750.
141. Latimer, W. M., *The Oxidation States of the Elements and their Potentials in Aqueous Solution*, Prentice-Hall, New York (2nd edn, 1952).
142. Lattman, E. A., Nockolds, C. E., Kretsinger, R. H., and Love, W. E., *J. molec. Biol.*, **60** (1971), 271.
143. Lemberg, R., and Legge J. W., *Hematin Compounds and Bile Pigments*, Interscience, New York (1949).
144. Love, W. E., Klock, P. A., Lattman, E. A., Padlan, E. A., Ward, K. B., and Hendrickson, W. A., *Cold Spring Harb. Symp. quant. Biol.*, **36** (1972), 349.
145. McCord, J. M., Keele, B. B., and Fridovich, I., *Proc. natn. Acad. Sci. U.S.A.* **68** (1971), 1024.
146. McKenna, C. E., Benemann, J. R., and Traylor, T. G., *Biochem. Biophys. Res. Commun.* **41** (1970), 1501.
147. Macnicol, P. K., *Arch. Biochem. Biophys.*, **117** (1966), 347.
148. Mangum, C., *Am. Scient.*, **58** (1970), 641.
149. Marvich, R. H., and Brintzinger, H. H., *J. Am. chem. Soc.*, **93** (1971), 2046.
150. Maskill, R., and Pratt, J. M., *J. chem. Soc. A* (1968), 1914.
151. Mazza, G., Charles, C., Bouchet, M., Ricard, J., and Raynand, J., *Biochim. Biophys. Acta.*, **167** (1968), 89.
152. Muirhead, H., and Greer, J., *Nature, Lond.*, **228** (1970), 516.
153. Murawski, K., Carta, S., Sorcini, M., Tentori, L., Vivaldi, G., Antonini, E., Brunori, M., Wyman, J., Bucci, E., and Rossi Fanelli, A., *Arch. Biochem. Biophys.*, **111** (1965) 197.
154. Nakamura, Y., Tohjo, M., and Shibata, K., *Arch. Biochem. Biophys.*, **102** (1963), 144.
155. Nakahara, A., and Wang, J. H., *J. Am. chem. Soc.*, **80** (1958), 6526.
156. Nason, A., Lee, K., Pan, S., Ketchum, P. A., Lamberti, A., and DeVries, J., *Proc. natn. Acad. Sci. U.S.A.*, **68** (1971), 3242.
157. Nathan, D. G., and Baehner, R. L., *Prog. Hemat.*, 7 (1971), 235.
158. Nechiporenko, G. N., Tabrina, G. M., Shilova, A. K., and Shilov, A. E., *Dokl. Akad. Nauk SSSR*, **164** (1965), 1062.
159. Nicholls, P., and Schonbaum, G. R., in *The Enzymes,* Vol. 8 (ed. D. Boyer, H. Lardy and K. Myrbäck), Academic Press, New York (2nd edn, 1963), p. 147.
160. Nobbs, C. L., *J. molec. Biol.*, **13** (1965), 325.
161. Nobbs, C. L., in *Hemes and Hemoproteins* (ed. B. Chance, R. W. Estabrook and T. Yonetani), Academic Press, New York (1966), p. 143.
162. Nobbs, C. L., Watson, H. C., and Kendrew, J. C., *Nature, Lond.*, **209** (1966), 339.
163. Ogura, Y., *Arch. Biochem. Biophys.*, **57** (1955), 288.
164. Ohme, R., and Gründemann, C., *Z. Chemie, Lpz.*, **10** (1970), 268.
165. Okazaki, T., Briehl, R. W., Wittenberg, J. B., and Wittenberg, B. A., *Biochim. Biophys. Acta*, **111** (1965), 496.
166. Okazaki, T., and Wittenberg, J. B., *Biochim. Biophys. Acta*, **111** (1965), 503.
167. Paul, K. G., *Acta chem. Scand.*, **13** (1959), 1239.
168. Paul, K. G., in *The Enzymes*, Vol. 8 (ed. D. Boyer, H. Lardy and K. Myrbäck), Academic Press, New York (2nd edn, 1963), p. 227.
169. Perutz, M. F., *Nature, Lond.*, **228** (1970), 726.
170. Perutz, M. F., and Lehmann, H., *Nature, Lond.*, **219** (1968), 902.
171. Perutz, M. F., and Matthews, F. S., *J. molec. Biol.,* **21** (1966), 199.
172. Perutz, M. F., Muirhead, H., Cox, J. M., and Goaman, L. C. G., *Nature, Lond.* **219** (1968), 131.
173. Perutz, M. F., and Ten Eyck, L. F., *Cold Spring Harb. Symp. quant. Biol.*, **36** (1972), 295.
174. Petty, M. A., *Bact. Rev.*, **25** (1961), 111.
175. Phelps, C., Antonini, E., and Brunori, M., *Biochem. J.*, **122** (1971), 79.
176. Phelps, C., Forlani, L., and Antonini, E., *Biochem. J.*, **124** (1971), 605.

177. Piette, L. H., Yamazaki, I., and Mason, H. S., *J. biol. Chem.*, **235** (1960), 2444.
178. Portsmouth, D., and Beal, E. A., *Eur. J. Biochem.*, **19** (1971), 479.
179. Pratt, J. M., *J. theor. Biol.*, **2** (1962), 251.
180. Pratt, J. M., Paper presented at the Chemical Society of London's Symposium: 'N₂, O₂ and hydrogen as ligands for metals', Leeds (1968).
181. Pratt, J. M., *Inorganic Chemistry of Vitamin B₁₂*, Academic Press, London (1972).
182. Pratt, J. M., and Craig, P. J., *Adv. organometal. Chem.*, **11** (1973), 331.
183. Pratt, J. M., and Wilkinson, P. J., to be published
184. Prince, R. H., and Stotter, D. A., *J. inorg. nucl. Chem.*, **35** (1973), 321.
185. Ricard, J., Mazza, G., and Williams, R. J. P., *Eur. J. Biochem.*, **28** (1972), 566.
186. Rodley, G. A., and Robinson, W. T., *Nature, Lond.*, **235** (1972), 438.
187. Roman, R., and Dunford, H. B., *Biochemistry*, **11** (1972), 2076.
188. Roman, R., Dunford, H. B., and Evett, M., *Can. J. Chem.*, **49** (1971), 3059.
189. Rossi Fanelli, A., and Antonini, E., *Arch. Biochem. Biophys.*, **77** (1958), 478.
190. Rossi Fanelli, A., and Antonini, E., *Nature, Lond.*, **186** (1960), 895.
191. Rossi Fanelli, A., and Antonini, E., *J. biol. Chem.*, **235** (1960), PC4.
192. Rossi Fanelli, A., Antonini, E., and Caputo, A., *Adv. Protein Chem.*, **19** (1964), 74.
193. Rossotti, F. J. C., in *Modern Co-ordination Chemistry* (ed. J. Lewis and R. G. Wilkins), Interscience, New York (1960), p. 1.
194. Roughton, F. J. W., in *Oxygen in the Animal Organism* (ed. F. Dickens and E. Neil), Pergamon, Oxford (1964), p. 5.
195. Sasazuki, T., Isomoto, A., and Nakajima, H., *J. molec. Biol.*, **65** (1972), 365.
196. Saunders, B. C., Holmes-Siedle, A. G., and Stark, B. P., *Peroxidase*, Butterworths, London (1964).
197. Schildknecht, H., *Angew. Chem. int. Edn*, **9** (1970), 1.
198. Schoenborn, B. O., Watson, H. C., and Kendrew, J. C., *Nature, Lond.*, **207** (1965), 28.
199. Schonbaum, G. R., and Lo, S., *J. biol. Chem.*, **247** (1972), 3353.
200. Schrauzer, G. N., and Lee, L. P. *J. Am. chem. Soc.*, **92** (1970), 1551.
201. Schroeder, W. A., Shelton, J. R., Shelton, J. B., Robberson, B., and Apell, G., *Arch. Biochem. Biophys.* **131** (1969), 653.
202. Scouloudi, H., *J. molec. Biol.*, **40** (1969), 35.
203. Shilov, A. E., Shilova, A. K., and Kvashina, E. F., *Kinet. Katal.*, **10** (1969), 1402.
204. Shulman, R. G., Ogawa, S., and Hopfield, J. J., *Cold Spring Harb. Symp. quant. Biol.*, **36** (1972), 337.
205. Sick, H., Gersonde, K., Thompson, J. C., Maurer, W., Haar, W., and Rüterjans, H., *Eur. J. Biochem.*, **29** (1972), 217.
206. Smith, D. W., and Williams, R. J. P., *Struct. Bond.*, **7** (1970), 1.
207. Spilburg, C. A., Hoffman, B. M., and Petering, D. H., *J. biol. Chem.*, **247** (1972), 4219.
208. Steen, J. B., in *Oxygen in the Animal Organism* (ed. F. Dickens and E. Neil), Pergamon, Oxford (1964), p. 621.
209. Stern, K. G., *Hoppe-Seyler's Z. physiol. Chem.*, **215** (1933), 35.
210. Strother, G. K., and Ackerman, E., *Biochim. Biophys. Acta*, **47** (1961), 317.
211. Stryer, L., Kendrew, J. C., and Watson, H. C., *J. molec. Biol.*, **8** (1964), 96.
212. Stynes, H. C., and Ibers, J. A., *J. Am. chem. Soc.*, **94** (1972), 5125.
213. Sudo, M., Ichikawa, M., Soma, M., Onishi, T., and Tamaru, K., *J. phys. Chem.*, **73** (1969), 1174.
214. Takano, T., Swanson, R., Kallai, O. B., and Dickerson, R. E., *Cold. Spring Harb. Symp. quant. Biol.*, **36** (1972), 397.
215. Teale, F. W. J., *Biochim. Biophys. Acta*, **35** (1959), 289.
216. Theorell, H., *Nature, Lond.*, **156** (1945), 474.
217. Thomas, J. A., Morris, D. R., and Hager, L. P., *J. biol. Chem.*, **245** (1970), 3129.
218. Thomas, J. A., Morris, D. R., and Hager, L. P., *J. biol. Chem.*, **245** (1970), 3135,
219. Tohjo, M., Nakamura, Y., Kurihara, K., Samejima, T., Hachimori, Y., and Shibata, K., *Arch. Biochem. Biophys.*, **99**, (1962), 222.
220. Treitel, I. M., Flood, M. T., Marsh, R. E., and Gray, H. B., *J. Am. chem. Soc.*, **91** (1969), 6512.
221. Ullrich, V., *Angew. Chem. int. Edn,* **11** (1972), 701.
222. Vallee, B. L., and Williams, R. J. P., *Proc. natn. Acad. Sci. U.S.A.,* **59** (1968), 498.

223. Volpin, M. E., and Shur, V. B., *Dokl. Akad. Nauk SSSR* **156** (1964), 1102.
224. Vonderschmitt, D., Bernauer, K., and Fallab, S., *Helv. chim. Acta*, **48** (1965), 951.
225. Walker, F. A., *J. Am. chem. Soc.*, **92** (1970), 4235.
226. Wang, J. H., *J. Am. chem. Soc.*, **77** (1955), 4715.
227. Wang, J. H., in *Haematin Enzymes* (ed. J. E. Falk, R. Lemberg, and R. K. Morton), Pergamon, Oxford (1961), p. 95.
228. Wang, J. H., in *Haematin Enzymes* (ed. J. E. Falk, R. Lemberg and R. K. Morton), Pergamon, Oxford (1961), p. 98.
229. Wang, J. H., in *Oxygenases* (ed. O. Hayaishi), Academic Press, New York (1962), p. 469.
230. Wang, J. H., *Acct chem. Res.*, **3** (1970), 90.
231. Watson, H. C., and Chance, B., in *Hemes and Hemoproteins* (ed. B. Chance, R. W. Estabrook and T. Yonetani), Academic Press, New York (1966), pp. 149–55.
232. Watson, H. C., and Nobbs, C. L., in *Biochemie des Sauerstoffs* (ed. B. Hess and H. Staudinger), Springer, Berlin (1968), p. 37.
233. Weber, J. H., and Busch, D. H., *Inorg. Chem.*, **4** (1965), 469.
234. Wilkins, R. G., *Adv. Chem.*, **100** (1971), 111.
235. Williams, R. J. P., *Fedn Proc. Fedn Am. Socs exp. Biol.*, **20** No. 3 (1961), 5.
236. Wilt, J. W., in *Free Radicals*, Vol. 1 (ed. J. K. Kochi), Wiley, New York, p. 333.
237. Winterhalter, K. H., Anderson, N. M., Amiconi, G., Antonini, E., and Brunori, M., *Eur. J. Biochem.*, **11** (1969), 435.
238. Wittenberg, B. A., Okazaki, T., and Wittenberg, J. B., *Biochim. Biophys. Acta*, **111** (1965), 485.
239. Wittenberg, J. B., Appleby, C. A., and Wittenberg, B. A., *J. biol. Chem.*, **247** (1972), 527.
240. Wittenberg, J. B., and Wittenberg, B. A., *Nature, Lond.*, **194** (1962), 106.
241. Wülker, W., Maier, W., and Bertau, B., *Z. Naturf.*, **24B** (1969), 110.
242. Wyman, J., *Adv. Protein Chem.*, **19** (1964), 224.
243. Yamazaki, I., Yokota, K., and Nakajima, R., in *Oxidases and Related Redox Systems* (ed. T. E. King, H. S. Mason and M. Morrison), Wiley, New York (1965), p. 485.
244. Yonetani, T., *J. biol. Chem.*, **242** (1967), 5008.
245. Yonetani, T., *Adv. Enzymol.*, **33** (1970), 309.
246. Yonetani, T., and Asakura, T., *J. biol. Chem.*, **243** (1968), 3996.
247. Yonetani, T., and Asakura, T., *Fedn Proc. Fedn Am. Socs exp. Biol.* **27** (1968), 526.
248. Yonetani, T., and Asakura, T., *J. biol. Chem.*, **244** (1969), 4580.
249. Yonetani, T., and Schleyer, H., *J. biol. Chem.*, **242** (1967), 1974.
250. von Zelewsky, A., *Helv. chim. Acta*, **55** (1972), 2941.

PART 3

The Biochemical Function of Molybdenum

F. L. BOWDEN

Department of Chemistry, University of Manchester, Institute of Science and Technology

11 INTRODUCTION

Molybdenum is unique among the elements of the second and third transition series of the Periodic Table in having an essential role in several biochemical reactions[1]. The most important of these concern the utilisation of inorganic nitrogen for the production of proteins, nucleic acids and other nitrogenous cell constituents[2-4]. Living organisms differ widely in their ability to synthesise the amino-acid precursors of these molecules and with respect to the forms of nitrogen they can utilise for this purpose. Higher animals are unable to synthesise certain amino acids—the essential amino acids—and must obtain them from an exogenous source. Furthermore, they cannot utilise the most abundant natural sources of inorganic nitrogen, namely soil nitrate and atmospheric nitrogen, in the synthesis of even the nonessential amino acids. Fortunately, plants and many micro-organisms are more versatile; they can make all the amino acids found in proteins starting from either nitrate or ammonia. The metabolic assimilation of nitrate into the form of ammonia, proceeds in two major steps: (i) reduction of nitrate to nitrite and (ii) reduction of nitrite to ammonia. The first of these reactions is catalysed by the molybdenum-containing enzyme, nitrate reductase. Ammonia is also produced by the bacterial fixation of molecular nitrogen. This is accomplished by free-living soil bacteria, for example *Clostridium pasteuranium* and, more importantly, by degenerate forms of other soil bacteria, for example *Rhizobium* in association with root nodules of leguminous plants. Both fixation processes are catalysed by another molybdenum-containing enzyme, nitrogenase. Thus molybdenum has a central role in the principal routes of nitrogen incorporation into plants and there-fore animals. It has been estimated that approximately 10^{10} tons of nitrogen are incorporated annually into plants[5]. In this respect alone, molybdenum must be considered one of the most important of the biologically active metals.

The importance of molybdenum to plants has been stressed, but it is also an essential constituent of several mammalian enzymes. Xanthine oxidase is the most extensively studied of these. It functions in the oxidation of purine bases to uric acid. Dual substrate specificities are exhibited by xanthine oxidase and the closely related enzyme aldehyde oxidase. Both enzymes catalyse the oxidation of a wide range of heterocyclic nitrogen compounds and aldehydes, and probably utilise oxygen as the physiological terminal electron acceptor. A third enzyme, xanthine dehydrogenase, has a very similar functionality pattern but appears to use NAD as electron acceptor. There are significant differences between the molybdenum electron paramagnetic resonance (e.p.r.) signals of these three enzymes; this could suggest that their differences arise, at least in part, from subtle variations in the ligand environment of the molybdenum prosthetic group. The most recent addition to the list of mammalian molybdenum-containing enzymes is sulphite oxidase. Its molybdenum was discovered accidentally during an e.p.r. investigation of the haem component[6]. The importance of these mammalian enzymes is not well established. However, there is one report of an infant who died at 23 months with neurological and other abnormalities apparently as a consequence of a lack of sulphite oxidase activity[7].

A redox role for molybdenum in biochemical reactions is to be anticipated from its physicochemical properties, since in keeping with the other redox-active ions its principal oxidation states, Mo(V) and Mo(VI), are readily interconvertible. Moreover, molybdenum has the unique feature among the biologically active metal ions of a range of oxidation states, Mo(III), Mo(V) and Mo(VI), which can be stabilised in an aqueous environment by ligands found in biological systems. It might, therefore, be able to participate in multi-electron transfer reactions such as may be involved in nitrogen fixation. Herein may lie the key to its selection as a biochemical redox ion.

Despite the clearly established importance of molybdenum as a biological trace element, very little is known concerning the detailed structure and function of the molybdenum-containing enzymes. The chief reasons for this are their low molybdenum contents and the structural complexities of the enzymes. They are of the many-headed enzyme type[8] in which four functions—reduction, oxidation, electron transfer and energy conservation—may be occurring simultaneously, with the molybdenum component as part of an electron-transport chain. The structural complexity of the enzymes often means that the response of the molybdenum centre to spectroscopic probes is overshadowed by those of other enzyme components, for example flavins and iron–sulphur moieties. E.P.R. spectroscopy has proved to be a technique that is not subject to this limitation. It has been used to monitor the rates of enzyme reactions and to establish the involvement of molybdenum in the catalytic phase of reaction. A comparison of the e.p.r. spectra of the enzymes with those of synthetic model compounds has been used to indicate the nature of the biological ligands binding to molybdenum. Selective oxidation and reduction using series of electron donors and acceptors, and studies of sequential inhibition have been used in biochemical investigations designed to establish its position in the electron-transport chain.

It is the purpose of part 3 of the book to summarise the results of these chemical and biochemical studies and to present the current picture of the biochemical role of molybdenum. Various individual aspects of the biochemistry of molybdenum have been reviewed elsewhere[2,7-16].

12 OCCURRENCE, GENERAL CONSTITUTION AND PROPERTIES OF MOLYBDENUM ENZYMES

Molybdenum is known to be an essential constituent of six enzymes: aldehyde oxidase, nitrate reductase, nitrogenase, sulphite oxidase, xanthine dehydrogenase and xanthine oxidase. Of these, four have a FAD[†] component and all six contain iron, either as a cytochrome or an iron–sulphur moiety. The enzymes have high molecular weights, in the range 10^5–10^6. In most of them there are two atoms of molybdenum per molecule of protein so that the molybdenum content is only about 10^{-2} per cent. This is one of the factors that has made the study of molybdenum-containing enzymes difficult. The occurrence of two molybdenum atoms per molecule of protein may be a reflection of the strong tendency of low molecular weight molybdenum compounds to exist as dimers, although it could be coincidental since there is no evidence for molybdenum–molybdenum inter-actions in any of the enzymes. The mammalian enzymes: aldehyde oxidase, sulphite oxidase, xanthine dehydrogenase and xanthine oxidase have been isolated from liver, but milk is the best source of xanthine oxidase. Bacteria are the chief sources of nitrate reductase and nitrogenase. The properties of a representative group of molybdenum-containing enzymes are summarised in table 21.

† FAD = flavin adenine dinucleotide.

FAD (VII)

TABLE 21 SOME MOLYBDENUM-CONTAINING ENZYMES

Enzyme	Source	Composition			Molecular weight × 10^{-3}	Substrates ¶	
						Oxidising	Reducing
aldehyde oxidase	rabbit liver	Mo(2)†	FAD(1)‡	Fe(4)§,||	300	O_2^-	aldehydes
nitrate reductase	Escherichia coli	Mo(1)	–	Fe(40)§	1000	NO_3^-	$FMNH_2$‡‡
	Neurospora crassa	Mo(1–2)	FAD††	Fe(?)‡‡	230	NO_3^-	NADPH¶¶
	Aspergillus nidulans	Mo(?)	FAD(?)	Fe(?)	200	NO_3^-	NADPH
nitrogenase	Azotobacter vinlandii	Mo(2)§§		Fe(16–18)|||	270	N_2	ferredoxin
	Clostridium pasteuranium	Mo(1)§§		Fe(11–14)|||	170	CH:CH	$S_2O_4^{2-}$
	Klebsiella pneumoniae	Mo(1)§§		Fe(10)|||	216	MeNC	viologen dyes
sulphite oxidase	bovine liver	Mo(2)		Fe(2)‡‡	110	O_2	SO_3^{2-}
xanthine dehydrogenase	chicken liver	Mo(2)	FAD(1)	Fe(4)§	300	NAD†††	purines / aldehydes
	Micrococcus lactilyticus	Mo(2)	FAD(1)	Fe(4)§	250	ferredoxin	purines / aldehydes
xanthine oxidase	cow's milk	Mo(2)	FAD(1)	Fe(4)§	275	O_2	purines / aldehydes

† numbers in parentheses indicate the number of molybdenum atoms per protein molecule.
‡ contents of FAD and other constituents are expressed per molybdenum atom owing to uncertainties in some of the molecular weights.
§ the iron is present as an iron–sulphur moiety.
|| aldehyde oxidase also contains coenzyme Q.
¶ substrates are typical. Most of the enzymes accept a wide range of oxidising and reducing substrates.
†† FAD is not established as a functional component of the enzyme; it dissociates with K_m about 2×10^{-8}.
‡‡ the iron is present as a cytochrome-b.
§§ composition and molecular-weight data refer to the molybdoferredoxin component.
||| also contains 'acid-labile' sulphide groups.
¶¶ NADPH = nicotinamide adenine dinucleotide phosphate, $R = PO_3H_2$.

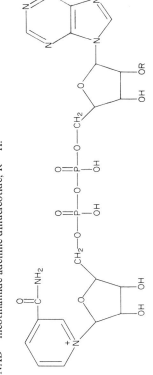

††† NAD = nicotinamide adenine dinucleotide, R = H.

NAD/NADP (VIII)

‡‡‡ FMNH₂.

FMNH₂ (IX)

13 PHYSICOCHEMICAL PROPERTIES OF MOLYBDENUM COMPOUNDS

This chapter is concerned with those properties of molybdenum compounds that can be used to provide information about the steric and electronic environment of the metal ion. Most attention has been concentrated on molybdenum compounds in aqueous solution although a recent report of the localisation of rat-liver sulphite oxidase in the intermembrane space of mitochondria (section 14.6) suggests that more emphasis should be placed on studies of nonaqueous systems. Of the molybdenum ions that are likely to be stable in a biological environment, Mo(III), Mo(IV) and Mo(V) have three, two and one d electron(s), respectively. They should therefore give rise to d–d electronic spectra, e.p.r. spectra and bulk magnetic susceptibility. These can in principle provide information about the nature of the ligands co-ordinated to the metal ion, its oxidation state and the stereochemistry of its ligand environment.

13.1 REDOX POTENTIALS AND OXIDATION STATES

Mo(VI) is the only oxidation state of molybdenum that is expected to be stable to air oxidation in aqueous solution. However, in the absence of air, oxidation states from Mo(III) to Mo(V) can be stabilised in aqueous solution by combination with suitable ligands, for example $[Mo^{III}Cl_6]^{3-}$ and $[Mo(CN)_8]^{4-}$. Complexes of Mo(V) and Mo(VI) are complexes of the oxocations MoO^{3+} and MoO^{2+}, respectively. Studies of these ions in aqueous media are complicated by their ready hydrolysis. Williams and Mitchell[18] showed that the MoO^{2+} ion is a typical class a[19] or hard[20] cation and that a change to the lower oxidation state in the MoO^{3+} ion introduces a significant amount of class b or soft character, since the MoO^{3+} ion forms stable complexes with sulphur ligands as well as with anionic oxygen ligands. The magnetic moments of the Mo(V) complexes are lower than anticipated for one unpaired electron per metal atom; this has been attributed to dimerisation via metal–metal bonding. Williams[21] has pointed out that the d electrons and orbitals of elements with few d electrons appear to be strongly exposed to ligands and in the case of Mo(V) compounds this leads to dimerisation. This tendency to dimer formation may account for the low-spin recovery of reduced xanthine oxidase in the e.p.r. spectra. An equilibrium between paramagnetic monomer and diamagnetic dimer has been proposed to explain the intensity of the e.p.r. signal that corresponds to only 37 per cent of the total molybdenum. The spin recovery is even lower ($<$ 1 per cent) with aqueous solutions of Mo(V) model complexes. This indicates a specific ability of the biological ligand system to stabilise monomeric Mo(V). Specific stabilisation effects can also control the balance between oxidation states. Such control is essential if the molybdenum enzymes are to function effectively as electron-transfer agents, since low activation-energy pathways for electron transfer between molecules are likely only when the redox potentials of the components of the electron-transfer process are correctly balanced. The results of Williams and Mitchell[18] show how the control of redox potentials can be achieved for molybdenum. They illustrate the specific stabilisation of Mo(IV) by cyanide ion, the enhanced stability of Mo(VI) relative to Mo(III) with hydroxide as ligand, and the similar stabilities of Mo(III) and Mo(V) with chloride and thiocyanate. Estimates[21] of the redox potentials for the Mo(VI)/Mo(V) and Mo(V)/Mo(III) couples at neutral pH of -0.2 to -0.4 V and -0.6 to -1.0 V, place the former close to the redox potentials of flavin systems (about 0.25 V) and the latter outside the range of potentials for biological systems. However, the ability of mercaptoacetic acid (E^0 about -0.30 V) to reduce Mo(V) to Mo(III) illustrates how the redox potential of a molybdenum couple can be brought into the biological range by the preferential stabilisation of a lower oxidation state (see also chapter 15).

Evidently, all oxidation states of molybdenum from Mo(III) to Mo(VI) inclusive can be stabilised by suitable combinations of ligands and therefore may be anticipated to occur in molybdenum enzymes. The lower oxidation states will be preferentially stabilised by ligands with some π-acceptor character. E.P.R. results indicate that thiol ligands may be biologically significant ligands of this type. In keeping with these comments, it is generally accepted that molybdenum cycles between the Mo(VI) and Mo(V) oxidation states during the catalytic stages of electron transport in reactions are catalysed by xanthine oxidase and nitrate reductase. There is evidence[22] for the involvement of Mo(IV) in the noncatalytic reduction phase of xanthine-oxidase activity; the potent inhibitory effect of alloxanthine on xanthine oxidase has been attributed to the ability of this purine base to form a stable complex with catalytically inactive Mo(IV)[23]. The extreme sensitivity of bacterial nitrogenase to oxygen, the inhibitory effect of carbon monoxide and the necessity to account for the transfer of six electrons to the nitrogen molecule could all be accommodated by the assumption that molybdenum in the reduced enzyme is in a low oxidation state, probably Mo(III).

13.2 ELECTRONIC SPECTRA

Absorption in the visible or near u.v. region in the spectra of transition-metal complexes may arise from d–d transitions of the metal cation, or from charge-transfer transitions either from ligand to metal or vice versa. Although d–d transitions are formally forbidden, absorption bands can gain intensity from intense charge-transfer bands, the gain being greater the lower the symmetry of the complex.

The electronic spectra of representative complexes of Mo(III), Mo(IV) and Mo(V) are summarised in table 22[24]. In most cases the d–d absorption bands are very weak and are often partly obscured by intense charge-transfer bands. For a

TABLE 22 ELECTRONIC SPECTROSCOPIC PARAMETERS OF SOME
REPRESENTATIVE MOLYBDENUM COMPLEXES

Mo(III)	$^2E_g, \, ^2T_{2g} \leftarrow \, ^4A_{2g}$[†]	$^2T_{2g} \leftarrow \, ^4A_{2g}$	$^4T_{2g} \leftarrow \, ^4A_{2g}$	$^4T_{1g} \leftarrow \, ^4A_{2g}$
$MoCl_6{}^{3-}$	9 650[‡]	14 800	19 200	24 000
$MoBr_6{}^{3-}$	9 500	14 500	18 300	23 200
			(33)[§]	
$MoI_6{}^{3-}$	9 200	13 800	16 650	20 050
$Mo(urea)_6{}^{3+}$			25 900	31 000
$Mo(acac)_3$			23 300	27 500
$MoCl_3(MeCN)_3$	8 400?	13 500	25 200	28 900
	8 750			

Mo(IV)	$^3T_{2g} \leftarrow \, ^3T_{1g}$	$^3T_{1g}(P) \leftarrow \, ^3T_{1g}$	Charge-transfer band
$MoCl_6{}^{2-}$	22 000	25 800	28 400
			33 900
			36 760
			41 300
$MoCl_4 \cdot 2PPh_3$	15 600	25 000	27 800
	20 000		
$MoCl_4 \cdot 2py$	21 200?	24 700	
$MoCl_4 \cdot 2Ph_3PO$	20 800?	27 000	
$MoCl_4(bipy)$	18 300	25 200	27 200 (1 200)
	(690)	(800)	33 100 (12 000)
			41 700 (14 000)

Mo(V)	$^2E \leftarrow \, ^3B_2$	$^2B_1 \leftarrow \, ^2B_2$
$MoOCl_5{}^{2-}$	13 800	23 000
$MoOBr_5{}^{2-}$	14 300	21 300
	(7)	(560)
$MoOCl_3 \cdot 2Ph_3PO$	13 500	22 300
	(19)	(11)
$MoOCl_3 \cdot 2MeCN$	13 700	19 000
$MoOCl_3(bipy)$	13 600	18 800
		23 000
$MoO(NCS)_5{}^{2-}$	12 800	19 200

† tentative assignments.
‡ absorption maxima (cm^{-1}).
§ figures in parentheses are molar absorption coefficients.

series of Mo(V) complexes the charge-transfer band moved to longer wavelengths
with increasing donor power of the ligand[21]; this parallels the behaviour of the
charge-transfer band in corresponding series of Fe(III) complexes and indicates
that as with these[25] the charge transfer is from ligand to metal. The strong interaction between both Mo(V) and Fe(III) and their respective ligands and the
common association of iron with molybdenum in enzymes is a strong indication
that these metals are principal components of an electron-transfer chain. Of the
molybdenum-containing enzymes, only xanthine oxidase has been studied in

detail of electronic spectroscopy. Absorptions due to d-d transitions of molybdenum are obscured by the more intense absorptions of the iron and flavin chromophores (see section 14.1). The absorption, circular dichroism and magnetic circular-dichroism spectra of xanthine oxidase and the Mo(V) model complex sodium cysteinato molybdenum(V) show little resemblance, again indicating the dominant influence of the iron and flavin components of the enzyme[26].

13.3 ELECTRON PARAMAGNETIC RESONANCE SPECTROSCOPY

Electron paramagnetic resonance spectroscopy has contributed more than any other technique to an understanding of molybdenum enzymes[27]. Molybdenum e.p.r. signals have characteristic g-values, hyperfine splittings and power-saturation characteristics that distinguish them from signals due to, say, flavin radicals and paramagnetic iron centres. The six-line hyperfine structure characteristic of molybdenum signals arises from the coupling between the unpaired electron and the nuclear spin of the isotope ^{95}Mo ($I = 5/2$), which constitutes 25 per cent of natural molybdenum. The other 75 per cent consists of the isotopes ^{94}Mo, ^{96}Mo and ^{98}Mo ($I = 0$). Isotope substitution has been used in order to enhance the contribution from the six-line hyperfine splitting or to establish whether a signal is due to molybdenum or some other paramagnetic species.

Mo(III) and Mo(V) are the only paramagnetic molybdenum species and since signals due to the former are likely to lose intensity by exchange broadening, it is probable that only signals due to Mo(V) have been observed so far. Whereas the appearance of an e.p.r. signal may indicate the presence of a Mo(V) species, the absence of a signal does not prove the absence of this oxidation state. For example, monomeric Mo(V) shows a strong tendency to form diamagnetic dimers. Weak signals in the spectra of molybdenum compounds have been attributed to small amounts of monomeric Mo(V) in equilibrium with diamagnetic dimer (see chapter 15). Such equilibria are strongly solvent dependent. An alternative source of signal is a paramagnetic dimer, that is a triplet with two unpaired electrons. The uncoupling of the electron spins leading to the conversion of the diamagnetic dimer into the paramagnetic dimer could occur by twisting about a bond between molybdenum and a bridging atom in a bridged dimer.

In summary, e.p.r. can be used: to detect paramagnetic Mo(V); to monitor changes in its concentration during catalytic reactions of the enzyme; to give information regarding the nature of the ligands binding to the metal and the symmetry of its environment and hence to distinguish between molybdenum ions in different chemical environments.

The striking similarity between the e.p.r. parameters of certain thiol complexes of Mo(V)[28] and those of molybdoflavoprotein enzymes (table 23) provides some of the strongest evidence yet available that enzymatic molybdenum is bound by sulphur ligands. In octahedral Mo(V) complexes no nonthiol ligand produces a g-value greater than 1.95, or a hyperfine coupling constant lower than 4.7 millitesla.

TABLE 23 E.P.R. PARAMETERS FOR SOME MOLYBDENUM-CONTAINING
ENZYMES AND REPRESENTATIVE Mo(V) COMPLEXES

Enzyme		g_{av}[†]	A (millitesla)[‡]	Ref.
aldehyde oxidase		1.97	6.1	60
nitrate reductase[§]		1.97	–	74
nitrogenase[‖]		1.97	–	85
sulphite oxidase		2.00, 1.968	6.25, 4.68	109
xanthine oxidase[¶]	γδ	1.977	3.4	28
	αβ	1.977	4.1	28

Ligand[††]			
HSCH$_2$COOH	1.978[‡‡]	3.8	28
	1.987[§§]	4.1	
	2.006[‖‖]	3.4	
$\overset{+}{N}H_3CH(CH_2SH)COO^-$	1.975	3.5	118
glutathione	1.951	3.2	119
HS(CH$_2$)$_2$SH	2.002	3.0	28
HS(CH$_2$)$_3$SH	1.993	3.5	28
HSCH$_2$CH(SH)CH$_2$OH	2.002	–	16
CF$_3$C(S):C(S)CF$_3$	2.0097	1.22	16
cyanide[¶¶]	1.99	3.5	16
EDTA	1.936	5.6	16
CH$_3$CH(OH)COOH	1.94	–	16
HOCH(CH$_3$)COOH	1.921	5.6	28
chloride	1.949	4.7	28

† parameters refer to molybdenum signal only.
‡ ^{95}Mo hyperfine splitting.
§ from *N. crassa*.
‖ molybdoferredoxin component.
¶ γδ corresponds to the Very Rapid Signal and αβ to the Rapid signal.
†† ligand given because of uncertainties in the formulae of some complexes.
‡‡ initial signal.
§§ signal after about 20 hours.
‖‖ intermediate signal, maximum intensity after about 5 hours.
¶¶ in Mo(CN)$_8^{3-}$.

An interpretation in terms of delocalisation of the unpaired electron into sulphur-ligand orbitals allows for the simultaneous reduction of the h.f.s. and the increase in *g* towards the free-electron value for the sulphur complexes. In view of these similarities in e.p.r. parameters, complexes of molybdenum with sulphur-containing ligands have been studied extensively as enzyme models (see chapter 15).

14 MOLYBDENUM-CONTAINING ENZYMES

14.1 XANTHINE OXIDASE

Purine bases such as adenine and guanine, arising from the degradation of nucleic acids are conveyed to the liver in the blood. Deamination produces hypoxanthine and xanthine, respectively, and these are ultimately oxidised to uric acid

$$\text{xanthine} + [O_2] \longrightarrow \text{uric acid (enol form)} + H_2O_2 \quad (14.1)$$

Xanthine oxidase is the enzyme that catalyses such oxidations. Uric acid levels in the blood tend to rise when cells are being destroyed and nucleoprotein liberated, for example in leukaemia and in some stages of pneumonia. Xanthine oxidase also catalyses the oxidation of aldehydes to carboxylic acids; as with purines, these oxidations are formally hydroxylations in which the hydroxyl function derives from the solvent. Although xanthine oxidase is widely distributed in mammals, milk and calf liver being particularly good sources, its biological significance is not clear. Since aldehydes do not appear in appreciable quantities in mammals and since the oxidation of purines by molecular oxygen is among the fastest reactions catalysed by xanthine oxidase, it is assumed that the catalysis of purine and aldehyde oxidations are the principal roles of the enzyme. However, the absence of xanthine-oxidase activity does not appear to have severe pathological consequences, at least in humans. A patient apparently totally lacking xanthine-oxidase activity suffered no more than a xanthine stone in the urine[29].

A combination of the ease of isolation and purification, and the relatively high stability of xanthine oxidase, have made it the most studied of the molybdenum enzymes. Molecular oxygen is the physiological terminal electron acceptor in oxidations catalysed by xanthine oxidase but cytochrome c, ferritin and methylene blue may also function in this capacity. The substrate specificity of the enzyme is also low. More than one hundred substances serve as substrates; these are principally purines or related heterocyclic compounds and aldehydes, but also include NADH and some quaternary nitrogen compounds, for example N-methylnicotinamide. Binding constants for aldehyde substrates are some one hundred times greater than those for purines.

Milk xanthine oxidase contains 2 Mo atoms, 2 FAD molecules and 8 Fe atoms per molecule of molecular weight 275 000[30]. Although oxygen is the most probable terminal electron acceptor in *in vivo* hydroxylation reactions catalysed by xanthine oxidase, [18]O radio-tracer studies by Mason and Onoprienko[31] have shown that the oxygen incorporated into the product molecule derives from the reaction medium and not from the oxygen molecule. Thus xanthine oxidase is an example of an oxygen-facultative electron-transfer oxidase[32].

Investigations of the ligand environments of the metal atoms in xanthine oxidase are difficult since neither Mo nor Fe[9,33] can be removed reversibly or replaced by metal ions that can act as electronic spectroscopic probes[34]. However, the native enzyme exhibits absorption in the visible region. Although there is some conflict in the literature concerning the chromophore responsible for the absorption, all workers agree that it is not the enzyme's molybdenum component. Xanthine oxidase substantially free from iron (0.3 g. atom iron/mole protein) was reported to have an absorption spectrum almost identical to that of the native enzyme[35]. However, similarities between the absorption, circular dichroism and optical rotatory dispersion spectra of xanthine oxidase and spinach ferredoxin (figure 36) were held to imply the existence of similar chromophores—namely iron–sulphur moieties—in the two proteins. The greater intensities of the absorptions from xanthine oxidase were attributed to the presence of eight atoms of iron in this molecule compared to only two in spinach ferredoxin. The molybdenum of methanol-inactivated xanthine oxidase has been shown by e.p.r. spectroscopy to be Mo(V); in the oxidised active enzyme it is probably Mo(VI). Differences in the optical properties of the active and inactive enzymes due to the change from Mo(VI) to Mo(V) might be expected. However, the absorption, circular dichroism and optical rotatory dispersion spectra of the oxidised and dithionite-reduced methanol-inactivated xanthine oxidase were found to be identical, indicating that molybdenum makes an insignificant contribution to the optical properties of the enzyme[36]. The 450 nm absorption of xanthine oxidase disappears on reduction by substrate. Early work by Morell[37] showed that this bleaching occurs in two phases, one rapid and one slow. The rate of the slow-phase bleaching was too low for it to be involved in the catalytic reaction. Morell suggested that the slow phase corresponds to a reaction of an inactive form of the enzyme; this has been in dispute[38-40] but the elegant e.p.r. work of Bray and Swann[17,22] largely supports Morell's original proposal.

The potential complexity of the electron-transport sequence in xanthine oxidase is illustrated in figure 37. Any of the three nonprotein components of xanthine oxidase can in principle act as a binding site for substrate or terminal electron acceptor. Indeed, with xanthine as substrate and phenozine methosulphate as electron acceptor there is a common binding site, namely molybdenum[40,41].

That molybdenum is the general substrate binding site, with a few exceptions such as NADH, is strongly supported by kinetic studies of the inhibition of xanthine oxidase by methanol[42,43]. These established the locus of methanol inhibition as the substrate binding site and furthermore that progressive inactivation of the enzyme by methanol occurred only in the presence of reducing substrate, indicating that inactivation occurs via binding of methanol to the reduced enzyme. Since neither the flavin nor the iron components of xanthine oxidase are reduced in the

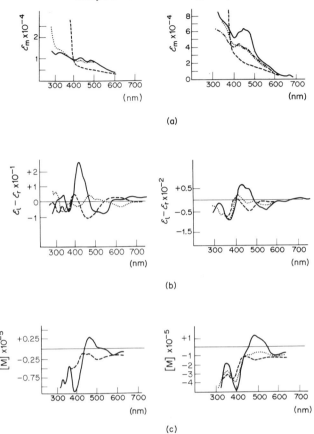

Figure 36 The optical properties of spinach ferredoxin (left) and xanthine oxidase (right): (a) absorption spectra, (b) circular dichroism, (c) optical rotatory dispersion (note differences in scale); ———, oxidised; . . . . , reduced with xanthine; - - - - - -, reduced with dithionite; —·—·—·—, oxidised flavin and molybdenum-free enzyme. From Garbett *et al.*[36]. The greater intensities of the xanthine oxidase spectra are attributed to the presence of eight iron atoms per molecule compared with only two in the case of the spinach ferredoxin.

presence of methanol, it seems likely that the methanol binding site is molybdenum in a low oxidation state. The involvement of molybdenum(V) in the catalytic cycle and, in particular, in substrate binding is very strongly supported by the rapid-freezing e.p.r. work of Bray and coworkers, which is the principal source of information on the details of the electron-transport chain of xanthine oxidase.

Resting or oxidised xanthine oxidase in the absence of substrate does not give an e.p.r. signal[44,45]. A complex pattern of signals appears in mild reduction with substrate and disappears again on stronger reduction. In initial e.p.r. experiments, the signals were identified with Mo(V), flavin semiquinone radical and iron, in order of appearance[44,46]. Subsequent work by Massey and coworkers has shown that the deflavoenzyme retains substrate dehydrogenase activity, but has lost its oxygen reductase activity[47]. This implicates flavin as the oxygen binding site. The function of the iron may be to store reducing equivalents or to transfer them between Mo and flavin. This latter role is indicated by e.p.r. experiments and stopped-flow monitoring of the visible-absorption spectrum, which show that the redox reaction of iron occurs within the time scale of the catalytic reaction[38,44,48]. The initial e.p.r. signals (figure 38) exhibit g values and hyperfine splittings typical of Mo(V) complexes, especially those with sulphur ligands; this provides the most convincing evidence that the e.p.r. signals are due to Mo and that the molybdenum is bound to sulphur ligands in the enzyme. Further support for this comes from

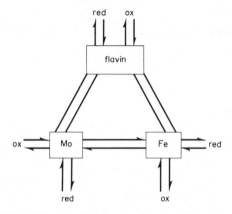

Figure 37 Possible routes for the transfer of oxidising and reducing equivalents between the components of xanthine oxidase and its oxidising (ox) and reducing (red) substrates.

the inhibition of xanthine oxidase by Cu(I) and thiol reagents. The individual components of the e.p.r. signals due to Mo in different chemical environments have been obtained by a careful choice of reducing conditions[17,22]. The signals and the conditions under which they are formed are shown in figure 38 and their parameters are summarised in table 24. The appearance of Mo(V) signals at a very early stage of the enzyme's catalytic cycle provides strong evidence for the binding of substrates to molybdenum. For example, with xanthine as substrate the maximum intensity of the Very Rapid signal is achieved in a time short compared with the turnover time[44]. Furthermore a substrate deuterated specifically on the carbon atom at which oxidation occurs (C-8) causes the Rapid signal to appear initially in its deuterium form[49,50]; within one second at 10°C the signal had changed to the hydrogen (doublet splitting) form indicating transfer of deuterium to the enzyme followed by exchange with water protons[50]. This process must be occurring at a site interacting with and therefore close to the Mo atom.

Every substrate for xanthine oxidase tested to date gives rise to a Rapid signal. This is a complex of signals representing as many as four individual e.p.r.-active species. The intensity of the Rapid signal increases with enzyme activity, as monitored by activity: absorbance ratios[51], and therefore it is due to the active enzyme. Reduction of xanthine oxidase by dithionate, formaldehyde or salicyl-

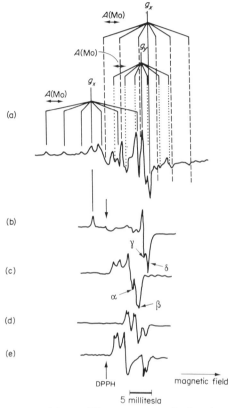

Figure 38 Molybdenum(V) e.p.r. spectra of reduced xanthine oxidase: (a) Very Rapid signal†, ^{95}Mo-enriched enzyme; enzyme reduced with xanthine at pH 10 for 10 ms; (b) Very Rapid signal†, native enzyme; reducing conditions as for a; (c) Rapid signal‡, native enzyme reduced with dithionite at pH 8.2 for 1 s; (d) Slow signal, enzyme reduced with dithionite at pH 8.2 for 30 min; or inactive enzyme; (e) inhibited signal, enzyme dialysed against aerated 1 M methanol containing salicyl-aldehyde for several days at 5°.

aldehyde produced identical Rapid signals, consisting of two components and indicating two inequivalent molybdenum(V) sites. Since neither dithionate nor its oxidation products were considered likely to bind to the enzyme, it was suggested

† Also referred to as γδ signal in early literature
‡ Also referred to as αβ signal in early literature

TABLE 24 E.P.R. PARAMETERS OF MOLYBDENUM(V) SPECIES FROM MILK XANTHINE OXIDASE
(HYPERFINE SPLITTINGS IN MILLITESLA)

Signal	g_{av}	$A(^{95}\text{Mo})$	$A(^1\text{H})$	$z(\parallel)$			$y(\perp)$			x		
				$A(^{95}\text{Mo})$	$A(^1\text{H})$	g	$A(^{95}\text{Mo})$	$A(^1\text{H})$	g	$A(^{95}\text{Mo})$	$A(^1\text{H})$	g
Very Rapid	1.977	3.4	–	4.1	–	2.025	2.4	–	1.956	3.7	–	1.951
Rapid, no complex { A	1.974	–	1.4	–	1.4	1.991	–	1.4, 0.3	1.968	–	1.4	1.963
{ B	1.973	–	1.4	–	1.4	1.991	–	1.4, 0.3	1.966	–	1.4	1.963
Rapid, complex { I	1.974	–	1.2	6.4	1.2	1.989	–	1.2, 0.4	1.969	–	1.2, 0	1.964
{ II	1.973	–	1.2, 1.2	–	1.2, 1.2	1.994	–	1.2, 1.2	1.968	–	1.2, 1.2	1.961
Slow	1.967	–	1.6	–	1.6	1.975	–	1.6	1.970	7.0	1.6	1.957
Inhibited	1.973	4.6	0.46	5.7	0.44	1.989	2.5	0.39	1.977	5.7	0.56	1.953

that these signals are due to uncomplexed active enzyme and that if the enzyme has two independent catalytic sites the chemical environments of the molybdenum atoms in these sites are not identical. With xanthine as reducing agent two sets of signals were observed, two signals corresponding to the reduced active enzyme

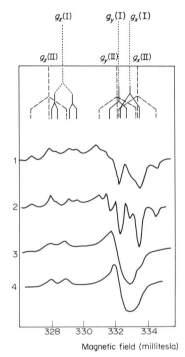

Figure 39 Rapid Mo(V) e.p.r. signals obtained on reducing xanthine oxidase at pH 10 with xanthine for 1 min at about 20 °C (15 moles xanthine per mole enzyme). Spectra are at 9.1 GHz.

1. Experimental spectrum, H_2O solvent; 2. computer simulation of 1 based on an interpretation involving two Mo(V) species—see text and table 24; 3. experimental spectrum, D_2O solvent; 4. computer simulation of 3.

The three triplets (dotted-stick spectra) are g_z, g_y and g_x of complex (II) showing splitting due to two equivalent protons. The doublet of doublets (solid-stick spectra) is g_z of complex (I), showing splitting by two nonequivalent protons. The other two doublets are g_y and g_x of complex (I), showing zero coupling to one of the protons. Redrawn from reference 17.

appearing at very low xanthine concentrations and two signals corresponding to complexes of xanthine and the reduced enzyme at higher xanthine concentration. This indicates more than one binding site for xanthine: once the molybdenum has been reoxidised via flavin and iron the bound xanthine can be oxidised. A comparison of the spectra obtained using H_2O and D_2O as solvent (figure 39) led Bray and Swann[17,22] to suggest the occurrence of interaction between the Mo(V) e.p.r.-active species and exchangeable protons. For one species the protons are equivalent and for the other they are nonequivalent. However, the precise nature of the complexes remains obscure. Further deuteration studies established that the Very Rapid signal species, which predominates at high pHs, does not interact with protons. The interpretation offered is that the Very Rapid- and Rapid-signal species are simply high- and low-pH forms of the same intermediate.

Quantitative studies on the Rapid signal have thrown some light on the sequence of oxidation-state changes of the molybdenum atom. Two types of experiment

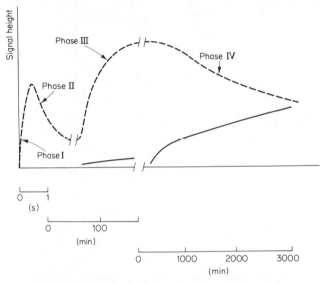

Figure 40 Phases in the reduction of xanthine oxidase at pH 8.2 by xanthine (20 moles of substrate per mole of enzyme). (Redrawn from ref. 17.)

were performed, one where the intensity of the Rapid signal was measured as a function of xanthine concentration at a fixed reaction time (1–2 mins) and another where the change in signal was monitored as a function of time at fixed xanthine concentrations. Maximum signal intensity occurs for about 4 moles xanthine per mole active enzyme; above this, intensity rapidly diminishes, reaching almost zero at about 10 moles xanthine per mole of enzyme. With an excess of xanthine ($\approx$ 20 moles per mole of enzyme) signals appear within milliseconds of substrate addition; they undergo a complex sequence of changes (figure 40), which is still incomplete after two days. A transient Very Rapid signal is followed by development of Phase I of the Rapid signal within about 200 ms. This fades during 1 s to about half the original intensity in Phase II and then increases to higher than the original intensity over about 2 h in Phase III. At longer reaction times (Phase IV) the Slow signal of the inactive enzyme[52,53] develops at the expense of the Rapid signal. Since only Phase I occurs within a time comparable to the enzyme turnover time it seems that only Phase I reflects changes occurring during the catalytic step.

The explanation put forward by Bray and Swann[17,22] of the various signal phases is that the resting enzyme contains Mo(VI), which is reduced initially to Mo(V) (Phase I), and then overreduced to Mo(IV), which accounts for the Phase II fading. This fading requires the presence of excess substrate and since the Mo(V) species is stable in Phase III, the Mo(IV) species must be being produced by a reducing agent that is only present while the substrate is being reduced. Thus substrate radicals are the prime candidates for this reducing agency. There are several lines of evidence supporting the formation of substrate radicals in reactions of xanthine oxidase. These include the auto-oxidation of sulphite by a system containing xanthine and xanthine oxidase[54] and the induction of chemiluminescence in luminol[†] by xanthine oxidase in the presence of xanthine and oxygen[55]. Such reactions are characteristic of the superoxide anion (O_2^-)[56]. This free radical of molecular oxygen has been implicated in xanthine-oxidase activity. The formation of substrate radicals necessitates a one-electron reduction of Mo(VI) rather than the alternative two-electron reduction mechanism leading to Mo(IV) and oxidised substrate. In the latter case Mo(V) could be produced by intramolecular redox processes since flavin and iron are reduced. However, the sequence of appearance of signals on reduction of the enzyme is Mo(V) → flavinsemiquinone radical → iron[44,46], suggesting that the two-electron mechanism is incorrect. The appearance of Phase III of the Rapid signal following Phases I and II initially indicates nonequilibrium conditions in the system. The rate of the development of the Phase III signal is proportional to the content of inactive enzyme. Kinetic evidence implicates the inactive enzyme in the rate-determining step, and Bray and Swann suggest that electron transfer occurs from Mo(IV) of active enzyme to iron–sulphur and flavin in the inactive form (figure 41). Support for this suggestion comes from the marked increase in iron–sulphur signals during Phase III for low-activity samples.

Further evidence for the occurrence of a Mo(IV) species in the reduction of xanthine oxidase comes from the work of Massey and his associates on the inactivation of the enzyme with pyrazolo-3,4-pyrimidine derivatives such as alloxanthine[57]. Alloxanthine (X) is the product of the hydroxylation of allopurinol (XI). Both allopurinol and xanthine reduce xanthine oxidase, the former

(X)
Alloxanthine

(XI)
Allopurinol

† luminol = 5-amino-2,3-dihydrophthalazine-1,4-dione.

to the level corresponding to the Rapid phase of xanthine-oxidase reduction; while over several hours xanthine causes further reduction to the level achieved by dithionite. Alloxanthine abolishes the slow reappearance of Mo(V) signals. This is consistent with reduction to Mo(IV), which is stabilised by alloxanthine but not by any of the oxidation products of xanthine. The further reduction observed with xanthine could be the reduction of Mo(VI) by Mo(IV). Addition of xanthine to enzyme previously incubated with alloxanthine causes rapid initial reaction, which ceases after about 2 minutes. The colour of the original enzyme returns on the admission of oxygen; indeed the spectrum of the reoxidised inhibited enzyme is identical to that of the native enzyme above 500 nm, indicating that the iron-sulphur chromophores are reoxidised. Furthermore, changes in the spectrum in

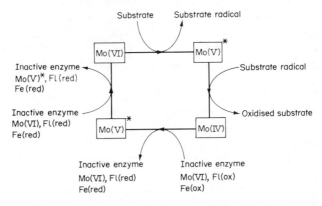

Figure 41 Representation of the scheme suggested by Bray and Swann for the reduction of xanthine oxidase by, for example, xanthine. Molybdenum e.p.r. signal-generating species are marked with an asterisk. In the active enzyme Mo is reduced before the flavin and iron components. In the inactive enzyme Mo is the last component to be reduced.

the region of 450 nm show that the flavin moiety is also rapidly reoxidised. However, the xanthine-oxygen reductase activity does not return, implicating the molybdenum as the binding site of the inhibitory alloxanthine. Interestingly, incubation of the alloxanthine-inhibited enzyme with the electron acceptors phenazine methosulphate or ferricyanide completely restores enzyme activity. This is consistent with these acceptors binding at molybdenum, which is oxidised and the alloxanthine liberated. The failure of Mo(VI) to bind alloxanthine is suggested by the insignificant spectral perturbation caused by its addition to the native enzyme. In contrast the binding of alloxanthine to the reduced enzyme causes marked changes in the spectra of the reduced enzyme and partially re-oxidised enzyme.

Attempts to determine the oxidation state of molybdenum in the alloxanthine-inhibited enzyme by titration with ferricyanide were complicated by the presence of an inactive form of the enzyme and also by the labile sulphydryl group[57]. This was blocked by phenylmercuric acetate. The amount of inactive enzyme was determined from the stoichiometry of the alloxanthine inhibition, 0.73 moles of inhibitor per mole of enzyme being required. Radioisotopic investigation of ^{14}C-alloxanthine binding gave the same result. Evidently alloxanthine binds only to complete functional-active sites. The difference between these and the non-functional sites is not known. The ferricyanide titration experiments, after correction for inactive enzyme, showed that two equivalents of electrons per mole of enzyme had been consumed and that the reoxidation Mo(IV) → Mo(VI) accompanies reactivation of the enzyme. This receives further support from the stoichiometry of the enzyme–allopurinol reaction. Three moles of alloxanthine are produced per mole of active site. From the absorption and e.p.r. spectroscopic results, it is evident that allopurinol reduces the iron–sulphur and flavin chromophores. Each iron–sulphur chromophore can accept one electron. Hence the iron-sulphur chromophores together with the flavin prosthetic group account for the production of 2 moles of alloxanthine. Reduction of Mo(VI) to Mo(IV) would then account for the formation of the third mole of product.

The picture of the mechanism of xanthine-oxidase activity, which has emerged chiefly from e.p.r. studies, is of an enzyme with two independent catalytic sites. Molybdenum is the binding site for reducing substrates, and during the enzymic reaction it is reduced from Mo(VI) to both Mo(V) and Mo(IV). Electron transfer to the terminal electron-donor flavin may be direct or via an intervening iron-sulphur moiety. Although flavin is the principal binding site for oxygen, the physiological oxidising substrate, other acceptors may bind elsewhere, for example phenazine methosulphate at molybdenum and ferricyanide at iron.

Many challenging problems remain unsolved regarding the detailed behaviour of xanthine oxidase. For example, it is quite possible that the differences between the active and inactive forms of the enzyme arise from a subtle change in the ligand environment of the molybdenum but the details remain obscure. Further, the hydroxylation reaction involves the formal transfer of two electrons and a proton from the substrate. This could occur in a concerted single step or in two separate steps. Since Mo(V) is implicated as the catalytically significant oxidation state of molybdenum and since there is strong presumptive evidence for the formation of substrate radicals, the two-step process seems more likely but the detailed mechanism is unknown.

Another problem concerns the proton-accepting group of the enzyme. The proton interactions observed in the e.p.r. spectra indicate that the proton-acceptor group is at least near to the molybdenum and may even be one of its ligands.

If, as seems very likely, conventional spectroscopic methods have yielded as much information as they can on xanthine oxidase, solutions of the problems outlined above, essential for a better understanding of the relation between structure and function in the enzyme, must await a determination of its structure by X-ray crystallography.

14.2 ALDEHYDE OXIDASE

There are some striking similarities between rabbit-liver aldehyde oxidase and milk xanthine oxidase. These include: the ability to catalyse the hydroxylation of aldehyde and N-heterocyclic substrates[58]; the utilisation of molecular oxygen, various dyes and cytochrome c as oxidising substrates; the identical content of FAD, Fe and Mo[59]; similarities in the behaviour towards methanol inhibition[43]; the production of e.p.r. signals characteristic of Mo(V) on reduction[60,61]; the generation of superoxide-anion radicals in the reduction of O_2[54]. However, there are also significant differences, which establish the separate identity of aldehyde oxidase, contrary to an earlier report[62] that it is a form of xanthine oxidase. Rajagopalan *et al.* found that in addition to its protein, Fe, Mo and FAD components, rabbit-liver aldehyde oxidase contains one or two molecules of coenzyme Q, or a similar quinone-type molecule, per molecule of enzyme[59,63]. They also observed inhibition of aldehyde oxidase by typical inhibitors, for example amytal and some steroidal hormones, of mitochondrial electron transport and on this basis suggested that the coenzyme Q component may be an integral part of the electron-transport system of the enzyme. Other differences between the two oxidase enzymes were illustrated by further studies.

When aldehyde oxidase is incubated with substrate, both FAD molecules are reduced; under comparable conditions, only one FAD of xanthine oxidase undergoes reduction. Xanthine is not a substrate for aldehyde oxidase, while purine is oxidised to 8-hydroxypurine rather than to uric acid, the product of xanthine-oxidase action on this substrate. These differences show that the principal factor determining the outcome of the enzymatic oxidation of purine and its derivatives is the environment of the substrate binding site rather than the electronic structure of the substrate[64]. Identical K_i values for the competitive inhibition of enzyme-substrate interaction with aldehyde and N-heterocyclic substrates and a wide range of inhibitors strongly favours a common binding site for these two classes of substrate. In view of the compelling evidence that reducing substrates are bound at molybdenum in xanthine oxidase it seems highly probable that this is also the substrate binding site of aldehyde oxidase. This receives powerful support from the striking similarity of the e.p.r. signals of inhibited milk xanthine oxidase and aldehyde oxidase[60,65].

A detailed study of the differential reductase behaviour exhibited by various combinations of aldehyde oxidase, inhibitor and electron acceptor led Rajagopalan and Handler[63] to propose the existence of a linear sequence of at least four electron carriers in the enzyme, with the reduction of molecular oxygen occurring terminally in the sequence. These four carriers are presumably Mo, coenzyme Q, FAD and Fe. Further evidence relating to the nature of the electron-transport chain was obtained from e.p.r. experiments. Aldehyde oxidase differs from xanthine oxidase in that the resting enzyme shows a weak Mo(V) e.p.r. signal ($g = 1.97$) corresponding to about 3 per cent of the total enzymic molybdenum. When the enzyme is incubated with excess substrate, the intensity of the signal increases but only to a level corresponding to about 15–20 per cent of the total Mo. Under comparable conditions, 24 per cent of the flavin is in the semiquinone form (g = 2.00) and 25 per cent of the iron in the reduced form[60, 61].

In the early stages of reoxidation of the enzyme by air, all the signals diminish in intensity; this is fastest for the iron and slowest for the Mo(V), with the radical signal at a stage between the other two. These results show that the iron component of aldehyde oxidase is closest to the oxygen binding site and molybdenum the farthest but it does not reveal the relative positions of the flavin and coenzyme Q components in the electron-transport chain. E.P.R. experiments on equilibrium mixtures of aldehyde oxidase and N-methylnicotinamide as reducing substrate have indicated that about 4 moles of substrate per mole of FAD are required to elicit the maximum Mo and Fe signals[60]. Since the substrates are two-electron donors, this corresponds to the uptake of approximately eight electrons per catalytically active unit. Furthermore the electrons from the first 1–2 moles of substrate appear to be consumed in reducing the iron before any *permanent* reduction of the molybdenum occurs. This is further support for iron as the site of electron egress from the enzyme.

Although xanthine oxidase and aldehyde oxidase have a number of common substrates, which bind to a molybdenum binding site, there must be subtle variations in this binding site to account for differences in the behaviour of the two enzymes. The presence of a sulphydryl function at the catalytic site of aldehyde oxidase is implicated by the potent inhibitory effect of cyanide and *p*-CMB[†61]. Indeed in the absence of substrate, aldehyde oxidase is inhibited at cyanide concentrations one order of magnitude less than those used for xanthine oxidase. Inhibition by *p*-CMB, however, only occurs in the presence of substrate; that is to say, it is a competitive inhibitor of aldehyde oxidase but it inhibits xanthine oxidase noncompetitively[66].

As was pointed out earlier the e.p.r. spectra of methanol-inhibited xanthine oxidase and aldehyde oxidase are very similar. However, the proton splittings are slightly larger for the latter enzyme.

All the available evidence suggests that the binding sites for reducing substrates in xanthine and aldehyde oxidases are very similar and it is likely that they differ by virtue of only a small variation in the ligand environment of the molybdenum atom. As with xanthine oxidase, a determination of the structure of aldehyde oxidase will be required to reveal the nature of this variation and the relative spatial distributions of the components of the electron-transport chain.

14.3 XANTHINE DEHYDROGENASE

Although the gross molecular properties of xanthine-dehydrogenase enzymes from a variety of sources closely resemble those of milk xanthine oxidase (table 21)

† *p*–CMB = *p*-chloromercuribenzoate.

their electron-acceptor specificities differ widely. Molecular oxygen is probably the physiological electron acceptor of the mammalian oxidase, but does not appear to act in this capacity to any significant extent for chicken-liver xanthine dehydrogenase or for bacterial dehydrogenases from *C. cylindrosporum* and *M. lactilyticus*. NAD seems to be the physiological acceptor for avian dehydrogenase[67], while ferredoxin can perform this function for the enzyme from *M. lactilyticus*[68] and possibly also for the clostridial enzyme. Xanthine dehydrogenase resembles aldehyde oxidase in converting purine solely to 8-hydroxypurine. One of the most interesting features of the bacterial dehydrogenase from *M. lactilyticus* is that the resting enzyme exhibits a large Mo(V) signal, which changes and even decreases slightly on reduction with substrate[43, 61]. One explanation for this is that the molybdenum remains as Mo(V) throughout the catalytic cycle. The slight decrease in signal could indicate some overreduction by substrate as in the case of xanthine oxidase. Thus the molybdenum in the *M. lactilyticus* enzyme could serve as a binding site for substrate and hydroxyl ions rather than as an electron-transfer agent. Alternatively, Mo(IV) may be the catalytic species. So far there is insufficient evidence to distinguish between these possibilities.

14.4 NITRATE REDUCTASE

Nitrate is the major source of nitrogen for most green plants and fungi. Biological processes by which soil nitrate is converted to ammonia for the synthesis of protein, nucleic acids and other cell constituents are known as assimilatory nitrate reductions. Nitrate may also be utilised in place of oxygen as the terminal electron acceptor in the anaerobic production of energy by certain bacterial species. The production of energy via this route is known as nitrate respiration or dissimilatory nitrate reduction. Some species of bacteria can carry out both processes, the relative amounts of the two nitrate-reducing systems depending on the growth conditions.

The assimilatory nitrate-reducing system from algae and higher plants has been shown to consist of two distinct enzymes, which participate in the sequential reduction of nitrate through nitrite to ammonia. A flavomolybdoprotein, NADH-nitrate reductase, catalyses the reduction of nitrate to nitrite while the reduction of nitrite to ammonia is catalysed by the iron–sulphur protein ferredoxin–nitrite reductase.

Nicholas and Nason[69] established that molybdenum is essential for the enzymatic reduction of nitrate to nitrite by the assimilatory nitrate reductase from *Neurospora* but not for the reduction of FAD by NADPH, since cyanide-inactivated, molybdenum-free and native enzyme preparations mediate this reaction equally well. Enzyme preparations containing molybdate catalysed the oxidation of $FMNH_2$ (structure (IX), page 211) while those containing reduced molybdenum formed from molybdate and $Na_2S_2O_4$-H_2 catalysed the enzymatic reduction of nitrate to nitrite. The enzyme also exhibits NADPH–cytochrome c reductase, FADH–nitrate reductase and reduced methyl viologen–nitrate reductase activities, indicating that it may be an enzyme complex of several tightly bound subunits.

More recently, Garrett and Nason[70, 71] showed that a b-type cytochrome, cytochrome b_{557} (*N. crassa*), concentrates in parallel with NADPH-nitrate reductase activity. Its reduction by NADPH shows the same flavin requirement as nitrate reductase activity, and it is insensitive to cyanide ion, although cyanide blocks its reoxidation by nitrate. On the basis of these results, Garrett and Nason proposed the following electron-transport sequence for assimilatory nitrate reductase with NADPH as electron donor

$$
\text{NADPH} \longrightarrow \text{FAD} \underset{\text{cytochrome c}}{\overset{\underset{\text{(metal?)}}{\text{cytochrome } b_{557} \longrightarrow \text{Mo} \longrightarrow \text{NO}_3^-}}{\bigg\langle}}
$$
(14.2)

Cytochrome c is included in this sequence because NADPH–nitrate reductase and NADPH–cytochrome c reductase are induced concomitantly in mycelia by nitrate[72]. These reductases have so far proved inseparable. The NADPH–cytochrome c reductase activity of the *Neurospora* enzyme is inhibited by *o*-phenanthroline and by 8-hydroxyquinoline but not by cyanide, azide or thiourea. NADPH–nitrate reductase activity is inhibited by all of these reagents. FADH– and reduced methyl viologen–nitrate reductase activities are inhibited by cyanide, azide and thiourea but not by the chelating ligands. Garrett and Nason suggested that since the iron of the porphyrin system in cytochrome b_{557} is unlikely to be influenced by metal-binding reagents a second metal component may be involved at an early stage in the electron-transport sequence, possibly near to the flavin. The inhibition of nitrate reductase by *p*-CMB and the restoration of enzymatic activity by sulphydryl reagents indicated the involvement of an SH group in the active site of the enzyme. The inhibitory effect of *p*-CMB is concentrated at the first stage of the electron-transport chain, and it was suggested[73] that the SH group is involved in binding the electron donor or oxidised flavin in the enzyme.

Deficiency and inhibitor studies showed that both iron and molybdenum are essential for enzymatic activity in the dissimilatory nitrate reductase from the denitrifying bacteria *Pseudomonas aeruginosa*. According to the difference spectrum of the dithionite-reduced enzyme, it contains cytochrome c. The cytochrome c requirement with NADH as electron donor was confirmed by Kinsky and McElroy[72] who showed that the NADPH–cytochrome c reductase and NADH nitrate reductase activities increase in parallel during purification of the enzyme. FAD is also a cofactor. A sample of the enzyme incubated with nitrate and NADH exhibited two e.p.r. signals, one at $g = 2.004$ and the other, a much weaker signal, at $g = 1.97$. The signals decayed with time of incubation but could be restored by adding more NADH or dithionite. They were tentatively assigned to flavin semi-quinone and Mo(V). These results coupled with the observation that nitrate-reductase activity was abolished in the absence of cytochrome c and restored on its addition, and that with NADH or reduced FAD as electron donor there was a cytochrome–c requirement but with Mp(V) or reduced cytochrome c there was not, led to the proposal of the following electron-transport sequence[74]

$$NADH \diagdown \quad FAD \diagdown \quad Fe(II) \diagdown \quad Mo(VI) \diagdown \quad NO_2^-$$

cytochrome c   (14.3)

$$NAD \diagup \quad FADH \diagup \quad Fe(III) \diagup \quad Mo(V) \diagup \quad NO_3^-$$

The cytochrome–c requirement of dissimilatory nitrate reductase and the inhibition of substrate reduction of nitrate or nitrite oxidation of cytochrome by inhibitors of the oxygen-based respiratory chain imply the involvement of similar, possibly common, pathways for the two processes. Indeed, there are several examples of competition between nitrate and oxygen respiration.

Recent e.p.r. studies of a dissimilatory nitrate reductase from another bacterial source *Micrococcus denitrificans*[75] are the first to be carried out on a homogeneous preparation of the enzyme. They support the earlier conclusions that molybdenum is located at the substrate binding site. The resting enzyme at 80 K exhibits an asymmetric e.p.r. signal ($g_\perp = 1.985$, $g_\parallel = 2.045$; see figure 42), which accounts for only 15 per cent of the total molybdenum content. This spin recovery is in the range observed for other molybdenum-iron-sulphur proteins. The Mo(V) signal is altered by its substrate, nitrate ion, the reduction product (nitrite ion) and the competitive inhibitor (azide ion) (figure 42). With nitrate, the $g_\perp$ component remains unchanged at $g = 1.985$, but $g_\parallel$ has been shifted to 2.090 and the signal shape has been altered. The signal shape is also affected by nitrite but the g values remain the same. The inhibitory azide ion causes a change in signal shape and also a shift of $g_\parallel$ to 2.077. In order to produce these changes, the small molecules must be bound directly to the molybdenum or at least close enough to it to significantly alter its ligand field. The initial Mo(V) signal of the enzyme is destroyed by sodium dithionite reduction; a new signal ($g_\perp = 1.999$, $g_\parallel = 2.023$) appears in its place. It was suggested that this is due to Mo(III). This is possible although the e.p.r. signal of the d^3 Mo(III) ion might be expected to lose intensity owing to exchange broadening. Since there does not appear to be an appreciable loss in intensity, either the ligand field is sufficiently asymmetric to diminish, or even abolish, exchange broadening, or, alternatively, the new signal is due to Mo(V) in a different ligand environment created by the reduction process. The Mo(V) signal is also abolished by enzymatic reduction with hydrogen and a hydrogenase preparation from *Desulfovibrio gigas*. However, in this case no new signal appears. In the absence of detailed kinetic data, no firm conclusions can be drawn regarding the nature of the catalytically active oxidation state of molybdenum in the bacterial dissimilatory nitrate-reductase enzymes, although the occurrence of an e.p.r. signal due to Mo(V) in the resting enzyme and its disappearance on enzymatic reduction implies a catalytic cycle involving Mo(V) and either Mo(IV) or Mo(III).

N. crassa has been reported to form both assimilatory and dissimilatory nitrate reductases when it is grown under aerobic[76] and anaerobic conditions[77], respectively.

The dissimilatory enzyme requires iron and molybdenum for its NADPH–nitrate-reductase activity and thus resembles the dissimilatory enzyme found in denitrifying bacteria. However, the most purified enzyme preparation was not inhibited by iron-chelating agents, contained less iron than less pure preparations and had no detectable cytochromes or flavin. NADPH did not function as an electron donor for the purified enzyme but benzyl viologen (N,N'-dibenzyl-4,4'-dipyridyl) did.

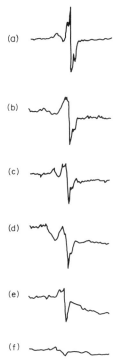

Figure 42 The e.p.r. spectrum of nitrate reductase at 80 K in 100 mM phosphate buffer (pH 7.4): (a) resting enzyme; (b) enzyme + nitrate ion; (c) enzyme + nitrite ion; (d) enzyme + azide ion; (e) enzyme reduced with dithionite; (f) enzyme reduced with hydrogen and hydrogenase.

These observations are consistent with the hypothesis[78] that two enzymes participate in NADPH–nitrate-reductase activity: one is an FAD-dependent NADPH diaphorase that can utilise a variety of oxidised compounds including cytochrome c as electron acceptor; the other is the terminal nitrate reductase proper, a molybdenum protein that can utilise reduced flavin nucleotides as electron donors.

More recently, Garrett and Nason[71] have reported that *N. crassa* cannot utilise nitrate as an electron acceptor in a terminal respiratory pathway or to form a respiratory type of nitrate-reductase enzyme. This surprising result, if confirmed, will raise some interesting questions relating to the nature of the nitrate binding site, commonly accepted to be molybdenum, and the nature of the relationship between the molybdenum prosthetic group and the other components of the electron-transport chain.

Studies of bacteria grown under molybdenum-deficient conditions showed that it is not essential for the synthesis of the protein components of nitrate reductase[79]. The whole reductase complex is synthesised but only the component exhibiting cytochrome-c reductase activity functions. It is not impossible to restore the nitrate-reductase activity by adding molybdenum in any common form to the molybdenum-deficient bacteria, indicating that the apoprotein is stabilised in a structure that forbids molybdenum binding to the active-site ligands.

Nason and coworkers[79,80] have established some interesting relationships between the subunits of nitrate reductase from *N. crassa,* which may have important implications for all the known molybdenum-containing enzymes. Combined cell-free extracts of nitrate-reductaseless mutant *nit-1* of *N. crassa* and either extracts of untreated uninduced wild-type organism or acid-treated molybdenum-containing enzymes were found to contain an assimilatory NADPH–nitrate reductase that is the same as the *Neurospora* wild-type enzyme. The *in vitro* restoration of NADPH–nitrate reductase activity causes a change in the nature of the cytochrome-c reductase activity, indicating that the enzyme is being formed by the association of subunits. Since the molybdenum-containing subunit can be supplied from a variety of sources including mammalian xanthine, aldehyde and sulphite oxidases, avian xanthine dehydrogenase and bacterial nitrogenases, the formation of a *Neurospora*-like enzyme must be determined by the subunit(s) provided by the *nit-1* mutant. Moreover, neither xanthine oxidase nor aldehyde oxidase have a cytochrome component so this too must be supplied by the *nit-1* mutant. In order to account for the ability of enzymes from various sources to supply the molyb-denum-containing subunit for the *in vitro* formation of assimilatory nitrate reduc-tase, Nason has proposed that the molybdenum is part of a relatively low molecular weight subunit, which can be considered as a cofactor and which is common to most if not all of the molybdenum-containing enzymes. Acid treatment of the enzymes is necessary to release this cofactor, which is then able to associate with the subunit(s) of the *nit-1* mutant to form the complete nitrate reductase. *Neuro-spora* extracts can supply the molybdenum subunit without prior acidification. This ready accessibility could be explained on the basis of the closely similar molybdenum binding sites in *nit-1* and wild-type extracts. An alternative possibility that different cofactors are involved but dissociate to provide a supply of molybdenum to *nit-1* does not seem very likely in view of the tenacity with which molybdenum enzymes bind the metal. Attempts to restore nitrate-reductase activity in *nit-1* by supplying molybdenum in the form of the model cysteine, cysteine ester and histidine complexes (XV, R = CO_2^-), (XII, R = Co_2Me) and (XXVII) (chapter 15) proved fruitless. This is not altogether surprising if the molybdenum in the enzyme is part of a mononuclear complex, since the stability of the dioxo bridges in (XV) and (XXVII) may preclude either the dissociation of the dimer or further complex formation with ligand groups on the protein. In this connection it is interesting to note that the $1:2$ molybdenum: cysteine complex (XII, $R = CO_2^-$) could not be obtained from the di-μ-oxo complex ($R = CO_2^-$) (see chapter 15). In the case of the mononuclear Mo(VI) complex (XII, $R = CO_2Me$) the ligands may be too tightly bound to allow the molybdenum to associate with the protein. A more realistic model would be provided by a mononuclear molybdenum complex containing one or more weakly held ligands. It would be interesting to see if monomeric molybdenum complexes produced *in situ* could restore the nitrate-reductase activity.

14.5 NITROGENASE

Nitrogenase is the enzyme responsible for the reductive fixation of atmospheric nitrogen to ammonia by bacteria and blue-green algae. This reduction is accomplished at atmospheric pressure, ambient temperature and in an essentially aqueous medium. In contrast, the industrial fixation of nitrogen by the Haber process requires high reaction temperatures and pressures. The ability of micro-organisms to reduce under mild conditions what has traditionally been regarded as a highly unreactive molecule, has aroused considerable interest in the nitrogenase enzyme. This arises not only from a desire to understand the fundamental biochemical processes involved in nitrogen fixation, but also to attempt to mimic the nitrogen-fixing ability of living organisms with synthetic catalysts and hence produce new routes to nitrogen-based compounds. Various aspects of the chemistry and biochemistry of nitrogen fixation have been reviewed elsewhere[3, 4, 81].

Nitrogen-fixing bacteria are of two kinds: symbiotic and asymbiotic (or free-living). Symbiotic bacteria are only able to fix nitrogen after they have infected the roots of certain plants, notably the legumes, and have changed to a degenerate form known as a bacterioid. These provide the plant with nitrogen in exchange for carbohydrate. Asymbiotic bacteria live and fix nitrogen in the soil but seem to be less important than the symbiotic type in terms of the relative amounts of nitrogen fixed. Despite this, most work has been done on the asymbiotic bacteria primarily because they are easier to grow and study.

Asymbiotic bacteria may be aerobic, for example *Azotobacter vinlandii*, anaerobic, for example *Clostridium pasteuranium*, or facultative, for example *Klebsiella pneumoniae*. However, all three types will only fix nitrogen in the strict absence of oxygen.

Nitrogenase from *Azotobacter* was the first enzyme for which molybdenum was shown to be essential[82]. The bacteria grow in the absence of molybdenum provided there is an exogenous source of ammoniacal nitrogen. With added molybdenum and no ammoniacal nitrogen, nitrogenase activity develops within a few hours. Iron and magnesium are also necessary for bacterial nitrogen fixation. Doubtless, the latter functions in the ATP-dependent reactions associated with nitrogen fixation. Nitrogenase has now been isolated from *C. pasteuranium*[83, 84], *A. vinlandii*[85-7] and *K. pneumoniae*[88, 89]. In each case it can be separated into two proteins; protein 1, fraction 1, molybdoferredoxin or the Mo–Fe protein and protein 2, fraction 2, azaferredoxin or the Fe protein, respectively. The Mo–Fe protein of *A. vinlandii* has been isolated in a crystalline form[85]. It has a molecular weight between 270 000 and 300 000 and contains molybdenum, iron, cysteine and labile sulphide in the ratio $1:20:20:15$. The Fe protein has a molecular weight of about 40 000 and contains two atoms of iron and two sulphide groups per molecule. The individual proteins do not show nitrogenase activity, but even Mo-Fe and Fe proteins from different species can be combined to produce systems that will fix nitrogen in the presence of an ATP-generating system and a reducing agent. As the difference between species increases, the efficiency of nitrogen fixation decreases. Thus a common site for nitrogen fixation in enzymes from different species is indicated; the difference in efficiency arising from the increasing incompatibility of the proteins as the species become more widely separated.

Turnover of nitrogenase enzyme involves the transfer of reducing equivalents from a reducing substrate—pyruvate in the case of *C. pasteuranium*, but hydrogen gas or dithionite will also serve—to an oxidising substrate. Nitrogen is the physiological oxidising substrate but all preparations of nitrogenase when supplied with a reducing agent and ATP also exhibit hydrogenase activity towards substrates with multiple bonds, for example acetylenes, isonitriles and cyanide ion (see also section 15.3). In the absence of substrate and to a lesser extent in its presence, hydrogen is evolved, and ATP is hydrolysed to ADP and inorganic phosphate. Hydrogen evolution is not inhibited by CO; nitrogen fixation is strongly and competitively inhibited by this molecule. Thus it seems that the sites of nitrogen fixation and hydrogen evolution in nitrogenase are quite distinct. The reduction of acetylene by the nitrogenase from *A. vinlandii* is ATP dependent like nitrogen fixation. In D_2O it is completely stereospecific giving entirely *cis*-$C_2H_2D_2$[90]. The apparent close similarity between the enzymic reduction of acetylene and nitrogen has led to the proposal that the first stage in nitrogen fixation is the two-electron reduction leading to di-imine[91]. So far there is no evidence for the formation of this species during enzymic nitrogen fixation. Acetylenes are only able to bind to metal atoms in a symmetrical side-on manner[92] and this accounts for the stereospecificity of the hydrogenation reaction. However, in all known complexes of molecular nitrogen, it is bound in an unsymmetrical linear fashion (see chapter 15). This difference suggests that it may be unwise to infer too much from comparisons between the behaviour of nitrogen and acetylenes.

The close similarity between the environments of the metal atoms in Mo–Fe proteins from different sources is highlighted by e.p.r. studies of the proteins from *C. pasteuranium*[93], *K. pneumoniae*[94] and *A. vinlandii*[95]. Each gives an e.p.r. spectrum comprising signals at $g = 2.00$, $g = 3.6$–3.7 and $g = 4.27$–4.3. The signal intensity in the case of the protein from *C. pasteuranium* is greatest in the presence of an excess of dithionite; the signals diminish in parallel following the addition of an oxidising substrate. The e.p.r. spectrum is simple in shape compared with that of the Fe protein. It has been rationalised in terms of an $S = 3/2$ system subject to both axial and rhombic zero-field splittings. Molybdenum(III) is an obvious candidate for the $S = 3/2$ species. The exchange broadening anticipated for a d^3 ion should be considerably reduced in a low-symmetry environment. In an attempt to implicate Mo(III) as the signal source, a culture of *C. pasteuranium* was grown in a medium supplied with ^{95}Mo. Chemical degradation of the protein and mass-spectrometric analysis of the degradation products indicated 76 per cent incorporation of ^{95}Mo. No difference could be detected between the e.p.r. spectra of the native and enriched molybdoferredoxin, although experimental conditions were such that a hyperfine interaction as low as 100 microtesla would have been detected. The e.p.r. signals are therefore probably due to iron. Since e.p.r. spectroscopy is the one spectroscopic technique that is capable of eliciting a specific molybdenum response, the interference from iron could be a serious obstacle to studies of the role and environment of molybdenum in nitrogenase enzymes.

14.6 SULPHITE OXIDASE

The catalysis of the nonenzymatic aerobic oxidation of sulphite by metal ions and free radicals, and its inhibition by scavengers of chain-propagating species have helped to establish this oxidation as a free-radical chain reaction. In contrast, the enzymatically catalysed oxidation of sulphite to sulphate by the enzyme sulphite oxidase, occurs in mammals, plants and bacteria, where there is a plentiful supply of substances that are scavengers of free radicals. Moreover, it occurs at concentrations of sulphite that are too low to support the effective propagation of the chain reaction.

Sulphite oxidase has been isolated in a high state of purity from mammalian and avian sources[96,97]. In mammals, its role appears to be in the oxidative detoxification of sulphur compounds, specifically in the oxidation of sulphur dioxide and sulphite to sulphate[98,99]. In the case of the enzyme from rat liver, its activity has been localised to the intermembrane space of mitochondria[100]. Evidence for the importance of sulphite oxidase to humans is the report of a child whose liver, kidney and brain were completely deficient in this enzyme. His urine contained abnormally large amounts of S-sulpho-L-cysteine, sulphite and thiosulphate, and virtually no sulphate. He died at 23 months with neurological and other abnormalities[7].

There is some confusion in the early literature on sulphite oxidase; reports of direct oxidase activity and the presence of a haem prosthetic group in the liver enzyme, and the ability of the bacterial sulphite oxidase to transmit electrons to oxygen have been both confirmed[101-3] and denied[104-6]. However, the more recent work of Cohen, Fridovich and Rajagopalan[109] has greatly clarified the situation, although the details of the mechanism whereby sulphite is oxidised by molecular oxygen are still unknown. Sulphite oxidase from bovine liver has been purified to about 75 per cent homogeneity[96]. It is capable of transmitting electrons

abstracted from sulphite to one-electron acceptors such as cytochrome c and ferricyanide and to two-electron acceptors such as molecular oxygen, 2,6-dichloro-
phenol, indophenol and methylene blue. Sulphite and other anions are non-
competitive inhibitors of the reduction of one-electron acceptors but are without
effect on the two-electron reduction. This indicates distinct enzyme sites for the
two processes. The existence on the enzyme of sites that are capable of accepting
electrons from sulphite and of subsequently transferring them to various acceptors
is also supported by the ping-pong kinetic behaviour of the sulphite oxidase-
oxygen reaction. The failure of superoxide dismutase[107] to influence the rate of
aerobic cytochrome reduction by sulphite oxidase shows that the reduction of
oxygen by the enzyme does not involve the O_2^- species as an intermediate.

Sulphite oxidase from bovine liver has a molecular weight of 115 000 and
consists of two subunits of molecular weight 55 000. The striking similarity
between the electronic absorption spectra of the reduced enzyme and reduced
cytochrome, the parallel enrichment of haem content and sulphite-oxidase
activity during purification of the enzyme and the identical migration of haem
and sulphite-oxidase activity during electrophoresis established haem as a prosthetic
group of the enzyme[108]. Functional congruity of haem and enzyme activity was
further demonstrated by the parallel loss of enzyme activity and sulphite-reducible
haem on thermal inactivation of sulphite oxidase. Analytical data indicates 2 haems
per molecule of sulphite oxidase. Sulphite reduces one haem rapidly and the other
much more slowly. Interestingly the haem is fully reduced when oxygen is accepting electrons from sulphite oxidase but it is fully oxidised when cytochrome c is
the acceptor. It was suggested that one-electron acceptors interact with sulphite
oxidase at a site that precedes the haem. Moreover, it was argued that if the
inhibition of one step inhibits the overall process it must be rate limiting, and
therefore transfer of electrons from some site on the enzyme to a one-electron
acceptor, for example ferricyanide, must be slower than transfer of electrons from
SO_3^{2-} to the site. Hence in the steady state the site must be in the reduced state.
The discovery of molybdenum(V) as this reduced site occurred[109] accidentally
during experiments designed to investigate the haem prosthetic group of the
enzyme. It is the first example of the analytical detection of molybdenum in an
enzyme by e.p.r. spectroscopy and further emphasises the value of this technique
in the study of molybdenum enzymes. Oxidised sulphite oxidase does not give an
e.p.r. signal, but on reduction with sulphite ion, a signal appears at $g = 1.97$ (see
figure 43). This signal is specifically elicited by sulphite but not by, for example,
substrates of xanthine oxidase, nor does sulphite generate an e.p.r. signal when
incubated with xanthine oxidase. Confirmation of the assignment of the $g = 1.97$
signal to Mo(V) came from the appearance of six-line hyperfine splitting due to
^{95}Mo $(I = 5/2)$ in the spectrum of sulphite oxidase in a D_2O run at high gain. The
Mo(V) signal and the specific activity of the enzyme increased in parallel over a
340-fold range of purification, thus demonstrating that the molybdenum is indeed
a prosthetic group of the enzyme. Further analysis established a Mo:haem ratio
of 1:1 and the absence of a flavin prosthetic group. On the basis of these results

and the differential inhibition of one- and two-electron transfer by anions, the following electron-transfer sequence was proposed for sulphite oxidase

$$SO_3^{2-} \rightarrow Mo \rightarrow haem \rightarrow O_2$$
$$\downarrow$$
cytochrome c   (14.4)
or $Fe(CN)_6$

Some 50 per cent of the total molybdenum of sulphite oxidase is e.p.r. active compared with about 37 per cent in the case of xanthine oxidase. This difference could arise from the greater stabilisation of the paramagnetic monomer in a mono-

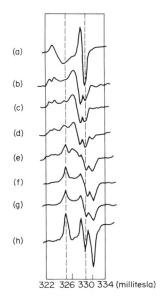

Figure 43 The sulphite-elicited Mo(V) e.p.r. spectrum of sulphite oxidase: (a) D_2O solvent, pD = 7.5; $g_{\parallel}$ = 2.00, $g_{\perp}$ = 1.968; (b) H_2O solvent, pH = 7.21; $a^H{}_{\parallel}$ = 1·1 millitesla, $a^H{}_{\perp}$ = 1.2 millitesla; (c) –(g) H_2O solvent; (c) pH 7.58; (d) pH 8.00; (e) pH 8.40; (f) pH 8.82; (g) pH 9.21; (h) D_2O solvent, pD = 9.6, g_z = 1.984, g_y = 1.961, g_x = 1.950.(Redrawn from ref. 109.)

mer–dimer equilibrium with sulphite oxidase or a weaker interaction between Mo(V) and another paramagnetic centre. Despite the greater signal: Mo ratio for sulphite oxidase, its e.p.r. spectra resemble those of xanthine oxidase in a number of ways. The spectra of both enzymes show a biphasic dependence on temperature: a transition in the environment of the molybdenum occurring at approximately −113°C. The spectra of both enzymes are pH dependent showing a doublet splitting, which vanishes at high pH or when D_2O is used as solvent (see figures 39 and 43). This doublet splitting is probably due to a dissociable proton. At lower pHs in the presence of the proton, the e.p.r. signal of sulphite oxidase is characteristic of an axially symmetric ligand field, but the symmetry is apparently reduced to rhombic on dissociation of the proton. The values of about 1 millitesla in the spectrum of the 'acid' form of sulphite oxidase (figure 43) suggest that the proton is not bound directly to the molybdenum but rather to a nearby group, for example a ligand. The changes in the symmetry of the ligand field around the molybdenum could be accounted for by successive protonation and deprotonation of one of the ligand groups. Since there is no evidence for the splitting of the Mo(V) by a nitrogen nucleus, it is possible that the proton is bound to either oxygen or sulphur. The pH dependence of the Mo(V) signal of sulphite oxidase provides an explanation for the differences between the spectra of sulphite oxidase reduced with sulphite after the addition of ferricyanide in the steady state and when all the ferricyanide has been reduced. The spectrum in the steady state is that of the 'acid' form of the enzyme. After reduction of all the ferricyanide the final spectrum is still in the 'acid' form but to a lesser extent than in the steady state. Possibly, substantial amounts of acid are produced in the oxidation of sulphite by ferricyanide.

Sulphite oxidase has been shown to be the simplest of the molybdenum enzymes with only two prosthetic-group components in its electron-transport chain, namely molybdenum and haem. E.P.R. spectroscopy and inhibition studies indicate that molybdenum is the binding site for the reducing substrate sulphite and is present as Mo(V) in the active enzyme. The two-electron oxidising substrate, molecular oxygen, is probably bound at the haem prosthetic group although there is no direct evidence for this. Certain similarities occur between the e.p.r. spectra of sulphite oxidase and xanthine oxidase, suggesting that the environment of the molybdenum in these two enzymes may be very similar. Further detailed e.p.r. studies on sulphite oxidase of the kind carried out for xanthine oxidase will be required to assess the extent of this similarity. Avian sulphite oxidase[97] is very similar to the bovine liver enzyme[96]. Anions inhibit the reactions of both enzymes with one-electron acceptors; but in addition they modify the sedimentation behaviour of the avian enzyme and also affect the Mo(V) e.p.r. signal. The reasons for these differences are not yet known.

As with the other molybdenum-containing enzymes, structural information on sulphite oxidase is sadly lacking. However, it must be one of the prime candidates for further study in view of its relative simplicity and its importance in human physiology.

15 MODEL STUDIES

15.1 GENERAL CONSIDERATIONS

Knowledge of the ligand environment of molybdenum in its enzymes is, to say the least, scanty. Such evidence as there is, based on comparisons between the e.p.r. parameters of, for example, xanthine oxidase and molybdenum complexes of sulphur ligands (see table 23, page 216), is thought to indicate molybdenum-sulphur binding. Most attention has been concentrated on cysteine as the obvious candidate for the role of a sulphur-containing biological ligand. Interest has also been shown in the potential alternative binding site for molybdenum, namely a flavin moiety, on the grounds that several of the molybdenum enzymes are metalo-flavoproteins.

This chapter deals with some general features of sulphur-ligand and flavin complexes of molybdenum as a preface to a discussion of individual model systems for some of the molybdenum enzymes.

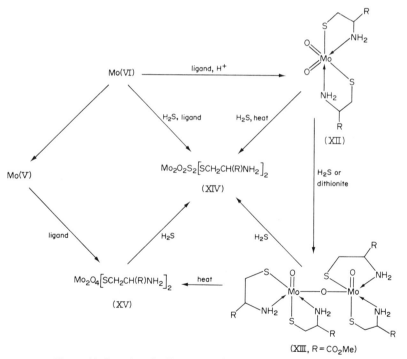

Figure 44 Reactions leading to cysteinate and cysteine-ester complexes of molybdenum. Charges on complex ions are not shown.

15.1.1 Sulphur-containing ligands

Complex formation has been observed between Mo(V) or Mo(VI) and a wide range of sulphur-containing amino acids, esters and carboxylic acids[110]. The Mo(V) complexes are oxygen- or sulphur-bridged diamagnetic dimers based on the MoO^{3+} ion. Complexes of the sulphur-containing amino acid, cysteine, seem to bear most closely on the molybdenum enzymes. Consequently they have received the greatest attention[111,112]. Reactions leading to molybdenum complexes of cysteine or its esters are summarised in figure 44. Structures have been assigned on the basis of analytical and infrared spectroscopic data (table 25). They have been confirmed by X-ray crystallography for (XIV, $R = CO_2Me$)[113], for (XVa, $R = CO_2Et$)[114] and for (XVb, $R = CO_2^-$)[115].

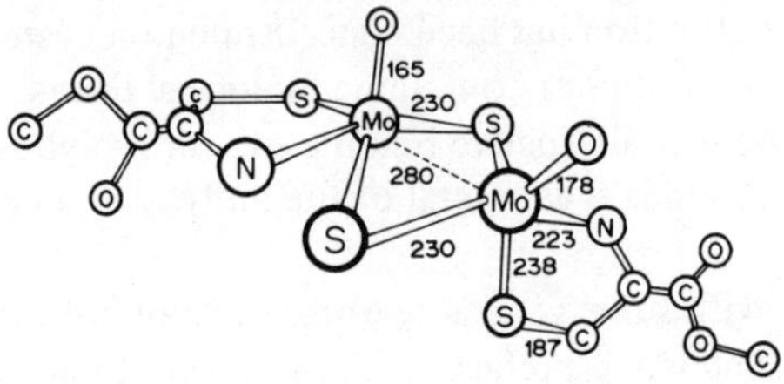

(XIV)

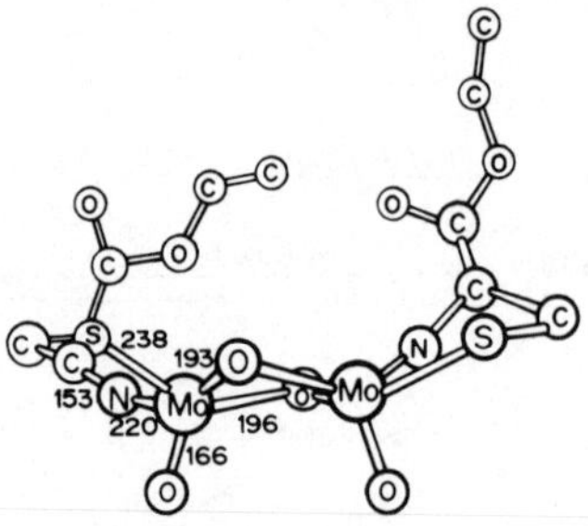

(XVa)

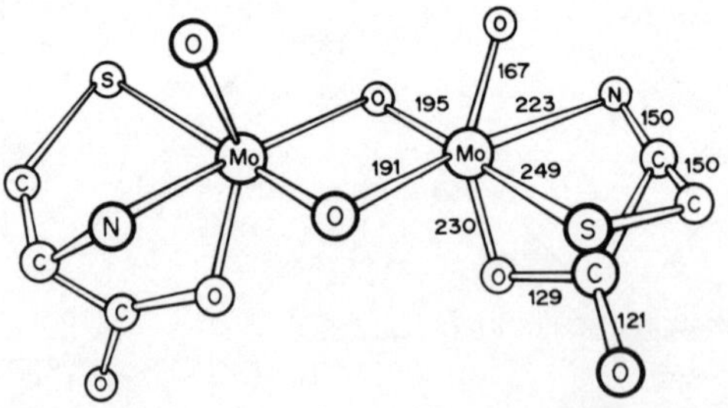

(XVb)

The Mo atoms in the dianion (XVb, $R = CO_2^-$) are in a distorted octahedral environment with the cysteine molecule acting as a tridentate ligand. The Mo—O bond *trans* to the sulphur atom is slightly longer than the Mo—O bond *trans* to

TABLE 25 INFRARED SPECTRA OF SOME MOLYBDENUM COMPLEXES OF CYSTEINE AND CYSTEINE ESTERS[†], [‡]

Complex	$\nu(N–H)$	$\nu_{as}(O–C–O)$	$\nu_s(O–C–O)$	$\delta(N–H)$	$\nu(Mo=O)$	$\nu(MoO_2Mo)$	Others
$HSCH_2 \cdot CH(NH_3^+) \cdot CO_2^-$	3250s	1590vs	1400s	1550s			2550s $\nu(S–H)$
$Na^+[HSCH_2 \cdot CH(NH_2) \cdot CO_2]^-$	3350m	1620vs	1415s	1590vs			2500m $\nu(S–H)$
$HSCH_2 \cdot CH(NH_3^+) \cdot CO_2H\ Cl^-$	3000vs	1737vs	1405s	1528vs			2500m $\nu(S–H)$
$Na_2[MoO_2\{SCH_2 \cdot CH(NH_2) \cdot CO_2\}_2]$, $HCONMe_2$	3230s	1610s	1419s	1556vs	922vs		
	3130s				892vs		
$Na_2[Mo_2O_4\{SCH_2 \cdot CH(NH_2)CO_2\}_2]$, $5H_2O$	3100s,b	1630vs	1390s	1590vs	955vs	735s	3430 $\nu(O–H)$
$Na_2[Mo_2O_2S_2\{SCH_2 \cdot CH(NH_2) \cdot CO_2\}_2]$, $3H_2O$	3200s,b	1630s	1392s	1588vs	948m		3500s $\nu(O–H)$
$HSCH_2 \cdot CH(NH_3^+) \cdot CO_2Me\ Cl^-$	2910vs	1742s		1544s			2500m $\nu(S–H)$
$MoO_2[SCH_2 \cdot CH(NH_2) \cdot CO_2Me]_2$	3295s	1736vs		1568m	912vs		
	3255s				884s		
	3160m						
$Mo_2O_3[SCH_2 \cdot CH(NH_2)CO_2Me]_4$	3200s	1725vs		1565m	932vs		
$Mo_2O_4[SCH_2 \cdot CH(NH_2) \cdot CO_2Me]_2$	3220s	1750vs		1574m	978vs	739vs	
	3185vs	1730vs		1550m	950m		
$Mo_2O_4S_2[SCH_2 \cdot CH(NH_2) \cdot CO_2Me]_2$	3260vs	1718vs		1582m	966vs		485m $\nu(Mo–S)$
	3140vs				946vs		
	3080vs				922m		
$Mo_2O_4[SCH_2 \cdot CH(NH_2) \cdot CO_2Et]_2$	3230vs	1738vs		1560vs	980vs	735vs	470s $\nu(Mo–S)$
	3180vs			1541s	945s		
	3100vs				935m		

† positions of main bands in cm^{-1} and assignments.
‡ spectra of free ligands shown for comparison.

nitrogen, 195 pm compared with 191 pm. Although the difference is not large, it may be a structural manifestation of the labilising effect of the cysteine ligand on the oxo-bridge. Such an effect has been proposed (section 15.2.2) to account for the greater rate of acetylene reduction with molybdenum–cysteine complexes compared with histidine complexes. In (XVa, R = CO_2Et) the molybdenum atoms have identical distorted trigonal bipyramidal environments. The Mo–S bond distance (238.5 pm) is significantly shorter than that (249 pm) in the dianion (XVb, R = CO_2^-). It has been suggested that a *trans* influence, this time due to a bridging oxygen in the dianion, causes the lengthening of the Mo–S bond. However, the different stereochemistries of the two molybdenum complexes render such arguments of doubtful value. The solid complexes (XII–XV) are diamagnetic but in several cases aqueous solutions give rise to weak e.p.r. signals indicating the presence of small amounts of paramagnetic species. Moreover, the strength of the e.p.r. signal, the metal : ligand ratios and even the outcome of reaction are pH dependent. Spence and coworkers[116] found that in the pH range 4–6 Mo(VI) forms complexes with L-cysteine with molybdenum : cysteine ratios of from 1:1 to 1:3. At pH 7.5 in a phosphate buffer there was no detectable complex formation; slow reduction to Mo(V) occurred with concomitant formation of a 1:2 Mo(V)–cysteine complex. Attempts to prepare the 1:2 complex by adding an excess of ligand to the 1:1 complex were unsuccessful, indicating that the two are not in equilibrium in solution.

Most interesting in connection with molybdenum enzymes, is the exhibition of a weak e.p.r. signal by aqueous solutions of Mo(V) and cysteine[117]. Although this signal accounts for less than 1 per cent of the total molybdenum–whereas with xanthine oxidase and sulphite oxidase up to 50 per cent of the molybdenum is e.p.r. active–it does indicate the possible existence of an equilibrium of the type

diamagnetic dimer ⇌ paramagnetic monomer

which has been proposed[61] to account for the incomplete spin recovery with the enzymes. Both the stability and intensity of the e.p.r. signal from the Mo(V)–cysteine solutions are critically dependent on concentration, pH and buffer. A weak unstable signal is produced at pH 6 with 10^{-2} M Mo(V) in 1 M phosphate buffer[117]. However, Huang and Haight[118] obtained a stable, well-defined signal (g_{zz} = 2.029, g_{yy} = 1.972, g_{xx} = 1.931; $A(^{95}Mo)$, zz = 5.4, yy = 2.4, xx = 3.4 millitesla) at pH 7–10 with 10^{-3} M Mo(V) in 0.2 M phosphate buffer. Even under these conditions, the signal accounted for only 2 per cent of the total molybdenum. Two colour changes occur when cysteine is added to a solution of Mo(V). The e.p.r. signal develops during the second change. Furthermore, its intensity is proportional to the square root of the concentration of molybdenum. These observations support the suggestion that Mo(V) and cysteine first form a diamagnetic dimeric species, which subsequently dissociates into a paramagnetic monomer. The inverse temperature dependence of the signal intensity indicates that the dissociation process is endothermic. The e.p.r. signals of the Mo(V)–cysteine solutions isotopically enriched with ^{98}Mo ($I = 0$) and ^{95}Mo ($I = 5/2$) confirm the presence of only one paramagnetic centre and rule out the possibility of an equilibrium between paramagnetic and diamagnetic dimers. That the e.p.r. signal of a Mo(V)–cysteine solution is due to Mo(III) formed by the reduction of Mo(V) by excess cysteine was excluded because of the marked differences between the experimental spectra and those of Mo(III) complexes with either cysteine or cystine. A mechanism (figure 45) for the dissociation of a dimeric cysteine complex involving cleavage of the oxygen bridge by OH⁻ ions is consistent with the pH dependence of the rate and extent of e.p.r. signal formation.

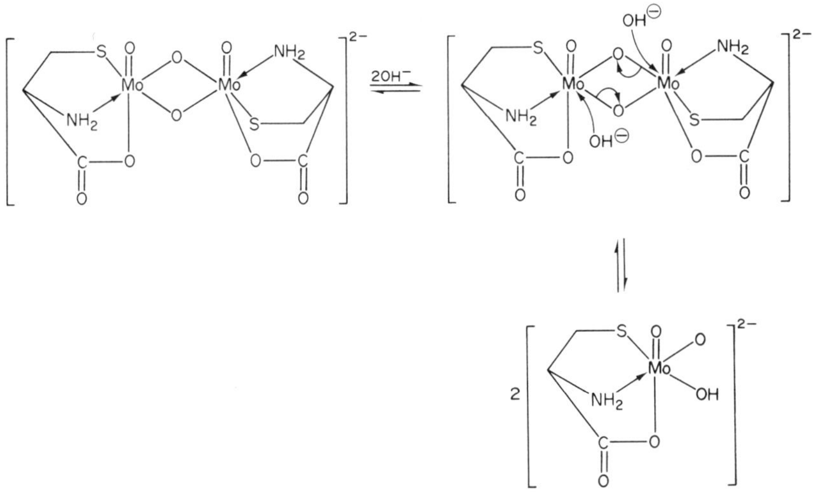

Figure 45 The formation of an e.p.r.-active Mo(V) complex by hydroxyl-ion attack at the oxo bridge of $[Mo_2O_4 (cysteinate)]^{2-}$.

Since e.p.r. signals could not be detected in solutions containing either monodentate sulphydryl or polydentate nonsulphydryl ligands, Huang and Haight concluded that the μ-dioxo bridge in the Mo(V)–cysteine dimer is labilised by the multidentate sulphydryl ligand. Furthermore, preliminary indications that the lability of the bridge is enhanced by increasing the chain length of the ligand were substantiated by their detailed study of the Mo(V)–glutathione system[119]. As with the Mo(V)–cysteine system, well-defined e.p.r. signals were only obtained under alkaline conditions; maximum signal intensity, representing 4 per cent of the total molybdenum, occurred at pH 9. However, with the glutathione system, the signal intensity was found to be proportional to $[Mo^V_2]$. Moreover, the room-temperature spectrum showed an eleven-line splitting characteristic of an interaction between an unpaired electron and two magnetically equivalent nuclei of spin 5/2. Thus the significant difference between the Mo(V)–cysteine and –glutathione systems is that the e.p.r.-active constituent of the latter is a paramagnetic binuclear Mo(V) species in equilibrium with a diamagnetic dimer. Since the pH and temperature dependence of signal intensity are similar for the two systems it was proposed that the paramagnetic glutathione complex is also formed by hydroxyl-ion attack at the μ-dioxo

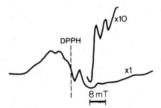

(XVI)

bridge. A diamagnetic binuclear complex $Na[Mo_2O_4(glutathione)H_2O]4H_2O$ was isolated from the Mo(V)–glutathione system. It was assigned the structure (XVI) on the basis of infrared spectroscopic and analytical data. Co-ordination of the NH group of the glycine moiety is assumed in order to complete the pseudo-octahedral co-ordination sphere of one of the molybdenum atoms. Further information regarding the nature of the paramagnetic binuclear species was

Figure 46 The first-derivative e.p.r. spectrum of isotopically enriched ^{95}Mo(V)$_2$–glutathione complex (XVII) at 77 K. From Huang and Haight[119].

obtained from an analysis of the 77 K spectra (figure 46) of frozen Mo(V)–glutathione solutions. The spectrum is complex and Huang and Haight have interpreted it in terms of two overlapping full-field transitions of a triplet-state binuclear complex. This overlap implies a low value of D, the zero-field splitting constant. An analysis of the spectrum, based on the simplifying assumption of axial symmetry, gave a value of $D = 83 \times 10^{-4}$ cm^{-1}. This was used to estimate a molybdenum–molybdenum separation in the dimer of 600 pm. On the basis of these results and model building, structure (XVIIa) is favoured for the paramagnetic dimer rather than the alternative (XVIIb).

(XVIIa)   (XVIIb)

The possibility that paramagnetism in Mo(V) compounds might arise from an equilibrium between paramagnetic and diamagnetic dimers

$$\text{dimer} \quad \rightleftharpoons \quad \text{dimer}$$

e.p.r. inactive, singlet   e.p.r. active, triplet

has been recognised previously[120]. However, complex (XVII) is the first reported example of a paramagnetic binuclear species with two molybdenum atoms co-ordinated to a biologically significant sulphur ligand. As such, it is of considerable interest as a model of molybdenum enzymes such as xanthine oxidase, which has two molybdenum atoms per molecule of enzyme and generates an e.p.r. signal on reduction by substrate. The absence of an eleven-line splitting pattern from the spectra of molybdenum enzymes does not necessarily imply that paramagnetic binuclear molybdenum species do not have any biological significance, since there are a number of ways in which electron–nuclear spin interaction in a binuclear complex can be destroyed. The simplest is separation of the two nuclei to moderately large distances. The zero-field splitting constant D is very sensitive to internuclear separation (see table 26), diminishing rapidly with increasing distance. D approaches zero in the region 700–800 pm. Such a separation could easily be achieved via a small change in the conformation of the protein in a molybdenum enzyme following the cleavage of a dioxo-bridge.

Mo(III) has been proposed as a possible site of nitrogen binding and activation in nitrogenase enzymes (section 10.2). The ability of sulphur ligands to stabilise

TABLE 26 METAL-METAL DISTANCES IN BINUCLEAR COMPLEXES FROM ZERO-FIELD SPLITTING PARAMETERS[†]

Complex	$10^4 \times D_{\text{expt}}$ (cm^{-1})	$10^4 \times D_{\text{pseudo}}$[‡] (cm^{-1})	R_{calc}[§] (pm)	R_{calc}[‖] (pm)	R_{expt}[¶] (pm)
$(\text{VO})_2(\text{d-tart})(\text{1-tart})^{4-}$	≈ -336	< 10	> 415	≈ 419	408
$\text{Cu}_2(\text{C}_5\text{H}_4\text{N}_5)_4 \cdot 4\text{H}_2\text{O}$	-1210 ± 50	130	284	292	295
$\text{Cu}_2(\text{d-tart})(\text{1-tart})^{4-}$	-572			385	340
				(330)	
$\text{Mo(V)}_2\text{–glutathione}$	83		600	600	610[††]
			$(\theta = 0°)$	$(\theta = 0°)$	$(\theta = 0°)$
				480	550–900[††]
				$(\theta = 90°)$	$(\theta = 90°)$

† from Huang and Haight[119].

‡ pseudodipolar contribution to D_{expt}.

§ calculated from $D_{\text{dd}} = (3/4)g^2\beta^2 \left\langle \dfrac{1 - 3\cos^2\theta}{r_{12}^3} \right\rangle_{\max}$ and
$R_{\text{calc}} = (0.325g^2|1 - 3\cos^2\theta|/D_{\text{dd}})^{1/2}$, where D_{dd} is the magnetic dipolar interaction between two electron spins separated by a distance r_{12}, and $D_{\text{dd}} = D_{\text{expt}} - D_{\text{pseudo}}$. H is taken to be along the Mo=O direction and thus $\theta \approx 0°$ or $\theta \approx 90°$.

‖ calculated as for §, but with $D_{\text{dd}} = D_{\text{expt}}$.

¶ from X-ray data.

†† estimated from models.

this oxidation state of molybdenum is demonstrated by the behaviour of the molybdenum(VI)–thioglycollic-acid system. Mo(VI) is quantitatively reduced to Mo(V) by thioglycollic acid. An equilibrium mixture containing about 10 per cent of a paramagnetic Mo(V) species ($g = 1.978$) is formed[28]. The g-value is time dependent, changing first to $g = 2.006$ and after about 20 hours to $g = 1.987$. The absorption spectrum of the final reaction mixture is identical to that of a mixture of $[MoCl_6]^{3-}$ and thioglycollic acid, indicating that the $g = 1.987$ species is a Mo(III) complex. A Mo(IV) intermediate in the reduction was proposed to account for the difference between the rate of disappearance of the Mo(V) complex and the rate of formation of the Mo(III) complex.

Mo(IV) complexes have been postulated as reactive intermediates in several other model reaction schemes but in only one case have the reactions of a known Mo(IV) complex been studied. The oxodithiocarbamate complex (XVIII), which is a potentially carbene-like d^2 species, has been investigated as a model of the catalytic site of nitrogenase. Complex (XVIII) reacts readily with azodiethyl-dicarboxylate, diethylacetylenedicarboxylate and tetracyanoethylene[121]. In each case seven-co-ordinate 1:1 adducts such as (XIX) are believed to be formed. The azodiethyldicarboxylate adduct undergoes rapid hydrolysis yielding bis(ethoxy-carbonyl)hydrazine and the *cis*-bis(oxo) complex (XX). This step could be con-sidered to model a possible second stage of nitrogen reduction by nitrogenase,

(XVIII)   (XIX)   (XX)

namely di-imine to hydrazine. However, even if di-imine is an intermediate in the enzymatic reduction of molecular nitrogen—and there is no evidence for this—it is not its reduction, presumably to hydrazine, which is likely to be the most difficult step in the overall reduction process, but the first stage, the reduction of nitrogen to di-imine, for which there is a large positive free-energy change. Thus the hydrolysis does not act as a model for the most significant catalytic phase of nitrogenase activity.

Other sulphur ligands that have attracted attention as possible models of biological molybdenum binding sites are dithiols, dithiolates and dithioketones. The dithiolate complexes (XXI) exhibit the high g values and small hyperfine splittings characteristic of xanthine oxidase. Toluene-3,4-dithiol forms an interest-ing sulphur-bridged binuclear complex (XXII) with molybdenum. The solid complex is diamagnetic but a solution in methylene chloride has a weak e.p.r. signal with $g = 1.99$ and $A(^{95}Mo) = 3.2$ millitesla, indicating a slight dissociation into para-magnetic monomers[122]. In contrast, complexes formed from the 8-substituted

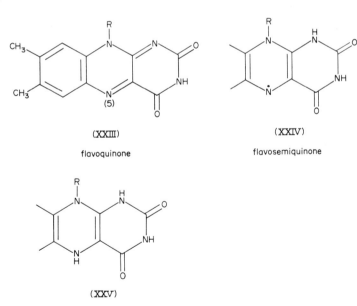

(XXI)   (XXII)

quinoline ligands 8X-Quin (X = QH, SH and NH_2) and Mo(V) are 100 per cent
e.p.r. active in dimethylformamide solution[123]. The formation constants of the
complexes decrease in the order of ligands SH > OH > NH_2, indicating the
preference of Mo(V) for sulphur ligands.

15.1.2 Flavin ligands

At least three of the molybdenum-containing enzymes—xanthine oxidase, aldehyde
oxidase and nitrate reductase—have flavin coenzymes. E.P.R. evidence[44,74,124] that
electron transfer occurs between molybdenum and the coenzyme has led to
speculation that the molybdenum is bound directly to the flavin molecule. Consequently the interaction between molybdenum and various flavins has been
studied as a possible model for the enzymatic electron-transport process. At pHs
in the normal physiological range, flavins can occur in three forms: the fully
oxidised form (flavoquinone, structure (XXIII)), the free-radical form (flavosemiquinone, structure (XXIV)) and the fully reduced form (flavohydroquinone, structure (XXV)). Monodentate co-ordination of the flavoquinone form through N-5 is
unlikely because of the extremely low basicity of this atom. Since the formation

(XXIII)

flavoquinone

(XXIV)

flavosemiquinone

(XXV)

flavohydroquinone

of the 4-iminol derivative requires a considerable expenditure of energy, stable oxinate-like complexes are only likely if there is extensive back-donation of electron density from the metal to the flavin ligand. This can occur for polarisable cations such as Cu(I) but will be negligible for Mo(V), a d^1 ion, and absent for Mo(VI). Co-ordination of the flavohydroquinone form through N-5 without displacement of the proton would result in folding of the flavin molecule with a concomitant loss in the resonance-stabilisation energy of the tricyclic system. The immeasurably low basicity of N-5 precludes complex formation by displacement of the proton. Only the flavosemiquinone form of the flavin molecule exhibits appreciable metal–ion affinity. It has a higher N-5 basicity, and a lower barrier to iminol formation than either of the other forms[125,126]. The results of studies by Spence and coworkers[127] of the reaction between lumiflavin-3-acetic acid and the Mo(VI)–Mo(V) redox system lend general support to the predictions outlined above. At pH 6 the equilibrium

$$FlH_2 + 2Mo(VI) \rightleftharpoons Fl + 2Mo(V) \tag{15.1}$$

lies to the right and at pH 8 to the left. At the lower pH the reaction is first order in molybdenum and first order in flavin. A transient red colour forms and e.p.r. signals at $g = 2.00$ and $g = 1.95$ can be detected. Neither the rates of formation nor the maximum concentrations of the coloured and e.p.r.-active species coincide. These observations were interpreted in terms of the formation of the flavosemiquinone radical ($g = 2.00$) and Mo(V) ($g = 1.95$), which combine to form the red e.p.r.-inactive oxinate-like complex (XXVI). In the reverse (pH 9) reaction, the

(XXVI)

enhancement of the flavosemiquinone e.p.r. signal by a factor of ten relative to molybdenum-free solutions of oxidised and reduced flavin was attributed to the formation of a Mo(VI)–flavosemiquinone complex. The scheme shown in figure 47 was suggested by Spence and coworkers to account for these observations[127]. Other examples of electron-transfer reactions involving molybdenum and flavins are the reduction of nitrate by reduced flavinmononucleotide catalysed by Mo(VI)[128], and the oxidation of the Mo(V)–cysteine complex (XV, R = CO_2^-) by FMN[129]. The former of these has been studied as a model of nitrate reduction by nitrate reductase (section 15.2.1).

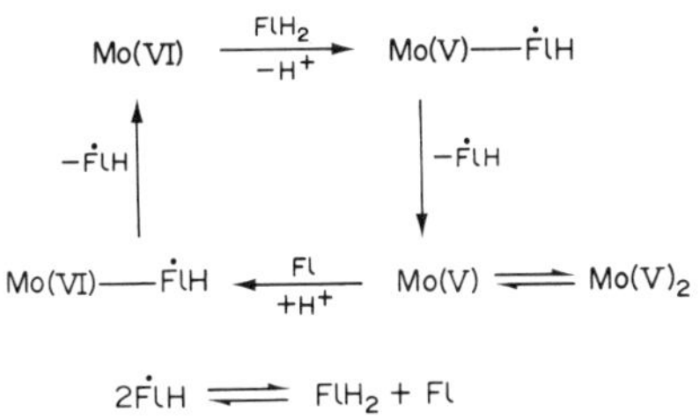

$$Mo(VI) \xrightarrow[-H^+]{F\iota H_2} Mo(V)—\dot{F}\iota H$$

Figure 47 A scheme for the interaction between molybdenum and flavin under acid and alkaline conditions.

Although these model studies have established the ability of molybdenum and flavins to participate in electron-transfer reactions, definitive evidence of a flavin binding site for molybdenum in its enzymes is still lacking. Indeed, with xanthine oxidase, the dissociation of the FAD coenzyme abolished the oxygen reductase activity of the enzyme but the substrate oxidase activity was retained. Since molybdenum is strongly implicated as the substrate binding site in this enzyme, it seems unlikely that the molybdenum and flavin components are combined in a single prosthetic group involving direct molybdenum–flavin interaction.

15.2 NITRATE REDUCTION

Investigations of model nitrate-reduction systems have thrown very little light on the enzymatic reduction of nitrate although they have established beyond question the ability of molybdenum ions to catalyse its chemical reduction. Early conclusions that the catalytically active species in the molybdenum-catalysed electro-chemical[130] and Sn(II) reduction[131] of nitrate is Mo(IV) have been questioned[132]. An alternative proposal that the active species is Mo(V) was substantiated by Guymon and Spence[133] who examined the reduction of nitrate by Mo(V) in the presence of a series of buffers: phosphate, borate, acetate, bicarbonate and tartrate. They found that only with tartrate did reduction occur. Moreover, the reaction was first order in tartrate and half order in total molybdenum, indicating the formation of the catalytically active species via dissociation of a dimeric tartrate complex. The following mechanism was proposed to account for the kinetic results

$$Mo(V)_2 \rightleftharpoons 2Mo(V)$$
$$2Mo(V) + 2NO_3^- + 4H^+ \rightarrow 2Mo(VI) + 2NO_2 + 2H_2O$$
$$2NO_2 \rightarrow NO^+ + NO_3^- \tag{15.2}$$
$$NO^+ + Mo(V) \rightarrow NO + Mo(VI)$$

It should be pointed out that neither NO_2 nor NO have been observed as products of the nitrate-reductase catalysed reduction of nitrate. In support of the findings regarding the formation of a paramagnetic Mo(V) monomer, Spence and Heydanek[134] observed a weak Mo(V) e.p.r. signal in solutions containing Mo(V) and tartrate.

Spectrophotometric investigation of these solutions indicated the existence of the dimer–monomer equilibrium

$$L_2Mo(=O)-O-Mo(=O)L_2 + 2L \rightleftharpoons 2MoL_3 \qquad (15.3)$$

$$L = \{CH(OH)COOH\}_3$$

with an estimated K_{eq} of 3.1×10^{-6} at pH 4.72 and 25°C.

Two model nitrate-reducing systems based on FMN have been investigated. Both of these afford NO as the principal reduction product. In the former, photochemical reduction of the FMN is thought to occur via hydrogen abstraction from the ribityl side chain[135]; this is followed by reduction of Mo(VI) to Mo(V) by the reduced flavin and reduction of the NO_3^- by Mo(V). Only very small amounts of NO_3^- are reduced by this system, but NO_3^- is quantitatively reduced to NO by chemically reduced FMN in tartrate or citrate buffer and in the presence of catalytic amounts of molybdenum[136]

$$3FMNH_2 + 2NO_3^- \xrightarrow{Mo(VI)} 3FMN + 2NO \qquad (15.4)$$

Kinetic studies of this reaction once again pointed to the involvement of a catalytically active Mo(V) species; this is stabilised by tartrate or citrate buffer but not by either phosphate or acetate buffers. NO_3^- reduction did not occur in the latter two buffers.

The model nitrate reductions demonstrate clearly the necessity for the stabilisation of some paramagnetic monomeric Mo(V) to achieve a significant degree of molybdenum catalysis. An e.p.r. signal attributable to Mo(V) arises when substrate is added to nitrate reductase. This has been taken to indicate that as in the models, Mo(V) is the catalytically important oxidation state. In the enzyme, the protein could readily provide the necessary stabilisation of monomeric Mo(V).

15.3 NITROGEN FIXATION

The evidence available on bacterial nitrogenase points to the activation of molecular nitrogen by metal ions. Molybdenum, nonhaem iron and acid-labile sulphide groups have been recognised as essential components of the bacterial nitrogen-fixing system, and indeed most attempts to produce chemical models have been based on the assumption that iron and molybdenum are the key elements in nitrogen fixation. The nitrogenase of *C. pasteuranium, A. vinlandii* and other bacterial nitrogen-fixing systems also exhibit hydrogenase activity towards other triply-bonded substrates. In the absence of added substrate and to a lesser extent in its presence hydrogen gas is evolved. Acetylene is reduced selectively to ethylene and isonitriles to alkanes and primary amines[90,137]. Since acetylene is also a competitive inhibitor of nitrogen fixation and is therefore thought to be bound at the active site, it has been introduced as a convenient substrate for assaying the nitrogenase activity of both native enzyme preparations and chemical model systems. However, its use as the *sole* criterion of nitrogen-fixing ability can lead to erronous conclusions about the nature of the enzyme[138]. The ultimate test of a nitrogenase model must be its ability to catalyse the reduction of molecular nitrogen at ambient temperature and pressure.

Attempts to model the active site of the molybdoferredoxin component of nitrogenase have centred on complexes of molybdenum with various sulphur ligands and $Na_2S_2O_4$ or $NaBH_4$ as reducing agent. Of some twenty-five other transition metals examined as potential model active-site centres[90] only iridium showed appreciable activity and this was rather less than 15 per cent of that shown by molybdenum. The best model system reduces acetylene at about 0.3 per cent of the enzymic rate. The molybdenum compounds tested included Na_2MoO_4, $MoOCl_3$, $MoCl_5$ and heteropolymolybdates. Of the sulphur ligands tested, dithioerythritol, 1-thioglycerol, 2-mercaptoethanol, cysteine and glutathione

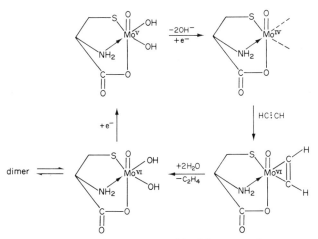

Figure 48 The mechanism proposed by Schrauzer and Schlesinger[137] for the reduction of acetylene catalysed by a molybdenum–cysteine complex.

showed greatest activity; maximum activity occurred with a 1:1 ratio of metal to ligand. Despite the encouraging ability of the model to mimic the hydrogenase activity of nitrogenase towards acetylene and isonitrile substrates, only trace yields of ammonia could be obtained from nitrogen even under high pressures. The initial rate of acetylene reduction by a model system comprising the molybdenum–cysteine complex (XV, R = CO_2^-) and $NaBH_4$ increases exponentially; this has been attributed to the production of a catalytically active species such as a Mo(V) monomer. Support for this comes from the e.p.r. studies of Huang and Haight[118] on molybdenum–cysteine complexes. In the pH range 6–10, Mo(V)–cysteine mixtures exhibit weak but well-defined e.p.r. signals, which increase with pH and account for a maximum of 2 per cent of the total molybdenum at pH 9–10. Samples enriched with ^{95}Mo ($I = 5/2$) exhibit the characteristic six-line hyperfine splitting, indicating that the e.p.r.-active species are monomeric Mo(V) complexes.

The slow decrease in signal intensity is due to the reduction of the Mo(V). Since both the rate and extent of signal formation increases with pH, the formation of monomeric Mo(V) species by nucleophilic attack of OH^- on the molybdenum-oxygen bridge was proposed (see figure 45); this was further supported by the dependence of signal strength on the square root of the total molybdenum concentration. According to the mechanistic scheme of figure 48 for acetylene reduction proposed by Schrauzer and Schlesinger[137], the initially formed monomer undergoes further reduction and loss of OH^- ligands to produce a co-ordinately unsaturated Mo(IV) species, which co-ordinates a molecule of acetylene. Cleavage of the molybdenum–carbon bonds by water then produces ethylene and the Mo(VI) analogue of the initial monomer; this is then reduced by the electron donor and the catalytic cycle is complete. The activating effect of ATP on this model system has been attributed to an enhancement of the leaving-group properties of the OH groups via phosphorylation. Further insight into the reduction process is provided by the effect of changes in the ligands bound to molybdenum on the reduction of acetylene. Histidine forms the oxygen-bridged binuclear complex (XXVIIa and b) with Mo(V)[139]. The carboxylate group is shown co-ordinated to the metal but there appears to be no direct evidence for this.

cis (**XXVIIa**)

trans (**XXVIIb**)

Acetylene is reduced to a 1:1 mixture of ethylene and ethane by the complex and $NaBH_4$. The reaction rate is only 5 per cent of the rate observed with the molybdenum–cysteine system. This can be rationalised on the basis of a much reduced dissociation of the complex into catalytically active monomers compared to the cysteine complex. Reduction of the undissociated dimers could provide a binuclear Mo(IV) species that is capable of reducing acetylene to ethane. However, the nitrogenase of *A. vinlandii* produces only 0.1 per cent of ethane and if the model results bear at all on the enzyme it seems likely that the molybdenum sites on the enzyme are not part of a bridged system.

As noted above, isonitriles can act as oxidising substrates for nitrogenase[140, 141, 142]. They are reduced to hydrocarbons derived from the isonitrile carbon atom and primary amine from the R–N moiety. Isonitriles bind to transition metals in an end-on manner; in this they resemble N_2. The reduction of co-ordinated methyliso-nitrile affords methane as a major product in a six-electron reduction rather than dimethylamine, the product of reduction of the unco-ordinated molecule. This

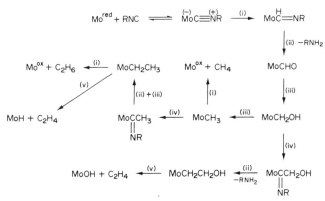

Figure 49 Possible pathways in the reduction of isonitriles to hydrocarbons catalysed by molybdenum complexes: (i) protonation; (ii) hydrolysis; (iii) reduction; (iv) 'insertion' of RNC; (v) elimination of olefin.

combination of properties renders isonitriles excellent substrates for use with both biological and model nitrogenases. Ethylene and ethane were the principal products of the reduction of isonitriles using a 1:1 molybdenum:cysteine complex as catalyst and excess $NaBH_4$ as reducing agent[137]. As with acetylenic substrates, the experimental data were consistent with a monomeric catalytically active molybdenum species. The reduction was weakly inhibited by N_2 and more strongly inhibited by CO. Experiments with $^{15}N_2$ established that the inhibitory N_2 was being reduced to ammonia and that the N_2 and RNC molecules must have a common binding site, namely the molybdenum atom. Moreover, N_2 and CO are also competitive inhibitors of isonitrile reduction by nitrogenase, demonstrating convincingly that molybdenum possesses the necessary properties for substrate binding and reduction. A mechanistic scheme for isonitrile reduction suggested by Schrauzer and Schlesinger[137] is shown in figure 49.

Isonitriles are reduced more slowly than acetylenes by a factor of 10^{-3}–10^{-4} for the model complexes but only by a factor of 10^{-1} for nitrogenase. This coupled with $K_M(CH_3NC)$ values that are greater for the model than for the enzyme by a factor of 10–100 indicate a substantially lower substrate affinity for the catalytically active model complex compared with the enzyme. As Schrauzer and Schlesinger[137] point out this probably reflects differences in the π-acceptor. σ-donor characteristics of the metal atoms in the model and enzyme, respectively. The binding of ligands such as N_2, CO, isocyanides and acetylenes, is expected to be sensitive to small changes in the donor–acceptor characteristics of the binding site. These can be varied in the model systems by the use of different ancillary ligands. In keeping with these ideas, Hill and Richards[143] found that a catalyst system based on $MoO_4{}^{2-}$, Fe(II) and 2-aminoethanethiol gave a thirtyfold improvement in the yield of $^{15}NH_3$ from $^{15}N_2$ compared with the system employing cysteine as the ligand.

The model studies described above show that the combination of molybdenum with a thiol ligand provides a highly efficient *homogeneous* catalyst system with the necessary qualities for substrate binding and reduction. This strongly implicates molybdenum as the substrate binding and reduction site in the nitrogenase enzyme. A recent suggestion that iron too can fulfil this role[138] is based on the observation that the incorporation of an iron source improves the catalytic efficiency of molybdenum systems[137, 143] and that Fe(II) complexes with reducing agents such as $NaBH_4$ can catalyse the reduction of acetylene to ethylene and ethane. However, it appears that the catalytic species here is the finely-divided metal; such a heterogeneous system cannot be considered a true model of the nitrogenase enzyme. The only comparative studies of the efficiencies of molybdenum and iron in substrate reduction[90] support the candidature of the former as the binding site and also the site of reduction of oxidising substrates. This leaves iron in the role of

(XXVIII)

the electron-transfer agent, forwarding electrons to the nitrogen-activating protein component of the nitrogenase complex. The mechanism of this process is unknown. One possibility is that metal–sulphur–metal bridges provide a low-energy pathway for the transport of electrons. The preparation and structure of the Mo–S–Fe complex (XXVIII)[144] are of interest in this connection; the redox properties of this molecule are currently under investigation.

15.4 TRANSITION-METAL COMPLEXES OF MOLECULAR NITROGEN

Attempts to model the active site of nitrogenase enzymes have been complemented by studies of the interaction of molecular nitrogen with transition-metal salts. Molecular nitrogen is traditionally regarded as a highly unreactive molecule[145]. The only reaction it is known to undergo, under conditions likely to obtain in a biological system, is the formation of complexes with transition-metal ions. Furthermore, the formation of the binuclear complex ion $[(NH_3)_5Ru(N_2)Ru(NH_3)_5]^{4+}$ from the aqua-ion $[Ru(NH_3)_5H_2O]^{2+}$[146] established the ability of the nitrogen molecule to compete successfully with water for a metal co-ordination site, and provided the first positive indication that a metal site might exist in nitrogenase at which substitution of water by molecular nitrogen could occur. If enzymatic nitrogen fixation does occur at a metal site, then the nitrogen molecule must be able to displace the product ammonia in order to sustain the catalytic cycle. Reaction 15.5 demonstrates that ammonia can indeed be displaced from a nitrogen-binding site[147]

$$CoH(NH_3)(PPh_3)_3 + N_2 \rightarrow CoH(N_2)(PPh_3)_3 + NH_3 \qquad (15.5)$$

Both of the metals iron and molybdenum, essential constituents of the nitrogen-fixing enzymes, form stable complexes with molecular nitrogen. This in itself has been taken to suggest that they are involved at least in the binding step of nitrogen fixation. Unfortunately, the complexes formed by molybdenum (see table 27) seem to bear little relation to biological systems since they are confined to molybdenum in its zero or +1 oxidation state and to nonbiological ligands, principally tertiary phosphines. Such combinations evidently satisfy the simultaneous demands of the nitrogen molecule for empty low-energy metal π-acceptor orbitals and occupied high-energy σ-donor orbitals. Although no X-ray crystallographic studies have been carried out on molybdenum complexes with molecular nitrogen it is to be expected that the nitrogen molecule is bound end-on to the metal as in figure 50[148]. The polarity induced in the co-ordinated molecule by this mode of binding causes the nitrogen–nitrogen stretching vibration to become infrared active. $\nu(N-N)$ for the molybdenum complexes (see table 26) lies in the range 1925–2220 cm^{-1} observed for N_2 complexes of Fe, Co, Ni, Ru, W, Re, Os and Ir.

In addition to establishing the ability of molecular nitrogen to bind to molybdenum at least in its lower oxidation states, studies of transition-metal complexes of molecular nitrogen have revealed an interesting bimetal–nitrogen interaction, which has provided a useful basis for discussions of models of biological nitrogen fixation. When molecular nitrogen is bound to certain metal–ligand combinations, it retains the ability to act as a donor to molecules such as $AlMe_3$, PF_5, $TiCl_3.3THF$, $CrCl_3.3THF$, $MoCl_4.2THF$ and $Pt_2Cl_4(PEt_3)_2$[149, 150]. The value of $\nu(N-N)$ in these binuclear adducts is up to 650 cm^{-1} lower than in the free nitrogen molecule, indicating a very substantial bond-weakening effect. Indeed in some examples $\nu(N-N)$ is reduced to the N=N level of, for example, azobenzene.

TABLE 27 MOLYBDENUM COMPLEXES OF MOLECULAR NITROGEN

Mononuclear complexes	Colour	ν(N–N) (cm^{-1})
trans-Mo(N$_2$)$_2$(diphos)$_2$	orange–yellow	2020, 1970 (Nujol) 1979 (benzene)
Mo(π-PhMe)(N$_2$)PPh$_3$)$_2$	orange	2005 (toluene)
Mo(N$_2$)Cl(diphos)$_2$	yellow	1970, 1950 sh(Nujol) 1975 (CH$_2$Cl$_2$)
trans-Mo(N$_2$)$_2$(Ph$_2$PCH : CHPPh$_2$)$_2$	–	–
trans-Mo(N$_2$)$_2$(Ph$_2$PMe)$_4$	orange	1926
cis-Mo(N$_2$)$_2$(PhMe$_2$P)$_4$	yellow	2010, 1937 (CsBr and benzene)
cis-Mo(N$_2$)$_2$(Bu^n$_3$P)$_4$	yellow	2065, 1980, 1940
Binuclear complexes		
[Mo(π-C$_6$H$_6$)(PPh$_3$)$_2$]$_2$N$_2$	maroon	1910 (Raman, solid)
(PMe$_2$Ph)$_4$Re(Cl)N$_2$MoOCl$_3$Et$_2$O	emerald green	1795
(PMe$_2$Ph)$_4$Re(Cl)N$_2$MoOCl$_3$. THF	emerald green	1795
(PMe$_2$Ph)$_4$Re(Cl)N$_2$MoOCl$_3$PMePh$_2$	emerald green	1810
(PMe$_2$Ph)$_4$Re(Cl)N$_2$MoOCl$_3$PEtPh$_2$	emerald green	1810
(PMe$_2$Ph)$_4$Re(Cl)N$_2$MoOCl$_3$PPh$_3$	emerald green	1805
Polynuclear complexes		
(PhMe$_2$P)$_4$Re(Cl)N$_2$Mo$_2$OCl$_5$Et$_2$O	bright blue	1680
(PhMe$_2$P)$_4$Re(Cl)N$_2$Mo$_2$OCl$_5$. THF	bright blue	1680
(PMe$_2$P)$_4$Re(Cl)N$_2$Mo$_2$OCl$_5$PMePh$_2$	purple	–

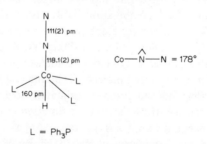

Figure 50 The structure of CoH(N$_2$)(Ph$_3$P)$_3$.

The type of acceptor molecule that produces the maximum bond-weakening effect has been distinguished by Chatt and coworkers[151] as one capable of accepting electron density from the bonding π-orbitals of nitrogen into vacant nonbonding orbitals. Complexes of transition metals with few d electrons satisfy this criterion, for example

$$\textit{trans-}(PMe_2P)_4ClRe(N_2)MoCl_4(Et_2O), \quad \nu(N–N) = 1795 \text{ cm}^{-1}$$
$$(diphos)_2ClMo(N_2)MoCl_4 . \text{THF}, \quad \nu(N–N) = 1770$$
$$1720 \text{ cm}^{-1}$$

With other acceptors, the nonbonding d-orbitals are either empty but energetically inaccessible (for example $AlMe_3$) or are half filled or filled (for example Cr(III) and Pt(II), respectively). The addition of these acceptors to the mononuclear nitrogen complexes hardly affects the N—N bond strength.

Complexes of molecular nitrogen have provided models for most of the steps in nitrogen fixation, namely binding and activation of the nitrogen molecule and product release, but not for the critical one, reduction to ammonia under conditions likely to obtain in enzymatic nitrogen fixation. Evidently, the N—N bond is not sufficiently weakened in any of the complexes prepared so far for it to be cleaved under mild aqueous reducing conditions. Chatt and his associates have

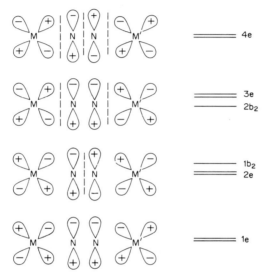

Figure 51 The simplified π-molecular-orbital scheme for a linear M—N—N—M' system. From Chatt *et al.*[152].

discussed the bonding in linear M—N—N—M' systems in terms of linear combinations of metal d and N_2 p orbitals[152] and have provided a rationalisation of this failure to achieve reduction of the co-ordinated nitrogen molecule. A simplified π molecular-orbital scheme for a C_{4v} M—N—N—M' system is shown in figure 51. The 1e and 4e orbitals are mainly π-bonding and π^*-antibonding N_2 orbitals with a small amount of metal orbital mixed in, depending on the electron affinity of the metal orbitals relative to those of nitrogen. The 2e and 3e orbitals are mainly metal d_{xz} and d_{yz} in character; $1b_2$ and $2b_2$ are mainly d_{xy} and contribute insignificantly to the bonding. The strength of the N—N bond is governed by the number of electrons in the e orbitals; the N_2 molecule provides four electrons to fill the 1e level and the partial metal character of this orbital results in a slight weakening of the N—N bond. All the known M—N—N—M' complexes have at least one d^n metal ion ($n = 4$ or 6); thus there are sufficient electrons to fill the 2e and

$1b_2$ molecular orbitals. Since the 2e orbital is antibonding for N_2 the N—N bond will be considerably weakened. However, if electrons are also available to occupy the $2b_2$ and 3e molecular orbitals, which are bonding for N_2, then the bond weakening will be counteracted. For example, in the $[(NH_3)_5 RuN_2 Ru(NH_3)_5]^{4+}$ ion[146], there are sufficient electrons to occupy all but the 4e molecular orbitals and the N—N bond strength is not much changed from that in the mononuclear complexes. When fewer electrons are available as with Ti(III) and Mo(IV), substantial weakening of the N—N bond via occupation of the $1b_2$ and 2e molecular orbitals occurs as evidenced by the low values of $\nu(N-N)$ in binuclear complexes containing these ions. However, even with these systems, there is some occupation of the $2b_2$ and 3e orbitals; any attempt at reduction would result in further occupation of π-bonding N—N orbitals with consequent strengthening of the N—N bond. Various possibilities have been suggested whereby such problems may be overcome.

One of these is an enforced departure of the M—N—N—M' system from linear C_4 symmetry to, say, C_2 as obtains in the easily reducible azobenzene. Such a situation would require a high degree of asymmetry in the π-bonding nature of the metal–ligand binding centre; it would be favoured by the provision of a highly unsymmetrical ligand-field environment by the protein of nitrogenase. Of the two metallic constituents of nitrogenase, molybdenum appears to be the most likely to adopt low-symmetry environments (see also section 15.1). A second possibility concerns the availability of nonbonding d-electrons. If the total number of these electrons provided by the two metals of the M—N—N—M' system is at least four and not more then eight, then the 1e and 2e and possibly also the $1b_2$ orbitals would be occupied and the weakening of the N—N bond would be at a maximum. It might be sufficient for a strong metal–nitrogen bond to be formed at the expense of the N—N bond. The strength of such a bond would need to be greater than about 460 kJ mol^{-1} (110 kcal/mole), and its formation would require the metal ion to be capable of a formal three-electron oxidation. Molybdenum in its lower oxidation states, namely Mo(II) and Mo(III), could satisfy all the criteria outlined above. The catalytic fixation of nitrogen would then involve cycling between Mo(III) and Mo(VI) or between Mo(II) and Mo(V). In view of the fact that Mo(II) is unknown in aqueous solution it seems that if molybdenum is involved at the active site of biological nitrogen fixation, and if fixation proceeds by way of the formation of Mo—N bonds, then it is the Mo(III)–Mo(VI) system that is more likely to be involved. On the basis of ideas such as these, Chatt[152] proposed a speculative scheme (see figure 52) for the reduction of molecular nitrogen. For the purposes of illustration the metal ions are shown only in formal octahedral co-ordination, although this is not necessary and may even hinder the reduction process. The function of iron in a scheme such as this may be to store and transfer electrons or to act as the initial binding site for the nitrogen molecule, which is then passed to the molybdenum atoms. This latter possibility seems less probable on the grounds that Fe(II) as a d^6 ion is likely to form too stable a complex with the nitrogen molecule; it is difficult to conceive the benefit that

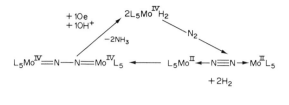

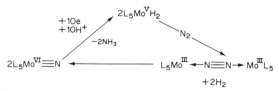

Figure 52 Speculative schemes for nitrogen reduction involving binuclear molybdenum binding sites.

could accrue from initial binding at an iron atom. In another scheme based on a binuclear binding site, this time involving iron, Bulen and coworkers[153] proposed that in the resting enzyme the protein conformation is such that the metal centres are not at the correct distance to bind nitrogen, but that a change in conformation can be brought about by ATP or H$^+$ ions. Another possibility is that changes in the metal–ligand distances that accompany reduction of the metal prior to nitrogen binding can bring about the desired conformational change. Once the metal sites

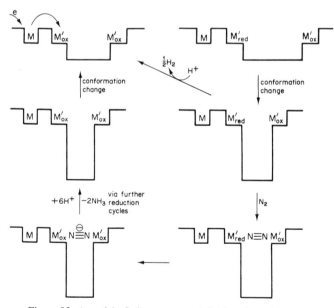

Figure 53 A model of nitrogenase activity based on the creation of a binuclear nitrogen binding site via a change in the conformation of the protein.

have been positioned correctly, the binuclear system can bind nitrogen and electrons can be transferred from the reduced metal ion(s). Cycles of reduction and protonation then lead to ammonia. This scheme (see figure 53) is equally applicable to molybdenum binding sites and iron as a component of an electron-transfer system.

16 CONCLUSION

Direct evidence of a role for molybdenum in substrate binding and of an involvement in the catalytic phase of enzyme activity, is available only for xanthine oxidase. The marked similarities between the e.p.r. parameters and substrate specificities of this enzyme and those of aldehyde oxidase suggest that the molybdenum is also present at the substrate binding site in the latter enzyme. Limited e.p.r. data also locate molybdenum at this site in nitrate reductase from *M. denitrificans* and in mammalian and avian sulphite oxidase. Studies of molybdenum complexes as models of the enzymes have centred on complexes with sulphur-containing ligands, especially cysteine, although there is only meagre evidence for sulphur-liganded molybdenum in the enzymes. A molybdenum-cysteine complex exhibits reductase activity towards several substrates that are also substrates for nitrogenase but fails to produce more than trace yields of ammonia from molecular nitrogen. The models also differ appreciably from the enzymes in their content of e.p.r.-active Mo(V). This is as high as about 50 per cent in the reduced oxidase enzymes, but does not exceed more than about 4 per cent in the models. Possibly, the conformation of the protein at the active sites of the enzymes results in the stabilisation of catalytically active, monomeric molybdenum species. But this, like the various proposed schemes of nitrogenase action, remain speculative in the absence of structural data.

REFERENCES

1. Hewitt, E. J., *Biol. Rev.*, **34** (1959), 333.
2. Nason, A., in *The Enzymes*, vol. 7 (eds P. D. Boyer, H. A. Lardy and K. Myrback), Academic Press, New York (1963), p. 587.
3. Postgate, J. R. (ed.), *The Chemistry and Biochemistry of Nitrogen Fixation*, Plenum, London (1971).
4. Hardy, R. W. F., Burns, R. C., and Parshall, G. W., *Adv. Chem. series Am. chem. Soc.*, **100** (1971), 219.
5. Beevers, L., and Hageman, R. H., *A. Rev. Pl. Physiol.*, **20** (1969), 495.
6. Cohen, H. J., Fridovich, I., and Rajagopalan, K. V., *J. biol. Chem.*, **246** (1971), 374.
7. Irreverre, F., Mudd, S. H., Heizer, W. D., and Laster, L., *Biochem. Med.*, **1** (1967), 187.
8. Williams, R. J. P., *RIC Rev.*, **1** (1968), 13.
9. Bray, R. C., in *The Enzymes*, vol. 7 (eds P. D. Boyer, H. A. Lardy and K. Myrback), Academic Press, New York (1963), p. 533.
10. Killeffer, D. H., and Linz, A., *Molybdenum Compounds*, Interscience, New York (1952).
11. McElroy, W. D., and Glass, B. (eds) *A Symposium on Inorganic Nitrogen Metabolism*, Johns Hopkins Press, Baltimore (1956).
12. Underwood, E. J., *Trace Elements in Human and Animal Nutrition*, Academic Press, New York (2nd ed., 1962), p. 100.
13. Bray, R. C., Chisholm, A. J., Hart, L. I., Meriwether, L. S., and Watts, D. C., in *Flavins and Flavoproteins* (ed., E. C. Slater), Elsevier, Amsterdam (1956), p. 117.
14. Beinert, H., Orme-Johnson, W. H., in *Magnetic Resonance in Biological Systems* (eds A. Ehrenberg, B. G. Malmstrom and T. Vanngard), Pergamon, Oxford (1967), p. 221.
15. Spence, J. T., *Z. Naturwiss. Med. Grundlagenforsch.*, **2** (1966), 267.
16. Spence, J. T., *Co-ord. Chem. Rev.*, **4** (1969), 475.
17. Bray, R. C., and Swann, J. C., *Struct. Bond.*, **11** (1972), 107.
18. Williams, R. J. P., and Mitchell, P. C. H., *J. chem. Soc.* (1960), 1912.
19. Ahrland, S., Chatt, J., and Davies, N. R., *Q. Rev., chem. Soc.*, **12** (1958), 265.
20. Pearson, R. G., *J. Am. chem. Soc.*, **85** (1963), 3533.
21. Williams, R. J. P., in *Advances in the Chemistry of Co-ordination Compounds* (ed. S. Kirschner), Macmillan, New York (1961), p. 65.
22. Swann, J. C., and Bray, R. C., *Eur. J. Biochem.*, **26** (1972), 407.
23. Massey, V., Komai, H., Palmer, G., and Elion, G. B., *J. biol. Chem.*, **245** (1970), 2837.
24. Lever, A. B. P., *Inorganic Electronic Spectroscopy*, Elsevier, Amsterdam (1968).
25. Tomkinson, J. C., and Williams, R. J. P., *J. chem. Soc.* (1970), 2070.
26. Bayer, E., Bacher, A., Krauss, P., Voelter, W., Barth, G., Bunnenberg Djerassi, C., *Eur. J. Biochem.*, **22** (1971), 580.
27. Ingram, D. J. E., *Biological and Biochemical Applications of Electron Spin Resonance* Hilger, London (1969).
28. Meriwether, L. S., Marzluff, W. F., and Hodgson, W. G., *Nature, Lond.*, **212** (1966), 465.
29. Dickinson, C. J., and Smellie, J. M., *Brit. Med. J.*, **102** (1959), 1217.
30. Andrews, P., Bray, R. C., Edwards, P., and Shooter, K. V., *Biochem. J.*, **93** (1964), 627.
31. Mason, H. S., and Onoprienko, I., unpublished results.
32. Mason, H. S., *Science, N.Y.*, **125** (1957), 1185.
33. Hart, L. I., McGartoll, M. A., Chapman, H. R., and Bray, R. C., *Biochem. J.*, **116** (1970), 851.
34. Dennard, A. E., and Williams, R. J. P., in *Transition Metal Chemistry*, vol. 2 (ed. R. L. Carlin) Dekker, New York (1966), p. 115.
35. Mozumi, M., Hayashikawa, R., and Piette, L. H., *Archs Biochem. Biophys.*, **119** (1967), 288.
36. Garbett, K., Gillard, R. D., Knowles, P. F., and Stangroom, J. E., *Nature, Lond.*, **215** (1967), 824.

37. Morell, D. B., *Biochem. J.*, **51** (1952), 657.
38. Massey, V., Brumby, P. E., Komai, H., and Palmer, G., *J. biol. Chem.*, **244** (1969), 1682.
39. Palmer, G., and Massey, V., *J. biol. Chem.*, **244** (1969), 2614.
40. Mackler, B., Mahler, H. R., and Green, D. E., *J. biol. Chem.*, **210** (1954), 149.
41. Spector, T., and Johns, D. G., *Biochem. Biophys. Res. Commun.*, **32** (1961), 1039.
42. Rajagopalan, K. V., and Handler, P., *J. biol. Chem.*, **239** (1964), 2027.
43. Aleman, V., Smith, S. T., Rajagopalan, K. V., and Handler, P., in *Flavins and Flavoproteins* (ed. E. C. Slater), Elsevier, Amsterdam (1966), p. 99.
44. Bray, R. C., Beinert, H., and Palmer, G., *J. biol. Chem.*, **239** (1964), 2667.
45. Bray, R. C., Malmstrom, B. G., and Vanngard, T., *Biochem. J.*, **73** (1959), 193.
46. Bray, R. C., Knowles, P. F., Meriwether, L. S., in *Magnetic Resonance in Biological Systems* (eds A. Ehrenberg, B. G. Malmstrom and T. Vanngard) Pergamon, Oxford (1967), p. 249.
47. Komai, H., Massey, V. and Palmer, G., *J. biol. Chem.*, **244** (1969), 1692.
48. Gutfreund, H., and Sturtevant, J. M., *Biochem. J.*, **73** (1959), 1.
49. Pick, F. M., Ph.D. Thesis, University of London (1971).
50. Bray, R. C., and Knowles, P. F., *Proc. R. Soc.*, **A. 302** (1968), 351.
51. McGartoll, M. A., Pick, F. M., Swann, J. C., and Bray, R. C., *Biochim. Biophys. Acta*, **212** (1970), 523.
52. Bray, R. C., and Vanngard, T., *Biochem. J.*, **114** (1969), 725.
53. Bray, R. C., Knowles, P. F., Pick, F. M., and Vanngard, T., *Biochem. J.*, **107** (1968), 601.
54. Fridovich, I., and Handler, P., *J. biol. Chem.*, **235** (1961), 1836.
55. Totter, J. R., Medina, V. J., and Scoseria, J. L., *J. biol. Chem.*, **235** (1960), 238.
56. Fridovich, I., *Acct chem. Res.*, **5** (1972), 321.
57. Massey, V., Komai, H., and Palmer, G., *J. biol. Chem.*, **245** (1970), 2837.
58. DeBernard, B., *Biochim. Biophys. Acta*, **23** (1957), 510.
59. Rajagopalan, K. V., Fridovich, I., and Handler, P., *J. biol. Chem.*, **237** (1962), 922.
60. Rajagopalan, K. V., Handler, P., Palmer, G., and Beinert, H., *J. biol. Chem.*, **243** (1968), 3784.
61. Rajagopalan, K. V., Handler, P., Palmer, G., and Beinert, G., *J. biol. Chem.*, **243** (1968), 3797.
62. Igo, R. P., and Mackler, B., *Biochim. Biophys. Acta*, **44** (1960), 310.
63. Rajagopalan, K. V., and Handler, P., *J. biol. Chem.*, **239** (1964), 2022.
64. Perault, A-M., Valdermoro, C., and Pullman, B., *J. theor. Biol.*, **2** (1961), 180.
65. Pick, F. M., McGartoll, M. A., and Bray, R. C., *Eur. J. Biochem.*, **18** (1971), 65.
66. Brumby, P. E., Miller, R. W., and Massey, V., *J. biol. Chem.*, **240** (1965), 2222.
67. Handler, P., *J. biol. Chem.*, **242** (1967), 4097.
68. Smith, S. T., Rajagopalan, K. V., and Handler, P., *J. biol. Chem.*, **242** (1967), 4108.
69. Nicholas, D. J. D., and Nason, A., *J. biol. Chem.*, **211** (1954), 183.
70. Garrett, R. H., and Nason, A., *Proc. natn. Acad. Sci. U.S.A.*, **58** (1967), 1603.
71. Garrett, R. H., and Nason, A., *J. biol. Chem.*, **244** (1969), 2870.
72. Kinsky, S. C., and McElroy, W. D., *Archs Biochem. Biophys.*, **73** (1958), 466.
73. Nason, A., and Evans, H. J., *J. biol. Chem.*, **202** (1953), 655.
74. Fewson, C. A., and Nicholas, D. J. D., *Biochim. Biophys. Acta*, **49** (1961), 335.
75. Forget, P., and Dervartanian, D. V., *Biochim. Biophys. Acta*, **256** (1972), 600.
76. Walker, G. W., and Nicholas, D. J. D., *Nature, Lond.*, **189** (1961), 141.
77. Fewson, C. A., and Nicholas, D. J. D., *Nature, Lond.*, **190** (1961), 2.
78. Subramanian, K. N., and Sorger, G. J., *Biochim. Biophys. Acta*, **256** (1972), 533.
79. Nason, A., Kuo-Yung Lee, Su-Shu Pan, Ketchum, P. A., Lamberti, A., and DeVries, J., *Proc. natn. Acad. Sci. U.S.A.*, **68** (1971), 3242.
80. Ketchum, P. A., Cambier, H. Y., Frazier, W. A., Mandansky, C. H., and Nason, A., *Proc. natn. Acad. Sci. U.S.A.*, **66** (1970), 1016.
81. Chatt, J., and Leigh, G. J., *Chem. Soc. Rev.*, **1** (1972), 121.
82. Bortels, H., *Arch. Mikrobiol.*, **1** (1930), 333.
83. Carnahan, J. E., Mortenson, L. E., and Castle, J. E., *Biochim. Biophys. Acta*, **44** (1960), 520.
84. Mortenson, L. E., *Meth. Enzym.*, **24B** (1972), 446.
85. Burns, R. C., and Hardy, R. W. F., *Meth. Enzym.*, **24B** (1972), 480.

86. Bulen, W. A., and Lecomte, J. R., *Proc. natn. Acad. Sci. U.S.A.*, **53** (1965), 532.
87. Burns, R. C., Holstein, R. D., and Hardy, R. W. F., *Biochem. Biophys. Res. Commun.*, **39** (1970), 90.
88. Kelly, M., and Lang, G., *Biochim. Biophys. Acta*, **223** (1970), 86.
89. Cook, K. A., D. Phil. Thesis, University of Sussex (1971).
90. Schrauzer, G. N., and Doemeny, P. A., *J. Am. chem. Soc.*, **93** (1971), 1608.
91. Dilworth, M. J., *Biochim. Biophys. Acta*, **127** (1966), 285.
92. Bowden, F. L., and Lever, A. B. P., *Organometal. Chem. Rev.*, **3** (1968), 227.
93. Dalton, H., Morris, J. A., Ward, M. A., and Mortenson, L. E., *Biochemistry*, **10** (1971), 2066.
94. Smith, B. E., Lowe, D. J., and Bray, R. C., *Biochem. J.*, **130** (1972), 641.
95. Bray, R. C., Holsten, R. D., and Hardy, R. W. F., *Biochem. Biophys. Res. Commun.*, **39** (1970), 90.
96. Cohen, H. J., and Fridovich, I., *J. biol. Chem.*, **246** (1971), 359.
97. Kessler, D. I., and Rajagopalan, K. V., *J. biol. Chem.*, **247** (1972), 6566.
98. Yokoyama, E., Yoder, R. E., and Frank, N. R., *Archs envir. Hlth*, **22** (1971), 389.
99. Bhaghat, B., and Lockett, M. F., *J. Pharm. Pharmac.*, **12** (1960), 690.
100. Cohen, H. J., Betcher-Lange, S., Kessler, D. L., and Rajagopalan, K. V., *J. biol. Chem.*, **247** (1972), 7759.
101. Fridovich, I., Farkas, W., and Handler, P., *J. biol. Chem.*, **236** (1961), 1841.
102. Hempfling, W. P., Trundinger, P. A., and Vishniac, W., *Arch. Mikrobiol.*, **59** (1967), 149.
103. Kodama, A., and Takeshi, M., *Pl. Cell Physiol.*, **9** (1968), 709, 725.
104. Howell, L. G., and Fridovich, I., *J. biol. Chem.*, **243** (1968), 5941.
105. Charles, A. M., *Archs Biochem. Biophys.*, **129** (1969), 124.
106. Lyric, R. M., and Suzuki, I., *Can. J. Biochem.*, **48** (1970), 334.
107. McCord, J. M., and Fridovich, I., *J. biol. Chem.*, **244** (1969), 6049.
108. Cohen, H., and Fridovich, I., *J. biol. Chem.*, **246** (1970), 367.
109. Cohen, H. J., Fridovich, I., and Rajagopalan, K. V., *J. biol. Chem.*, **246** (1971), 374.
110. McAuliffe, C. A., and Murray, S. G., *Inorg. Chim. Acta Rev.*, **6** (1972), 105.
111. Kay, A., and Mitchell, P. C. H., *J. chem. Soc. (A)* (1970), 2421.
112. Spivak, B., and Dori, Z., *Chem. Commun.* (1970), 1716.
113. Drew, M. G. B., and Kay, A., *J. chem. Soc. (A)* (1971), 1851.
114. Drew, M. G. B., and Kay, A., *J. chem. Soc. (A)* (1971) 1846.
115. Knox, J. R., and Prout, C. K., *Acta crystallogr.*, **25B** (1969), 1857.
116. Spence, J. T., and Chang, H., *Inorg. Chem.*, **2** (1963), 319.
117. Martin, J. M., and Spence, J. T., in Cais, M. (ed.), *Progress in Coordination Chemistry*, Elsevier, Amsterdam (1968), 492.
118. Huang, T. G., and Haight, G. P., *J. Am. chem. Soc.*, **92** (1970), 2336.
119. Huang, T. G., and Haight, G. P., *J. Am. chem. Soc.*, **93** (1971), 611.
120. Blake, A. B., Cotton, F. A., and Wood, J. S., *J. Am. chem. Soc.*, **86** (1964), 3024.
121. Schneider, P. W., Bravard, D. C., McDonald, J. W., and Newton, W. E., *J. Am. chem. Soc.*, **94** (1972), 8640.
122. Butcher, A., and Mitchell, P. C. H., *Chem. Commun.* (1967), 176.
123. Spence, J. T., and Lee, G. R., *Inorg. Chem.*, **11** (1972), 2354.
124. Handler, P., Rajagopalan, K. V., and Aleman, V., *Fedn Proc. Fedn Am. Socs exp. Biol.*, **23** (1964), 30.
125. Hemmerich, H., *Helv. chim. Acta*, **47** (1964), 464.
126. Ehrenberg, A., and Hemmerich, H., in *Biological Oxidations* (ed. T. P. Singer), Interscience, New York (1968), p. 239.
127. Spence, J. T., Heydanek, M., and Hemmerich, P., in *Magnetic Resonance in Biological Systems* (eds A. Ehrenberg, B. G. Malmstrom and T. Vanngard), Pergamon, Oxford (1967), p. 269.
128. Spence, J. T., *Archs. Biochem. Biophys.*, **137** (1970), 287.
129. Spence, J. T., and Kroneck, P. *Inorg. nucl. Chem. Lett.*, **9** (1973), 177.
130. Haight, G. P., *Acta chem. Scand.*, **15** (1961), 2012.
131. Haight, G. P., Mohilinier, P., and Katz, A., *Acta chem. Scand.*, **16** (1962), 221.
132. Kolthoff, I. M., and Hodara, I., *J. electroanal. Chem.*, **5** (1963), 2.
133. Guymon, E. P., and Spence, J. T., *J. phys. Chem.*, **70** (1966), 1964.

134. Spence, J. T., and Heydanek, M., *Inorg. Chem.,* 6 (1967), 1489.
135. Spence, J. T., and Frank, J. H., *J. Am. chem. Soc.,* 85 (1963), 116.
136. Schrauzer, G. N., Doemeny, P. A., Kiefer, G. W., and Frazier, R. H., *J. Am. chem. Soc.,* 95 (1972), 3604.
137. Schrauzer, G. N., and Schlesinger, G., *J. Am. chem. Soc.,* 92 (1970), 1808.
138. Newton, W. E., Corbin, J. L., Schneider, P. W., and Bulen, W. A., *J. Am. chem. Soc.,* 93 (1971), 268.
139. Melby, L. R., *Inorg. Chem.,* 8 (1969), 1539.
140. Kelly, M., Postgate, J. R., and Richards, R. L., *Biochem. J.,* 102 (1967), 1.
141. Kelly, M., *Biochim. Biophys. Acta,* 171 (1969), 9.
142. Kelly, M., *Biochim. Biophys. Acta,* 191 (1969), 527.
143. Hill, R. E. E., and Richards, R. L., *Nature, Lond.,* 233 (1971), 114.
144. Cameron, T. S., and Prout, C. K., *Acta crystallogr.,* 28B (1972), 453.
145. Leigh, G. J., in *The Chemistry and Biochemistry of Nitrogen Fixation* (ed. J. R. Postgate), Plenum, London (1971), p. 1.
146. Harrison, D. E., and Taube, H., *J. Am. chem. Soc.,* 89 (1967), 5706.
147. Yammamoto, A., Kitazume, S., Pu, L. S., and Ikeda, S., *J. Am. chem. Soc.,* 89 (1967), 3071.
148. Davies, B. R., Payne, N. C., and Ibers, J. A., *Inorg. Chem.,* 8 (1969), 2719.
149. Chatt, J., Dilworth, J. R., Leigh, G. J., and Richards, R. L., *Chem. Commun.* (1970), 955.
150. Atkinson, L. K., Mawby, A. H., and Smith, D. C., *Chem. Commun.,* (1971), 157.
151. Chatt, J., Dilworth, J. R., Richards, R. L., and Sanders, J. R., *Nature, Lond.,* 224 (1969), 5225.
152. Chatt, J., Fay, R. C., and Richards, R. L., *J. chem. Soc. (A)* (1971), 702.
153. Bulen, W. A., Lecomte, J. R., Burns, R. C., and Hinkson, J. in *Non-Haem Iron Proteins: Role in Energy Conversion* (ed. A. San Pietro), Antioch Press, Yellow Springs (1965), p. 261.

PART 4
Polynuclear Iron(III) Proteins

JOHN WEBB [†]

Division of Clinical Immunology and Rheumatology, Department of Medicine, University of Alabama, Birmingham, Alabama

† Current address: Research School of Chemistry, P.O. Box 4, Canberra, A.C.T. 2600, Australia.

17 INTRODUCTION

·The essential role of iron for the human organism was recognised in antiquity, whence legend (and a recent review[1]) records that a glass of wine and rust was sufficient to cure sexual impotence and to benefit those who, unfortunately, could not 'cohabit properly'. A complete understanding of these remarkable phenomena still escapes contemporary biochemistry, and even bioinorganic chemistry, yet, of all the essential trace metals, iron and its biological functions are probably the best understood. Many of these functions depend on the aqueous chemistry of iron(III), which is now known to be dominated by hydrolysis and polymerisation, with the formation of polynuclear iron(III) species, that is species that contain at least two iron atoms interacting via some bonding arrangement. This review discusses the results of structural studies, using a variety of physical methods, on a number of proteins that bind a large number ($\geqslant 50$) of iron(III) ions in a polynuclear fashion. Many of these studies have been carried out since 1969, when this research field was the subject of an excellent review[1].

Several proteins that bind more than one mole of iron per mole of protein are not included in this review. Transferrin binds two iron(III) ions but they appear to be sufficiently distant for no polynuclear interaction to occur[2-4]. Hemerythrin, which has been shown to contain antiferromagnetically coupled iron(III) ions[5-7], is considered elsewhere. Nonhaem iron–sulphur proteins, for example ferrodoxins[8], will also not be considered here.

Part 4 is organised in the following way. The presently available information on the hydrolytic polymerisation of iron(III) is summarised briefly in chapter 18. Chapter 19 considers the strengths and weaknesses of various experimental techniques used in studies of polynuclear iron(III) proteins, with particular emphasis on their application to model iron(III) compounds. Following consideration in chapter 20 of the several biological roles of polynuclear iron(III), chapter 21 discusses the structural results obtained for a number of biological systems. Finally a biological perspective of these studies is presented in chapter 22.

18 HYDROLYTIC POLYMERISATION OF IRON(III)

Potentiometric and other studies by a number of investigators have implicated a binuclear complex, $[Fe_2(OH)_2]^{4+}$, as the predominant hydrolysis species in solutions of iron(III) even at pHs less than 2[9–13]. Continued hydrolysis of iron(III) solutions readily leads to the formation of various poorly characterised polymeric precipitates and, eventually, to the formation of the insoluble brown polymerisation product, iron(III) hydroxide ($K_{sp} = 10^{-38.7}$, 25°, 3 M NaClO$_4$[14]. Although this familiar brown colloid has usually been assumed to be amorphous and ill defined, hydrolysates that are of small particle size (<10 nm diameter) but of sufficient crystallinity to give X-ray powder patterns have been isolated recently[15,16]. The iron(III) oxide and iron(III) oxyhydroxide systems contain several crystalline phases: α-, β-, γ-Fe$_2$O$_3$, Fe$_3$O$_4$, and α-, β-, γ-, δ-FeOOH. Their structures and interconversions have been extensively studied[17,18].

Inclusion of chelating agents in hydrolysed iron(III) solutions has led to the isolation of stable dimeric complexes. The proposed dimeric unit[19–21] of the chelate-free system, $[Fe(OH)_2Fe]^{4+}$, has been observed in the crystal structure of a dimeric species isolated from the iron(III)–picolinic-acid system[22]. Many dimers are analogous to that formed with HEDTA (N′-(2-hydroxyethyl)ethylenediamine-N,N,N′-triacetic acid), the crystal structure of which shows an almost linear (165°) oxo-bridged Fe—O—Fe structural unit linking the two subunits of the molecule together[23].

Gel filtration and ultrafiltration of an iron(III) nitrate solution to which two moles of OH⁻/Fe had been added resulted in the isolation of a remarkably homogeneous polymer, of molecular weight 150 000, incorporating about 1200 iron atoms[24,25]. It appears spherical in the electron microscope with a diameter of 7 nm. An iron(III)–citrate polymer also appears spherical, 7 nm in diameter, with a molecular weight of 2.1×10^5 [26,27]. A polymer of molecular weight 0.65×10^5 has been isolated by alcohol-induced precipitation from iron(III)–fructose solutions[28].

19 EXPERIMENTAL TECHNIQUES: THEORY AND APPLICATION TO IRON(III) MODEL COMPOUNDS

Particularly useful techniques for identifying the presence of polynuclear iron(III) ions are those that are sensitive to the characteristic feature of the cluster—the interaction between the different sets of 3d electrons. In this regard the magnetic moment, its temperature dependence and the e.p.r. spectrum are particularly important probes. Several other techniques have been used to elucidate structural details of the co-ordination geometry, the bonding between iron atoms in the cluster and between the cluster and the protein. These include electronic-absorption spectroscopy within the d^5 configuration, Mössbauer and vibrational spectroscopies and X-ray diffraction.

19.1 MAGNETIC BEHAVIOUR

19.1.1 Monomeric iron(III)
The familiar splittings of the d orbitals by the ligand field created by octahedral, tetrahedral and tetragonal arrays of ligand atoms are shown in figure 54. The figure also shows the possible spin configurations of the d^5 iron(III) ion in the various symmetries. In both octahedral (O_h) and tetrahedral (T_d) symmetries the high-spin case corresponds to five unpaired electrons and the low-spin case to one. With an electron-spin quantum number s of $1/2$ the total spin number S in the two cases is $5/2$ and $1/2$, respectively. The situation with three unpaired electrons, $S = 3/2$, which is not attainable in O_h or T_d symmetry has been characterised in only a few cases, for example in a series of monohalogen bis(N,N-dimethyldithiocarbamato) iron(III) complexes of general formula $[Fe(III)X . (S_2CNR_2)_2]$[29].

The choice between the $5/2$ and $1/2$ spin states in these symmetries is determined by the balance between the strength of the ligand field ($10Dq$) and the magnitude of the Coulomb and exchange energies for the d electrons. A ligand field of the order of 30 000 cm^{-1} attainable by strong-field ligands such as cyanide, dipyridine and o-phenanthroline, is required to form the $1/2$ ground state.

An intermediate situation occurs when the ligand-field strength is close to that required for the transition from the high-spin to the low-spin configuration. When the energy difference between these two configurations is comparable to kT (at 300 K, $kT = 208$ cm^{-1}), there is an equilibrium mixture of high-spin and low-spin states, resulting in a temperature-dependent magnetic moment that is a weighted average of the two components. One of the few established examples of this equilibrium situation with iron(III) occurs in the compound Fe(III) (diethyldithio-carbamate)$_3$[29,30].

In general, iron(III) has its greatest affinity for oxygen ligands. However, chelating ligands with oxygen-donor and nitrogen-donor atoms are well known (for example, EDTA, HEDTA) as are sulphur ligands (for example, ferredoxins[8], dithiolene chelates[31]) and strong-field ligands binding through nitrogen only (for example, dipyridine). The many amino-acid side chains of proteins that can bind through oxygen or nitrogen have a ligand-field strength too small ($<15\,000$ cm^{-1}) to induce either spin pairing or a high-spin–low-spin equilibrium.

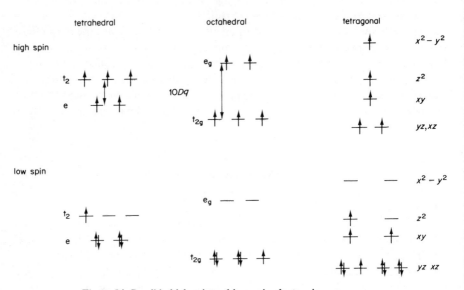

Figure 54 Possible high-spin and low-spin electronic configurations for a d^5 ion in ligand fields of tetrahedral, octahedral and tetragonal symmetries.

The magnetic moment μ_s of a single electron is simply related to the electron-spin quantum number. For an ion of total spin quantum number S describing the spin angular momentum, the magnetic moment μ for the ion is given (in Bohr magnetons, B.M.) by

$$\mu = g[S(S + 1)]^{1/2} \tag{19.1}$$

where g is the gyromagnetic ratio = 2.00023 for the free electron.

For the single-electron case, $S = s$ and μ is 1.73 B.M. For $S = 5/2$, μ is 5.92 B.M., and for $S = 3/2$, μ is 3.87 B.M. These are the values of the 'spin-only' magnetic moments, since no account is made of the orbital contribution to the moment from spin–orbit coupling. For the $S = 5/2$ case, the ground state of the complex $^6A_1$ is derived from the ^6S ground state of the free ion. For such a ground state $L = 0$ and no orbital contribution is possible. Thus, Na$_3$[FeF$_6$] has a magnetic moment per iron of 5.85 B.M.[32]. However, the moment for K$_3$[Fe(CN)$_6$] of 2.25 B.M. is increased from the spin-only value of 1.73 B.M. by an orbital contribution allowed in the $^2T_2$ ground state[32].

Magnetic moments are determined from the measured magnetic susceptibility χ of the sample in a magnetic field. The magnetic moment is calculated from the gram susceptibility χ_g by applying a diamagnetic correction to the molar susceptibility χ_M ($= \chi_g \times$ M.W.), giving χ_M^{corr}, which is used in the following equation

$$\mu_{\text{eff}} = 2.828\, (\chi_M^{\text{corr}} T)^{1/2} \qquad (19.2)$$

The temperature dependence of the magnetic susceptibility follows either the Curie ($\chi \propto 1/T$) or the Curie–Weiss law ($\chi \propto 1/(T - \theta)$). The Weiss constant θ, the intercept on the T axis of a plot of χ^{-1} against T, has often been taken as a measure of the degree of intermolecular interactions, yet a finite θ can result from other causes, for example spin–orbit coupling on the $^3T_{1g}$ ground state of the monomeric $K_3[Mn(CN)_6]$[32].

The diamagnetic correction accounts for the contribution to the susceptibility from the closed-shell electrons in the metal and ligand. Although the contribution from the organic component of a metal complex is small, $\approx -0.5 \times 10^{-6}$ CGS units per gram, a molar contribution from a protein of molecular weight 50 000 is $\approx -2.5 \times 10^{-2}$ CGS units. This is comparable to the paramagnetic molar susceptibility of $Fe(NO_3)_3 \cdot 9H_2O$ at 293 K of $+1.52 \times 10^{-2}$ CGS units[33]. Since diamagnetic and paramagnetic contributions to the observed susceptibility of a protein that binds only one or several metal ions can be of the same order of magnitude but of opposite sign, the net susceptibility is small and measurements of the paramagnetic component can be subject to significant errors.

Magnetic-susceptibility studies are usually carried out on solid samples, yet, particularly for metal-binding proteins, it is important to determine the solution susceptibility. This can be measured conveniently using an n.m.r. technique[34]. The solution under investigation is placed in a capillary tube, coaxially situated inside a normal n.m.r. tube containing the solvent identical to the test solution in all respects save the paramagnetic ions. For protein studies the reference solution ideally contains the metal-free protein in a concentration equimolar to that in the sample solution. Alternatively, the diamagnetism of the metal-free protein can be measured separately. Both tubes contain a standard proton reference, for example *tert*-butanol (TBA) for aqueous solutions. The paramagnetic ion changes the bulk susceptibility of the sample solution, changing the position of the TBA methyl resonance from that in the reference solution. From this shift the paramagnetic susceptibility can be calculated.

Superconducting magnets generate a magnetic field along the long axis of the sample rather than transverse to this axis as in the conventional configuration. This different magnetic polarisation changes the bulk-susceptibility shift to twice in magnitude and opposite in sign from that expected for conventional spectrometers[35]. This effect, together with the additional dispersion of chemical shifts at the higher n.m.r. frequency of superconducting units, makes them particularly suitable for paramagnetic susceptibility determinations.

19.1.2 Polynuclear iron(III)

The strength of the interaction between the sets of d electrons on different atoms of spins S_1 and S_2 can be expressed by the exchange-coupling constant J in the hamiltonian $\mathcal{H}'$ for the interaction[36]

$$\mathcal{H}' = -2JS_1S_2 \tag{19.3}$$

For $J > 0$, the interaction is ferromagnetic and for $J < 0$, antiferromagnetic. A few instances of ferromagnetic coupling are known for polynuclear clusters that are magnetically isolated from other clusters by diamagnetic matter[37]. None of these involve iron(III) ions. For a polynuclear cluster of iron(III) the magnetic interaction is antiferromagnetic, decreasing the magnetic moment per iron from the spin-only value. The strength of the antiferromagnetic coupling determines the extent of this decrease. For the polynuclear clusters of interest the antiferromagnetic coupling does not take place directly through space but rather indirectly through an intervening bridging unit, for example $-O-$ or $-OH-$, that is via a superexchange coupling mechanism.

In addition to lowering the magnetic moment per iron from the spin-only value, antiferromagnetic coupling changes the temperature-dependent behaviour of the moment. At any temperature the order resulting from magnetic effects competes with the disorder from thermal agitation. The magnetic susceptibility attains a maximum at the Néel temperature T_N. Although the susceptibility increases with decreasing temperature down to T_N, a plot of χ^{-1} against T is nonlinear. The antiferromagnetically lowered magnetic moment characteristically decreases with decreasing temperature. It is this behaviour that is the *most reliable criterion* for identifying polynuclear iron(III) species of moderate size (say, <100 iron(III) ions coupled together).

Several complexes containing the dimeric oxo-bridged iron(III) system, [Fe(III)$-$O$-$Fe(III)] have been characterised by detailed magnetic-susceptibility measurements[38-40]. In all cases the temperature dependence can be fitted quite well using the spin–spin interaction model described above. The resulting energy levels include a diamagnetic ground state and thermally accessible paramagnetic excited states. Strong antiferromagnetic coupling may lower the ground state sufficiently to allow the dimeric complex to become diamagnetic at some temperatures. The strongly coupled ($J \approx 100$ cm^{-1}) dimeric systems, [(Fe(III) HEDTA)$_2$O]$^{2-}$ and [Fe(III)EDTA)$_2$O]$^{4-}$ behave in this way, becoming diamagnetic at ≈ 30 K. The results for all such oxo-bridged systems known can be fitted equally well by the $S_1 = 5/2$, $S_2 = 5/2$ or $S_1 = 3/2$, $S_2 = 3/2$ models, but not the $S_1 = 1/2$, $S_2 = 1/2$ model. The extra spin levels of the 5/2, 5/2 model are not sufficiently populated at 300 K to discriminate between the two working models. Electronic spectral results to be discussed later are consistent only with the 5/2, 5/2 model, and indicate that this high-spin ligand-field model incorporating spin–spin interaction offers a more appropriate description[40] of the dimeric systems than does the molecular-orbital approach that had been reported earlier[41,42].

Mathematical expressions for the temperature dependence of these three models are well known[32]. In such calculations g is assumed equal to 2.0, consistent with $^6A_1$ iron(III) ions, and the temperature-independent paramagnetism $N\alpha$ is assumed to be zero. The upper limit for $N\alpha$ of 130×10^{-6} CGS units per mole obtained[40] from the $[(Fe(III)EDTA)_2O]^{4-}$ dimer study is less than 1 per cent of the diamagnetic correction for a 50 000 molecular-weight protein.

This theory of antiferromagnetically coupled clusters has been applied quite successfully to the magnetic behaviour of trimeric[43] and tetrameric systems[44,45]. Extensions of the theory to the case of a linear chain of up to 10 identical ions and to an infinite chain have been reported[46]. The limits of the spin–spin interaction model become apparent as the number of interacting ions becomes five or more[37,47]. Except for arrays of high symmetry, the proliferation of different Js prevents an unambiguous determination of the set of energy levels for the system.

Physically extended systems or lattices of magnetically interacting iron(III) ions are well known. The interaction can be ferromagnetic or antiferromagnetic. Examples of both are found among the various mineral phases of the iron-oxyhydroxide system. The theoretical treatment of this kind of magnetic behaviour is different from the interaction hamiltonian discussed above. The concepts of magnetic domains and molecular fields are useful in approaching this problem, but the phenomena are not completely understood[48].

19.1.3 Small particles: superparamagnetism

The magnetic behaviour of ferromagnetic and antiferromagnetic materials changes dramatically when the particles are small (<10 nm diameter) yet contain up to several thousand iron(III) ions. Néel has pointed out[49] that for particles of these dimensions the anisotropy energy associated with the possible orientations of the direction of ferromagnetism or antiferromagnetism can become of the order of kT. The orientation Δ has sufficient thermal energy to undergo a pseudo-Brownian rotation with a frequency f that can be expressed as

$$f = f_0 \exp(-KV/kT) \tag{19.4}$$

where
K = anisotropy energy per unit volume;

V = particle volume.

KV is, of course, just an expression of the energy barrier between possible orientations of Δ. As the temperature drops below the blocking temperature T_b, the magnetisation becomes thermally frozen and Δ becomes stable, that is definitely oriented. This fluctuation of Δ can be used to account for the unusual behaviour of small magnetic particles[50,51]. In any particle of diameter less than 10 nm the ions are not well organised into lattice planes, and the magnetic ions can be considered to constitute a single magnetic domain. The two sublattices of this 'sponge'-like species contain a different number of magnetic ions. This difference can be approximated by $M^{1/2}$, where M is the total number of magnetic ions in the particle. The particle magnetic moment is then $(M\mu\beta)^{\frac{1}{2}}$, where $\mu\beta$ is the magnetic moment of the ion. It is oriented with and follows the fluctuations of Δ. Using this assumption Néel showed that the magnetic susceptibility of a collection of N such particles containing ions of total spin S is

$$\chi = \frac{Ng^2 S^2 \mu\beta^2}{3kT} \tag{19.5}$$

where β is the Bohr magneton. The equivalent formula for an uncoupled system is

$$\chi = \frac{Ng^2 S(S + 1)\mu\beta^2}{3kT} \tag{19.6}$$

Consequently in the temperature range above T_b but below T_N the susceptibility of a collection of such antiferromagnetically ordered fine particles will obey the Curie law but with a constant of proportionality below that for the uncoupled system. This constant is that for the whole particle, rather than for the single ion— hence the term *superparamagnetism*. The experimental observations on fine particles of many systems, for example α-Fe_2O_3, α- and β-$FeOOH$, are in accord with this theory[52-5]. The impact of particle size on the magnetic behaviour can be appreciated from the observation that while grains of α- and β-$FeOOH$ less than 5 nm in diameter nearly obey the Curie law, the susceptibility of a sample in the massive state is almost temperature invariant.

19.2 ELECTRON PARAMAGNETIC-RESONANCE SPECTROSCOPY

19.2.1 Polynuclear iron(III)

A technique useful in identifying certain polynuclear clusters of iron(III) is electron paramagnetic-resonance (e.p.r.) spectroscopy. The traditional interest in e.p.r. studies of transition metals has been in molecular properties rather than the co-operative phenomena of coupled systems. In fact, dilution of solids and solutions is often used to reduce the exchange interactions, however weak. Although most dimeric systems that have been intensively studied involve copper(II)[56,57], not iron(III), the e.p.r. spectrum of single crystals containing the $(FeHEDTA)_2O$ unit has been reported[58]. The major features of the spectrum apparently result from a thermally populated $S = 2$ state and are consistent with the antiferromagnetically coupled 5/2, 5/2 model described for this system above. The line widths are not extensively broadened by magnetic dipole–dipole interaction, since even at room temperature most molecules occupy the diamagnetic ground state. Few e.p.r. studies of poly-nuclear clusters containing more than two paramagnetic ions have been reported[59].

When the number of iron atoms increases beyond the two of the dimeric system to twenty or thirty atoms, the magnetic dipole–dipole interaction within the cluster is expected to dominate the e.p.r. spectrum. In such situations e.p.r. can be an important diagnostic technique for identifying polynuclear iron(III). Other explan-ations for line broadening can usually be rejected by the temperature indepen-dence of the spectrum. Slow tumbling of the large protein molecules broadens reasonance lines, but a temperature increase sharpens them. Broadening due to spin–lattice relaxation effects is also temperature dependent, but it decreases with decreasing temperature. By contrast dipole–dipole broadening results from the physical proximity of magnetic dipoles in the coupled system and unless some significant structural changes occur it remains effectively independent of temperature.

19.2.2 Small particles

When the polynuclear iron(III) cluster attains the dimensions suitable for superparamagnetism, the e.p.r. spectrum broadens even further. In an analogous way to the magnetic-susceptibility results, the e.p.r. absorptions can be attributed to the total magnetic moment of the grain[49,51]. The additional line broadening observed at temperatures below the Néel temperature can be accounted for by the fluctuations of Δ. The temperature dependence of the line-width δH has been described by the equation

$$\delta H = A(KV) \exp (KV/kT) \tag{19.7}$$

where A is a constant dependent on the grain size. This behaviour has been observed by a number of investigators for fine particles of Fe_2O_3 and Cr_2O_3[60,61].

19.3 ELECTRONIC-ABSORPTION SPECTROSCOPY

19.3.1 Monomeric iron(III): octahedral and tetrahedral co-ordination
Magnetic susceptibility and e.p.r. measurements are valuable probes of the polynuclear interactions between iron(III) ions, but give no information concerning the co-ordination geometry of a $^6A_1$ iron(III) ion. The splitting of the energies of the five d orbitals under the influence of a ligand field of octahedral, tetrahedral and tetragonal symmetry has been presented earlier (figure 54). The electronic-absorption spectrum of the unusual five-co-ordinate compounds of ground state $S = 3/2$ is dominated by intense absorptions that are presumably charge transfer in origin, and the ligand-field transitions cannot be definitely identified[29]. These transitions are expected to be quite different in energy and intensity from those in the octahedral and tetrahedral cases. Although the temperature dependence of the magnetic susceptibility in the Fe–O–Fe dimeric systems can be accounted for by antiferromagnetic coupling between either two $S = 3/2$ or two $S = 5/2$ ground-state ions, the $S = 3/2$ ground state is excluded from octahedral and tetrahedral symmetries. The ligands of interest in polynuclear iron proteins are weak-field ligands unable to induce the formation of the $S = 1/2$ ground state. Hence, the subsequent discussion is restricted to the $S = 5/2$ ground state, which has been found to be the only one to fit the ligand-field bands in the smallest polynuclear iron systems, the Fe–O–Fe dimers[40]. The similarity of these bands in monomeric and dimeric six-co-ordinate iron(III) complexes is consistent with the relative energies of the antiferromagnetic spin–spin coupling ($J \approx 100$ cm^{-1}) and the ligand-field transitions ($\gtrsim 10\ 000$ cm^{-1}).

The ligand field created by four ligand atoms arranged in tetrahedral symmetry is predicted by ligand-field theory and by simple electrostatic considerations to be less ($\times 4/9$) than that of six ligands in octahedral symmetry. In general this means that a tetrahedral array of ligands will induce ligand-field transitions at lower energy than would an equivalent octahedral array. However, for the iron(III) electronic configuration d^5, this is not the case. The energy E of the first one-electron transition from the ground-state configuration $(t_2)^3(e)^2$ to the first excited-state configuration $(t_2)^4(e)^1$ can be simply related to the strength of the ligand field ($10Dq$) and to the pairing energy P

$$E = P - 10Dq \qquad (19.8)$$

As $10Dq$ increases, E decreases; that is, the energy required for the first one-electron transition in the d^5 configuration is less in a ligand field of octahedral symmetry than in one of tetrahedral symmetry.

This simple one-electron transition between d orbitals corresponds to several transitions between electronic states and hence absorption bands in the metal complexes. The octahedral and tetrahedral ligand fields (the cubic fields) produce the same ligand-field energy-level scheme from the free-ion levels of the d^5 ion. The free-ion ground state ^6S is not split by a cubic field; it assumes the $^6A_1$ symmetry label. The first excited state of the free ion ^4G splits into two T states, $^4T_1$, $^4T_2$, and into a degenerate pair, $^4A_1$, ^4E. The next free-ion term ^4D gives rise to $^4T_2$ and ^4E states. The first four ligand-field absorption transitions for the d^5 configuration are the spin-forbidden sextet-to-quartet transitions

$$^6A_1 \rightarrow {}^4T_1; \quad {}^6A_1 \rightarrow {}^4T_2; \quad {}^6A_1 \rightarrow {}^4A_1, {}^4E; \quad {}^6A_1 \rightarrow {}^4T_2$$

The calculated behaviour[62] of the four lowest energy levels relative to the $^6A_1$ ground state is shown as a function of the strength of the ligand field in figure 55. In octahedral iron(III) these transitions are both spin and parity forbidden. The d–p mixing allowed in tetrahedral symmetry relaxes this parity requirement, leading to bands of low intensity ($\epsilon < 1$) that are, however, more intense than those of the octahedral case ($\epsilon \approx 0.05$–0.01).

Reference spectra of iron(III) co-ordinated to oxygen-donor ligands in both octahedral and tetrahedral geometries have been studied fairly extensively. Although the ligand-field spectrum of $[Fe(H_2O)_6]^{3+}$ has long been obscured by the absorptions of the hydrolysis and polymerisation products of iron(III), the ligand-field bands for this species have now been observed in large single crystals of iron(III) ammonium sulphate[63] and in perchloric-acid solutions of iron(III) perchlorate[64]. The spectrum of iron(III) in a field of oxygen- and nitrogen-donor ligands has been studied in detail for the EDTA and HEDTA chelates[40]. Previous spectroscopic studies of tetrahedral iron(III) have included $[FeCl_4]$[65,66], 8-tungstoferrate(III) ion[67], and β-NaAl$_{1-x}$Fe$_x$O$_2$[68]. In these cases the ligand-field bands are either split in a complex fashion or are somewhat obscured by intense low-energy absorptions, even in spectra recorded at liquid-nitrogen temperature. A carefully studied spectrum of a $[Fe(III)O_4]$ system that contains many ligand-field absorptions is that of an iron(III)-doped sample of orthoclase feldspar, ideally $KAlSi_3O_8$[69,70].

The reference spectra show clearly that the first two spin-forbidden transitions, $^6A_1 \rightarrow {}^4T_2$ and $^6A_2 \rightarrow {}^4T_1$, occur at much lower energies (12 600; 18 200 cm^{-1}) in the [Fe(III)O$_6$] and in the [Fe(III)O$_4$N$_2$] (11 300; 18 400 cm^{-1}) complexes than in the tetrahedral system. The iron(III)-doped orthoclase feldspar sample shows an extremely broad weak system in the region of 20 000 cm^{-1} and bands at 22 500, 23 900 and 26 500 cm^{-1}. Although there is some uncertainty between two possible assignments in this spectrum[71,72], the calculated value of 10Dq in either case (5000–7000 cm^{-1}) is substantially lower than the 13 700 cm^{-1} value calculated from the

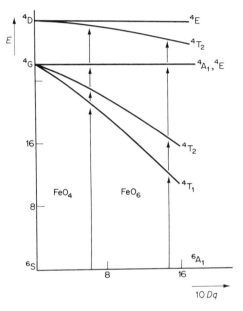

Figure 55 Dependence on the strength of the ligand field (10Dq) of the energies of the first four ligand-field transitions for d^5 iron(III) complexes in tetrahedral and octahedral co-ordination.

octahedral reference spectra[72]. An important point of this spectral comparison is that octahedral iron(III) invariably absorbs in the region 800–100 nm, where tetrahedral iron(III) does not, allowing in many cases an unequivocal determination of the presence or absence of octahedral iron(III).

19.3.2 Polynuclear iron(III)

These ligand-field bands have also been observed at similar energies in compounds containing octahedral iron(III) ions antiferromagnetically coupled together. For example, the basic iron(III) sulphate FeOHSO$_4$, the structure of which has been shown to contain [Fe(III)O$_6$] octahedra coupled together into chains by bridging OH and SO$_4$ groups[73], exhibits ligand-field bands at 10 600, 20 000 and 23 400

cm^{-1}[164] in the near i.r.-visible spectral regions. Comparable data confirming the presence of these d^5 ligand-field transitions in the spectra of polynuclear iron(III) species have also been reported for several crystalline phases of Fe_2O_3 and $FeOOH$[63,64], and in the oxo-bridged iron(III) dimers of EDTA and HEDTA[40].

In both tetrahedral and octahedral geometries the molar absorption coefficients of the ligand-field bands originating from the monomeric species are low, but the octahedral absorptions are generally lower than the corresponding tetrahedral ones. The intensity of these bands is increased in antiferromagnetically coupled systems by factors of 10–50 over the corresponding uncoupled systems, including both lattice antiferromagnetism (in the manganese(II) halides, for example $MnCl_2$)[74-7] and intermolecular antiferromagnetism, for example $[(FeEDTA)_2O]^{4-}$ and $[(FeHEDTA)_2O]^{2-}$[40,70].

19.4 MÖSSBAUER SPECTROSCOPY

19.4.1 Polynuclear iron(III)

Small perturbations of the nuclear energy levels of iron by the surrounding electrons are detected by Mössbauer spectroscopy (the recoilless emission and absorption of low-energy γ-radiation). The 14.36 keV nuclear transition of interest occurs between the $I = 3/2$ and $I = 1/2$ levels of the Mössbauer nuclide ^{57}Fe, where I is the nuclear-spin quantum number. A typical requirement for Mössbauer work is 1–2 micromoles of ^{57}Fe, which has a natural abundance of 2.19 per cent. For a protein of 50 000 molecular weight that binds one iron atom, and without isotopic enrichment, this amount of ^{57}Fe is contained in 2.5 g. The polynuclear iron proteins of this review are characteristically richer in iron than this and are quite suitable for Mösbauer spectroscopic studies. The four possible interactions between the ^{57}Fe nuclide and the electronic surroundings—isomer shift; quadrupole splitting; nuclear magnetic hyperfine interactions; the nuclear Zeeman interactions—have been widely studied. Their application to the iron proteins of the haem and iron–sulphur classes has been recently reviewed[78].

The isomer shift and the quadrupole splitting can be related to the electronic configuration and stereochemistry of the iron nucleus. Only high-spin iron(II) is unambiguously determined in this way. Analysis of the Mössbauer data for other states requires a detailed consideration of all four of the above parameters[79,80]. The reported Mössbauer data for a number of monomeric and dimeric iron(III) compounds show little difference in the isomer shifts between the two kinds of compounds, but the quadrupole splitting is usually greater in the dimeric case[5,39,40,81] The Mössbauer spectra of compounds with iron(III) occupying both octahedral and tetrahedral sites is expected to be broadened and perhaps resolvable into two sets of lines. Examples of such two-site materials include the ferromagnetic iron garnets $5Fe_2O_3 . 3M_2O_3$ where M is a rare earth or yttrium[42], and γ-Fe_2O_3[82]. For yttrium and dysprosium iron garnets, two hyperfine spectra are readily observed. However, the two hyperfine spectra of the structurally well-characterised γ-Fe_2O_3 can only be resolved by the application of a strong external field (1.35 MA m^{-1} $\equiv$ 17 kOe). The hyperfine splitting at zero applied field differs only slightly between the two sites. Spectra of the several other phases of Fe_2O_3 and FeOOH have been reported[79,83]. In any one compound the hyperfine field at the octahedral sites is greater than that at the tetrahedral sites, but the variation of each among several compounds is considerable. Similar variations are observed for each site symmetry when the ligand atoms change[84].

19.4.2 Small particles

In studies of polynuclear iron clusters, an important characteristic Mössbauer time is the mean lifetime of the nuclear excited state, 1.4×10^{-7} s for ^{57}Fe. When the anisotropy energy of very small particles ($\approx$10 nm diameter) becomes comparable to kT the orientation of the magnetic field at the nucleus fluctuates with a relaxation time τ, which is an exponential function of the particle volume V and the reciprocal temperature T.

$$\tau = \tau_0 \exp\left(KV/kT\right) \tag{19.9}$$

where K is an anisotropy constant.

By lowering the temperature the relaxation time becomes comparable to or larger than the mean lifetime of the nuclear excited state and a magnetic hyperfine spectrum is observed. A distribution in particle size results in a finite temperature range over which the hyperfine spectrum is established and the usual quadrupole split doublet disappears. This behaviour is characteristic of superparamagnetic particles[51,85,86].

19.5 X-RAY DIFFRACTION

Conventional X-ray diffraction study of single crystals has produced valuable structural information on the monomeric[87] and dimeric[23] iron(III) species that are important model compounds for polynuclear iron(III) proteins. Diffraction studies on these proteins, however, have been restricted in the main to studies of X-ray (and some electron) Debye–Scherrer diffraction patterns, and of low-angle X-ray solution scattering.

19.5.1 Debye–Scherrer patterns

The pattern of X-ray diffraction lines produced by an aggregate of finely ground crystallites corresponds to reflections from lattice planes that scatter in phase according to the usual Bragg equation[84]

$$n\lambda = 2d \sin \theta \tag{19.10}$$

Sharp diffraction maxima result from in-phase diffraction by thousands of parallel planes, posing a high degree of precision on the Bragg equation. As the crystallite size decreases, the number of co-operating planes decreases and the diffraction condition relaxes. As a result the diffraction lines broaden. A completely amorphous substance produces merely a broad halo at a small diffraction angle. In fact, line broadening can be used to estimate particle size in the range below $\approx$200 nm[88,89]. When only a few lines are observed, and their positions are poorly defined, it becomes extremely difficult to derive the cell dimensions unambiguously.

19.5.2 Low-angle solution scattering

In low-angle X-ray scattering studies[90] a monochromatic and highly collimated X-ray beam is directed onto a solution and measurements made of the intensity of that portion of the X-rays scattered at small angles ($\theta \lesssim 5°$) to the beam direction. The scattering curve of relative intensity against θ will usually contain several inflections superimposed on the side of a curve that rapidly decreases as θ increases from zero to 5° . To amplify these inflections, which reflect the shape and dimensions of the scattering unit as well as the short-range order between them, this curve is converted by a Fourier transformation into a radial distribution function (r.d.f.) of charge distribution plotted against the separation between the scattering centres.

For a solution containing particles of small dimensions, $\approx$5–50 nm, the r.d.f. can provide information on the particle shape, for example a measure of the effective radius of gyration. Such studies have been reported for some spherical viruses and a number of proteins, for example ovalbumin[90]. To obtain information about the internal composition of a particle such as a discrete polynuclear iron cluster in solution, the scattering curve of the solvent alone is determined and applied as a correction to the scattering curve of the solution of particles. A Fourier transformation of this difference function is then performed and the curve analysed to derive the interatomic separations for the scattering units. Analysis of the areas under these peaks is used to estimate the number of nearest neighbours around a scattering atom.

In this procedure the correction due to the solvent scattering must take into account the changes in this scattering caused by the inclusion of the solute. The scattering reflects the short-range order in the solvent, which can be seriously perturbed by the inclusion of a solute. Solute concentrations of several molar are commonly employed. Analysis of the peak areas to give the co-ordination number around a scattering centre depends in some measure on the mathematical procedure used. Thus, depending on the analytical method, the computer co-ordination number of argon in the comparatively simple liquid-argon system can vary from five to seven[91].

19.6 VIBRATIONAL SPECTROSCOPY

Characteristic vibrational energies of the important polynuclear [Fe—O—Fe] and [Fe—OH—Fe] structural units have been identified in inorganic complexes[92-5]. The asymmetric stretch of Fe—O—Fe is expected at $\approx$850 cm^{-1}; the symmetric stretch has been reported at $\approx$230 cm^{-1}. A weak band at 950 cm^{-1}, which moves to lower energy on deuteration has been identified as arising from the $[Fe(OH)_2Fe]^{4+}$ unit[22]. The search for these vibrations in a macromolecule is hampered by the preponderance in this spectral region of the carbon–carbon stretching, carbon–hydrogen and nitrogen–hydrogen rocking vibrations from the polypeptide chain. Oxo- and hydroxo-bridging groups, which couple together the several iron(III) ions in polynuclear clusters, are expected to adopt a variety of geometries. The i.r. absorptions of these various structural components may not correspond in a simple fashion with those of the model compounds. However, the spectral region below $1000 \ cm^{-1}$ where the metal–ligand vibrations occur offers promise of being an empirical 'fingerprint' for polynuclear iron clusters.

Comparative studies between the metal-free and the metal-bound protein are complicated by the unpredictable effects on the protein vibrational energies of the conformational changes induced by metal binding. The implications of small particle size for vibrational spectroscopy[96] have not received the theoretical analysis that the magnetic behaviour has, but the lack of a well-defined extended lattice in small particles is expected to broaden the absorptions, making comparative studies difficult.

20 BIOLOGICAL ROLES OF POLYNUCLEAR IRON(III)

Although iron metabolism has been extensively studied throughout the history of medicine, the importance of chelation and polymerisation of iron(III) to an understanding of iron biochemistry has been realised only recently. In contrast to the metabolism of other metals, most of the body iron is continuously recycled. Since very little iron is lost via urine, faeces and sweat, control of the iron level present occurs by regulation of uptake, not of excretion. Environmental iron that is presented to organisms as insoluble iron(III) hydroxide is biologically unavailable. A primary role of dietary chelating agents is to maintain iron(III) in a low molecular weight soluble form that can be utilised for absorption across the intestinal mucosa. Macromolecular chelates of iron(III), which compete with low molecular weight chelates for dietary iron and are resistant to the proteolytic enzymes in the gut, are expected to inhibit iron absorption. This inhibition becomes increasingly important as the molar ratio of iron to protein increases. In metalloproteins containing many iron(III) ions it is not surprising to find polynuclear iron(III) species. The first two examples of polynuclear iron(III) discussed in chapter 21, phosvitin and gastroferrin, have been implicated[97,98] in the regulation and inhibition of iron absorption. The egg protein phosvitin has been suggested also to act as an iron carrier during the development of the chick embryo[99]. Furthermore, gastroferrin possesses blood-group activity, at a level similar to that of purified blood-group substance from ovarian cysts, and blood-group active substances have been reported to bind iron(III)[98].

By virtue of their biological function, storage forms of iron are expected to be characterised by a high iron: protein ratio, and consequently by polynuclear interaction among the bound iron(III) ions. The inner iron core of the storage protein ferritin (section 21.3) can attain a molecular weight comparable to that of the polypeptide chains of the protein[100]. This storage iron appears as a discrete compact sphere of polymeric iron(III), which is maintained in a soluble biologically available form by a surrounding layer of polypeptide. The second storage form of iron considered in chapter 21, haemosiderin, also stores iron in a spatially compact form[101]. The general morphology of ferritin appears again in the final example of biological polynuclear iron, iron–dextran[102], the primary function of which seems to be to provide iron for an organism.

21 STRUCTURAL STUDIES OF BIOLOGICAL POLYNUCLEAR IRON(III)

21.1 PHOSVITIN

Several reports[97,103-5] have appeared concerning the interaction of iron with phosvitin, a phosphoglycoprotein of molecular weight 35 000 containing 6.5 per cent carbohydrate, which accounts for 60 per cent of the protein-bound phosphate in egg yolk[106-8]. In the presence of oxygen and phosvitin, iron(II) chloride is rapidly oxidised to a bound form of iron(III), with accompanying release of phosphoryl groups from the protein[103-5]. This iron(III) phosvitin, containing about 9 per cent iron by weight has not been characterised with respect to the structure of the bound iron. Nutritional reports[109] indicating that dietary eggs inhibit iron absorption stimulated further investigations of the interaction between iron(III) and this protein[97]. To mimic the nutritional forms of phosvitin commonly available in the diet, iron(III) (as the nitrilotriacetic acid chelate) was presented to phosvitin both before and after treatment of the protein with boiling water. These procedures led to the isolation of two metalloproteins, green and brown iron(III) phosvitin, respectively, which bind about 7 per cent iron by weight to a final phosphorus:iron ratio of 2. Both derivatives have been shown to contain polynuclear iron(III) but in sites of different co-ordination geometries.

21.1.1 Green Iron(III) Phosvitin

Evidence for the polynuclear nature of the bound iron(III) comes from the temperature dependence of the magnetic moment, and from a broad e.p.r. resonance, which is temperature independent. The room temperature $\mu_{eff/Fe}$ of 4.45 B.M. (4.56 B.M. in solution), which is substantially reduced from the high-spin ($S = 5/2$) value of 5.92 B.M., decreases with increasing temperature (curve I, figure 56), indicating some antiferromagnetic interaction. Curve II in figure 56 illustrates the temperature dependence of $\mu_{eff/Fe}$ for two $S = 5/2$ ions antiferromagnetically coupled together ($J = -12$ cm^{-1}), calculated by the usual spin-spin interaction model, assuming $g = 2.0$, $N\alpha = 0$[36]. When the value of $-J$ is increased (corresponding to stronger coupling) or decreased (weaker coupling), acceptable agreement between the observed and calculated behaviour of the magnetic moment still cannot be obtained. Although some of the iron(III) ions are undoubtedly antiferromagnetically coupled, a model consisting of an array of dimeric units does not account quantitatively for the data. As might be anticipated from this treatment and from the availability of the several parameters employed, this magnetic behaviour can be accounted for by a combination of high-spin monomers (25 per cent) and spin-spin coupled dimers (75 per cent, $J = -25$ cm^{-1}, $\mu_{eff/Fe,300\,K} = 3.88$ B.M.). However, such a structural model of a mixture of monomers and dimers is inconsistent with the e.p.r. data. The broad e.p.r.

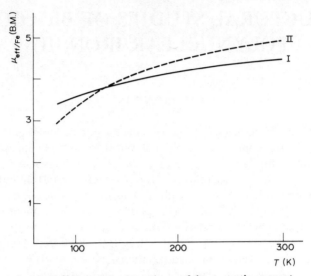

Figure 56 Temperature dependence of the magnetic moment
per iron ($\mu_{eff/Fe}$) over the range 300–85 K for green iron(III)
phosvitin (curve I). Curve II is calculated for a ($S = 5/2, 5/2$)
dimeric system with $J = -12$ cm^{-1}.

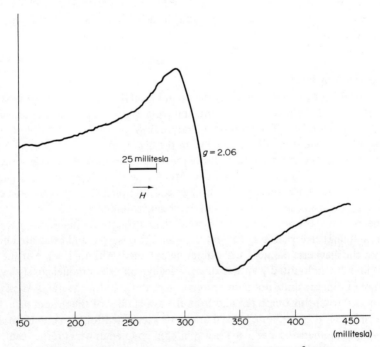

Figure 57 Electron paramagnetic-resonance spectrum of green
iron(III) phosvitin at $\approx$120 K.

spectrum (figure 57) is temperature independent, with a peak-to-peak width of 50 millitesla at both ambient temperature and 120 K. Although the e.p.r. resonances of any monomeric iron(III) present could be obscured by the broad polynuclear resonance, this broadening far exceeds that observed for dimeric and trimeric systems[56-9], indicating that the vast majority of the fifty iron(III) ions is bound in fairly large polynuclear clusters.

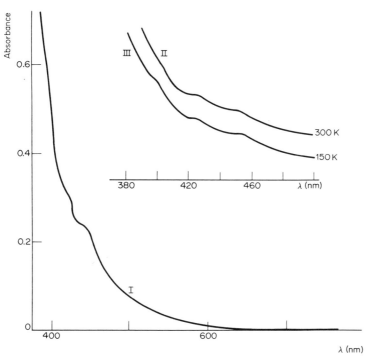

Figure 58 Visible and near i.r. absorption spectra of green iron(III) phosvitin at ambient temperature (curve I) and in a 1:1 (v/v) ethylene glycol:buffer mixture at ambient (curve II) and low temperature, $\approx$150 K (curve III).

The absorption spectrum in the visible and near i.r. regions shown in figure 58 exhibits two distinct shoulders (22 400 and 23 500 cm^{-1}) in aqueous solution at room temperature. On cooling a 1:1 ethylene glycol–buffer mixture to $\approx$ 150 K, where a stable clear glass forms, a third shoulder is observed at 25 000 cm^{-1}. These ligand-field bands have been assigned in increasing energy as $^6A_1 \rightarrow {}^4T_1, {}^4T_2$, ($^4A_1, {}^4E$), giving $10Dq = 5200$ cm^{-1}. This value of $10Dq$, which is far below the value for octahedral iron(III) of 13 700 cm^{-1} is comparable to that obtained for tetrahedral iron(III) in orthoclase feldspar[72]. Moreover, the close similarity between the spectra of green iron(III) phosvitin and orthoclase feldspar[69,70] strongly suggests that the binding sites in the protein are tetrahedral. The absence of peaks in the region 800–1000 nm rules out the possibility that significant numbers of the iron(III) sites are octahedral. The intensities of the first two electronic-absorption bands in green iron(III) phosvitin are almost an order of magnitude greater than those measured for analogous bands in iron(III)-doped orthoclase feldspar. Such intensity enhancement of spin-forbidden bands is consistent with the presence of antiferromagnetically coupled iron(III) centres[40,74-7].

The Mössbauer spectrum, which consists of a simple resonance doublet with a quadrupole splitting of $420 \, \mu m \, s^{-1}$ and an isomer shift, relative to ^{57}Co in platinum of $210 \, \mu m \, s^{-1}$ [110], provides additional evidence for the presence of a single (presumably tetrahedral) type of iron(III) binding site.

The complex i.r. spectra of phosvitin and green iron(III) phosvitin showed few changes in the oxo- and hydroxo-bridged region ($800–1000 \, cm^{-1}$) and in the phosphorus–oxygen stretching region around $1200 \, cm^{-1}$ [95]. However, the high concentration of phosvitin in serine phosphate residues (120/mole protein), many of which are arranged in continuous sequences of up to eight residues[107,111-12], suggests that the P=O groups bind to the iron(III) in polynuclear clusters[102]. The structural model for tetrahedral binding of iron(III) to β-pleated sheet arrays of serine phosphate side chains[113,114] that has been proposed[97] is shown in figure 59.

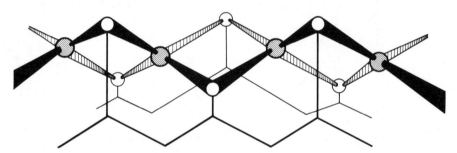

Figure 59 Proposed co-ordination structure for polynuclear iron(III) in green iron(III) phosvitin. Shaded spheres are iron atoms, open spheres oxygen atoms.

21.1.2 Brown Iron(III) phosvitin

The temperature-dependent $\mu_{eff/Fe}$ (1.56 B.M. at 300 K; 1.18 B.M. at 85 K) is depressed below that of high-spin monomeric iron(III) and of the polynuclear iron(III) in green iron(III) phosvitin. The e.p.r. signal is even broader—80 millitesla peak-to-peak—than that of the green form. Taken together, the magnetic data indicate that the iron(III) in brown iron(III) phosvitin is also bound in some type of polynuclear cluster[97]. The only ligand-field band ($12 \, 200 \, cm^{-1}$) observed in the visible–near i.r. spectrum has been assigned to the $^6A_1 \rightarrow {}^4T_1$ transition of octahedral iron(III). Although no bands are observed that may be assigned to tetrahedrally co-ordinated iron(III), such sites could easily escape detection because of the strong background absorption in the $20 \, 000–25 \, 000 \, cm^{-1}$ region.

It is interesting to note that, while both iron(III) phosvitins contain equivalent amounts of phosphorus and bind fifty iron(III) ions in a polynuclear fashion, the dominant co-ordination geometry present in each form is different. It is probable that the unusual tetrahedral co-ordination is favoured by the protein conformation, which serves to protect the metal ions from hydrolysis in the aqueous environment at physiological pH (7.4). This protection is evidently not available to the iron(III) ions found in brown iron(III) phosvitin. Presumably the heating of the protein prior to presentation of the metal has opened the iron(III) binding sites to hydrolytic polymerisation, thereby leading to octahedral co-ordination.

21.2 GASTROFERRIN

Gastroferrin is an iron(III)-binding glycoprotein, which can be isolated from human gastric juice and from pig gastric mucin[98, 115-16]. The glycoprotein is 85 per cent carbohydrate and 15 per cent polypeptide by weight, with a molecular weight estimated by a variety of techniques as 2.6×10^5.

Spectral and magnetic studies of iron(III) gastroferrin preparations that bound nearly 200 iron(III) ions—about 70 per cent of the glycoprotein's capacity—have been reported[117]. The room-temperature magnetic moment per iron for freeze-dried preparations of iron(III) gastroferrin is 3.45 (±0.05) B.M., consistent with the solution value of 3.5 B.M. The moment increased in an almost linear fashion with decreasing temperature, rising to 3.65 (±0.05) B.M. at 90 K. A preliminary field-strength study indicated the presence of a weak magnetic remanence. The e.p.r. spectrum at liquid-nitrogen temperature consists of a broad resonance (60 millitesla peak-to-peak) from $g = 6.61$ to $g = 4.05$ with a much weaker band at $g = 2.3$. Both the e.p.r. and magnetic-moment data suggest the presence of polynuclear iron.

The visible–near i.r. absorption spectrum of a solution of iron(III) gastroferrin (figure 60) contains three weak shoulders at 10 400, 16 200 and 20 400 cm^{-1} on the side of a strong ultraviolet absorption that tails into the visible. The energies of these spectral bands are consistent only with iron bound to gastroferrin as iron(III) in octahedral co-ordination. They are assigned to the first three spin-forbidden transitions from the $^6A_1$ ground state of the d^5 electronic configuration: $\rightarrow {}^4T_1$, $\rightarrow {}^4T_2$ and $\rightarrow [{}^4A_1, {}^4E]$, respectively. As expected for polynuclear iron the band intensities ($E \approx 2$–5) are increased from those of monomeric octahedral iron(III). There is no evidence of spectral bands attributable to tetrahedral iron(III). While a few tetrahedral sites could escape spectral detection, most of the iron(III) ions are undoubtedly bound in octahedral co-ordination to gastroferrin.

The Mössbauer spectrum of a frozen solution of iron(III) gastroferrin at 77 K consists[110] of a well-formed doublet characterised by a quadrupole splitting of 710 $\mu m\ s^{-1}$ and an isomer shift, relative to ^{57}Co in platinum, of 190 $\mu m\ s^{-1}$. These values are consistent with high-spin or antiferromagnetically coupled iron(III). There is no evidence for any hyperfine splitting or line-broadening at this temperature, indicating that any intrinsic ordering (magnetic remanence) present is small.

A structural model proposed for iron(III) binding to gastroferrin consists of

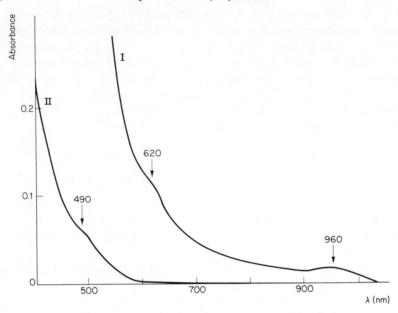

Figure 60 Visible and near i.r. absorption spectra of iron(III) gastroferrin at ambient temperature in a 10-cm cell (curve I) and a 1-cm cell (curve II).

[Fe(III)O$_6$] octahedra coupled together in a large polynuclear cluster and bound predominantly to hydroxyl oxygens of the large number of carbohydrate residues present[117]. Both small molecular weight and polymeric iron(III)–carbohydrate complexes containing polynuclear iron(III) have been reported[28,118−20].

21.3 FERRITIN

21.3.1 Protein core

The structure of the iron-storage protein ferritin has been the subject of a considerable number of recent investigations[100,121,122], yet it remains a matter of some controversy. In morphology, ferritin appears as a spherical core, 7 nm in diameter, of iron(III) oxy-hydroxy-phosphate, which is coated by twenty-four identical protein subunits of molecular weight 18 500[123,124]. The core, of approximate molecular composition (FeOOH)$_8$. FeO . PO$_4$H$_2$, attains a maximum molecular weight of ≈400 000[122,125,126]. It has been suggested that the subunit structure of the core reported in many electron-microscopic studies[127−31] is largely artifactual, but the 'true' appearance of the cores under the electron microscope remains controversial[132]. The apparent substructure observed[132−6] in ferritin molecules containing significantly incomplete cores has been attributed to asymmetric growth of the core[135].

The magnetic moment per iron of 3.8 B.M. at room temperature[125,126,137-9] indicates antiferromagnetic interaction among the iron(III) ions present. At lower temperatures the magnetic susceptibility increases to a maximum at 20 (±3) K. This behaviour has been attributed to superexchange between iron(III) and oxygen atoms, presumably analogous to the behaviour in the dimeric Fe—O—Fe systems[137]. However, a comprehensive study of the dependence of the magnetisation on the applied field, up to 6.4 GA m^{-1} (80 000 kOe), indicates that the explanation for the magnetic behaviour is rather more complex, requiring the presence of superparamagnetic and superantiferromagnetic terms. In this analysis[138,139] the magnetic moment per iron is 5.08 B.M. and the behaviour is in general agreement with the predictions of the Néel theory for superparamagnetic particles that are antiferromagnetically ordered[49].

The superparamagnetic nature of the ferritin core is reflected in the e.p.r. spectrum[140]. At room temperature the spectrum consists of an extremely broad resonance (>100 millitesla peak-to-peak) centred around $g = 2$ with another weaker, broad resonance near $g = 4$. Below about 200 K the Néel temperature of ferritin, this resonance at $g = 2$ broadens even further and decreases in intensity, consistent with superparamagnetic behaviour. The origin of the weak resonance at $g = 4$, which alone can be observed at 77 K, is not certain.

The near i.r. spectrum of ferritin contains the ligand-field band at 900 nm, indicative of the presence of octahedral iron(III)[64,141]. The near i.r. spectra of several mineral systems known to contain octahedral iron(III) ions coupled together in a crystalline array also show this band[63,64]. The spectra between 700 and 1300 nm of three of these minerals, $FeOHSO_4$[73], α-FeOOH and γ-FeOOH[17-18], are shown in figure 61, together with the spectra of ferritin and a synthetic model for the ferritin core, an iron(III) nitrate polymer[24], which is discussed further in a later section. The visible region of the ferritin spectrum is dominated by an intense absorption originating in the u.v. Consequently no other ligand fields are observed, and the possibility that some tetrahedrally bound iron(III) is present cannot be excluded. The Mössbauer spectrum of ferritin[86,139,142] consists of a quadrupole split (740 ± 40 μm s^{-1}) doublet with an isomer shift relative to stainless steel of 470 ± 50 μm s^{-1}. This doublet gradually decomposes below 70 K as the magnetic hyperfine structure is established. The spectrum appears symmetrical, suggesting the presence of only one type of co-ordination site for iron(III). The internal field established at low temperatures is close (39.2 MA m^{-1} $\equiv$ 493 kOe) to that of the calibrant, haematite (41.0 MA m^{-1} $\equiv$ 515 kOe), which contains octahedral iron(III). This temperature-dependent behaviour can be attributed to the superparamagnetic nature of the protein cores, which behave in the Mössbauer experiments as particles of variable size less than 10 nm in diameter.

Low-angle X-ray scattering studies[143] have been carried out on ferritin fractions of varying iron content. The solvent (53 per cent sucrose) was chosen to match the electron-scattering properties of the protein and thus amplify the resolution of the scattering due to the ferritin cores. The best agreement between the radial-distribution function and that calculated according to various models was given by a 'single-particle' model. There was no evidence in fractions containing fairly complete cores for any subunit structure. However, in fractions of lower iron content the average dimensions calculated for the core particle became smaller and more irregular. X-ray and electron Debye–Scherrer diffraction patterns produced by the ferritin core are characterised by a number of comparatively weak, diffuse lines. From the line breadths the size of the core crystallites has been

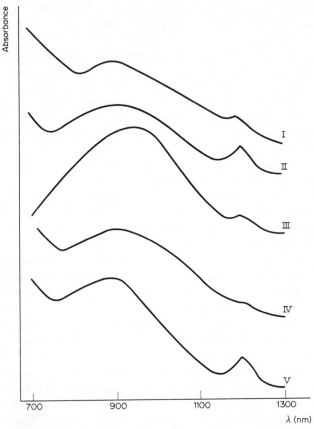

Figure 61 Near i.r. spectra at ambient temperature of ferritin (curve I), iron(III) nitrate polymer (curve II), FeOHSO₄ (curve III), α-FeOOH (curve IV) and γ-FeOOH (curve V).

estimated[144] at 7.5 ± 3.0 nm. The diffraction data indicate that the core is composed of either a single crystallite, or a very small number of crystallites, the shape and orientation to the protein layer of which are variable[145-7]. The lines of the patterns differ in number, spacings and intensities from those produced by the iron(III) oxide and iron(III) oxyhydroxide crystalline phases.

A unit cell proposed for the core is hexagonal (a = 294 pm, c = 940 pm) with iron atoms distributed at random among the available octahedral and tetrahedral sites[148]. A larger hexagonal cell (a = 1179 pm, c = 990 pm) has also been suggested[147]. Structural models (discussed below) for the core have been indexed using a hexagonal cell (a = 508 pm, c = 940 pm) with iron atoms in octahedral sites[16] and a cubic cell (a = 837 pm), for which no arrangement of iron atoms was proposed[15].

Extensive structural studies on three hydrated iron(III)-oxy-hydroxides, isolated by slightly different procedures from iron(III) nitrate solutions, have been reported; all of them have been variously proposed as structural models for the ferritin core. For convenience they are referred to by the term used in the original report, namely polymer[24], hydrolysate[16] and gel[15]. As the discussion shows, they are closely related materials.

Infrared spectra of the polymer, hydrolysate, isolated ferritin cores (micelles) and colloidal iron(III) hydroxide are shown in figure 62. In contrast to spectral data for well-defined crystalline mineral phases, for example α-FeOOH, γ-FeOOH and FeOHSO$_4$[63,64], the absorption bands are broader, particularly around 3500 cm^{-1}, where OH absorbs, and below 1000 cm^{-1}, where absorptions characteristic of metal-ligand bonds are expected. The spectrum of a freeze-dried preparation of ferritin is very similar to that of the isolated cores in both these spectral regions. Ferritin crystallised from cadmium sulphate (5 per cent) exhibits generally sharper absorption bands than does the powder preparation. An additional component can be resolved in the 3500-cm^{-1} envelope.

The low-energy ($<$1000 cm^{-1}) spectrum of ferritin, which consists of broad overlapping absorption bands, also differs appreciably from that of the oxides and oxyhydroxides of aluminium(III) and manganese(III) and of several iron(III) phosphate minerals[63]. Phosphate is present in ferritin at the level of one phosphate per nine iron atoms, but is released when the iron cores are freed from protein by a variety of treatments. Since the X-ray diffraction pattern and the low-energy i.r. spectrum of the isolated cores are essentially unchanged from those of the intact protein, it appears that phosphate plays a relatively insignificant structural role in the core. It is probable that the phosphate is largely surface bound, interacting both with the iron core and the surrounding protein coat.

21.3.2 Iron(III) nitrate polymer

The first synthetic structural model for the ferritin core is a homogeneous polymer of composition Fe$_4$O$_3$(OH)$_4$(NO$_3$)$_2$(H$_2$O)$_{1.4}$ readily isolated[24,25] from iron(III) nitrate solutions (0.3 M) to which two equivalents of base per mole of iron were added, giving a final pH of 2. The polymer appears in the electron microscope as a sphere of 7-nm diameter, strikingly similar to the ferritin core. The magnetic moment per iron is 3.1 B.M. and the magnetic susceptibility attains a maximum at 8 K. The magnetisation at 1.14 MA m^{-1} (14.3 kOe) is linear from 300–40 K. Weak ligand-field bands observed in the visible-near i.r. spectra were first assigned to spin-forbidden transitions of tetrahedral iron(III)[86]. Following the report of the iron(III)-doped orthoclase feldspar spectrum these bands have been reassigned to transitions of octahedral iron(III)[64,70,141]. The Mössbauer spectra closely resemble those of ferritin from 300–5 K[86]. The i.r. spectrum in the region of metal–ligand vibrations below 1000 cm^{-1} contains a pattern of broad overlapping bands that differs appreciably from that of ferritin and the iron-mineral phases[63,64,141]. On the basis of a small-angle X-ray scattering study a structural model consisting of iron(III) ions exclusively tetrahedrally co-ordinated has been proposed[86]. Difficulties inherent in

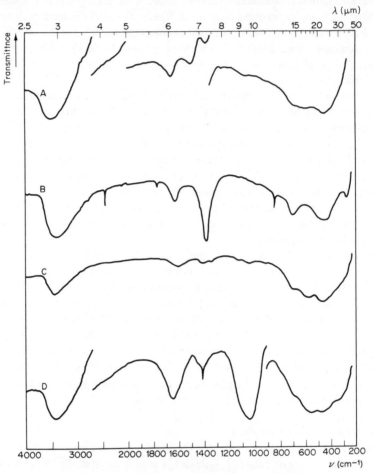

Figure 62 I.R. spectra at ambient temperature of air-dried iron(III) hydroxide (spectrum A), iron(III) nitrate polymer (spectrum B), iron(III) nitrate hydrolysate (spectrum C) and isolated ferritin micelles/cores (spectrum D).

such studies of concentrated solutions (3 M) of a polyelectrolyte have been noted earlier in this review. The spectroscopic data is persuasive evidence for the presence of significant numbers of octahedral iron(III) ions in the polymer.

Powder preparations of the polymer produce X-ray diffraction patterns that, with the exception of a strong 302 pm line in the polymer, closely resemble those of ferritin cores. By analogy to the α-FeOOH structure, the two strongest lines (250 and 150 pm) can be assigned[149] to higher-order manifestations of a unit cell based on oxygen octahedra 500 pm wide arranged in layers 300 pm high. The 300 pm line is then assigned to a first-order reflection from the interlayer spacing[141].

The energies of the nitrate absorptions in the i.r. spectrum indicate that the nitrate present is only slightly distorted from D_{3h} symmetry. Most, if not all, of the nitrate can be removed by treatment with nitron acetate[24]. This chemical and spectral evidence suggests that the nitrate is predominantly surface bound to the iron(III) oxyhydroxide polymer, analogous to the probable binding role of phosphate in ferritin.

21.3.3 Iron(III) nitrate hydrolysate
Hydrolysis of iron(III) nitrate solutions (0.06 M) at 80°C led to the isolation of a hydrolysate ($Fe_2O_3 . 1.8H_2O$) of small particle size (<10 nm). This material has also been proposed as a structural model for the ferritin core[16]. The magnetic moment per iron at room temperature is 3.9 B.M. Unlike ferritin and the iron(III) nitrate polymer the moment is markedly temperature and field dependent, even at comparatively low (<800 kA m^{-1}(10 kOe)) field strengths[64].

The visible and near i.r. spectra contain ligand-field bands at 11 100, 15 650 and 20 600 cm^{-1} assigned to iron(III) ions in octahedral co-ordination[64,141], consistent with the X-ray diffraction data[16]. The proposed unit cell is closely related to haematite, α-Fe_2O_3, which contains iron(III) ions in octahedral sites. Superparamagnetic behaviour is observed in the Mössbauer spectrum at low temperatures (figure 63). Unlike the iron(III) nitrate polymer, the hydrolysate closely resembles ferritin in the i.r. spectral region below 1000 cm^{-1} and in the predominant features of the X-ray diffraction pattern.

21.3.4 Iron(III) nitrate gel
An iron(III) oxide hydrate of formula [$Fe_2O_3 . 1.2H_2O$] has been isolated[15] from 1 M iron(III) nitrate solutions after ammonia-induced hydrolysis. The gel, which appears in the electron microscope as clusters of spheres (3 nm in diameter, molecular weight about 10^5 and containing about 1000 iron atoms), is sufficiently crystalline to produce an X-ray diffraction pattern that closely resembles that of ferritin and the iron(III) nitrate hydrolysate. A cubic unit cell (a = 837 pm) has been proposed. The magnetic susceptibility does not follow the Curie–Weiss law and, moreover, is field-strength dependent. The Mössbauer spectrum at 140 K shows only quadrupole splitting, but at liquid helium temperature the superparamagnetic six-line hyperfine spectrum is observed. On dehydration at 250–350 K the particle size increases to at least 10 nm, the α-Fe_2O_3 phase is established and the superparamagnetic character disappears[150].

21.4 HAEMOSIDERIN

Haemosiderin, a second major storage form of iron ($\approx$25 per cent iron by weight), can be isolated[101] as water-insoluble granules that contain protein, including ferritin, carbohydrate and lipid, in widely varying amounts[151-3]. Although ferritin and haemosiderin are physically and chemically distinct substances, structural investigations of haemosiderin iron (which, despite the name, is nonhaem iron) show it to be closely similar to the iron in ferritin. Reported values of the magnetic moment at room temperature[64,154] show considerable variation (±0.3 B.M.) around the value for ferritin (3.8 B.M.). The moment is both temperature and field-strength de-

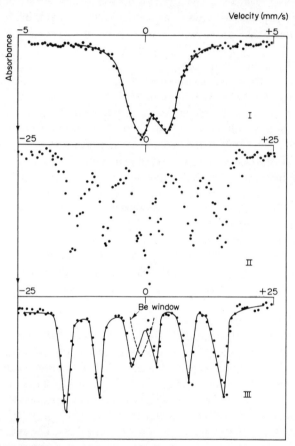

Figure 63 Mössbauer spectra of iron(III) nitrate hydrolysate at ambient temperature (curve I), ≈40 K (curve II) and 5 K (curve III).

pendent[64]. No e.p.r. data have been reported. The only ligand-field band observed is the 900 nm band in the near i.r., indicative of the presence of octahedral iron(III)[64]. Haemosiderin and ferritin are indistinguishable by the Mössbauer spectral parameters of isomer shift, quadrupole splitting and magnetic hyperfine field in the six-line low-temperature spectrum[155], by the i.r. spectrum in the region below 1000 cm^{-1} [64] and by the X-ray diffraction pattern[155]. The earlier diffraction studies[156] suggesting that haemosiderin could be either amorphous or a mixture of known phases of FeOOH have not been confirmed. The electron microscope reveals that haemosiderin is slightly smaller (10-20 per cent) than the ferritin core; this is reflected in the increased line widths of the X-ray diffraction pattern[155].

21.5 IRON–DEXTRAN

Iron–dextran is a colloid formed from iron(III) chloride and an alkali-modified dextran (a bacterial polysaccharide), which has found wide clinical use in parenteral administration of iron[157-9]. It contains up to 28 per cent iron by weight. The material appears in the electron microscope as a small (3-nm diameter) electron-dense iron-containing sphere coated by an electron-transparent sheath, presumably dextran. It has been included in this review because of this apparent structural similarity to ferritin, which also extends to the X-ray diffraction pattern and the Mössbauer spectrum at 300 and 77 K[102]. Although the i.r. spectrum has been reported only over a narrow range (1200-600 cm^{-1}) it closely resembles that of ferritin in the important 1000-700 cm^{-1} region. Other known similar therapeutic agents have been less well characterised but appear to resemble the iron–dextran colloid at least in gross morphology under the electron microscope[160-2].

22 BIOLOGICAL PERSPECTIVE

The detailed structural studies of several biological examples of polynuclear iron(III) discussed in the preceding chapter can be correlated with the biological roles of polynuclear iron(III) (see chapter 22). Phosvitin is a major dietary constituent of avian egg yolk, the inhibition of iron absorption of which is associated with resistance to digestive proteolytic enzymes[106] and with binding of about fifty moles of iron/mole protein in a polynuclear fashion[97]. Nutritional studies have shown that dietary phytic acid (inositolhexaphosphoric acid) and phosphates greatly depress iron uptake[1], presumably by forming highly insoluble iron(III) phosphate compounds. Binding of iron by phosvitin—primarily through the phosphate groups of the serine phosphate side chains—does not lead to precipitation but still effectively makes the iron unavailable for absorption by maintaining it in a macromolecular complex. However, scant information is available on the competition for dietary iron among the variety of potential ligands and chelates available in the diet.

Gastroferrin is an endogenous macromolecular ligand secreted in the digestive tract that binds about two hundred moles of iron per mole of protein, also in a polynuclear array[117]. Consequently its role as a potential regulating agent in both normal and abnormal control of iron absorption is under investigation[108-10]. Although the protein also possesses blood-group activity, the biological implications of the simultaneous occurrence of iron(III) binding and blood-group activity are not known.

Ferritin functions as an efficient storage compound by binding iron atoms into a sphere of polymeric iron(III) that is much more compact than the iron(III) nitrate polymer and hydrolysate since it can accommodate in the same volume several times the number of iron atoms as the structural models. The iron(III) nitrate gel, however, appears to be close in density to the ferritin core. The cores are prevented from aggregating by a surrounding layer of polypeptide. By contrast haemosiderin appears to consist largely of aggregates of ferritin cores formed when the ferritin subunits are unavailable. However, haemosiderin iron is known to be formed directly, not via ferritin, particularly in iron-storage pathologies. Although the major function of iron-dextran is to provide iron rather than to store it, the structural features of a polymeric iron core solubilised by a protective organic coating that can be metabolised, seems appropriate for this role.

Sizable polynuclear clusters of iron(III) are now known to be widespread in biological systems. These clusters, which vary in iron content from the fifty iron atoms in phosvitin and two hundred in gastroferrin, to over four thousand in a complete ferritin core, involve antiferromagnetic coupling among the bound iron(III) ions. The polynuclear nature of these iron(III) species correlates well with their particular biological effect, for example inhibition of iron absorption and storage of iron. It is anticipated that other systems binding various amounts of polynuclear iron(III) will be discovered.

REFERENCES

1. Spiro, T. G., and Saltman, P., *Struct. Bond.*, **6** (1969), 116.
2. Feeney, R. E., and Komatsu, St. K., *Struct. Bond.*, **1** (1966), 149.
3. Aasa, R., Malmstrom, B. G., Saltman, P., and Vanngard, T., *Biochim. biophys. Acta*, **75** (1963), 203.
4. Aasa, R., and Aisen, P., *J. biol. Chem.*, **243** (1968), 2399.
5. Okamura, M. Y., Klotz, I. M., Johnson, C. E., Winter, M. R. C., and Williams, R. J. P. *Biochemistry*, **8** (1969), 1951.
6. Moss, T. H., Moleski, C., and York, J. L. *Biochemistry*, **10** (1971), 840.
7. Dawson, J. W., Gray, H. B., Hoenig, H. E., Rossman, G. R., Schredder, J. M., and Wang, R.-H. *Biochemistry*, **11** (1972), 461.
8. Malkin, R., and Rabinowitz, J. C., *A. Rev. Biochem.*, **36** (1967), 113.
9. Hedstrom, B. O. A., *Ark. Kemi*, **6** (1953), 1.
10. Sillen, L. G., *Q. Rev. chem. Soc.*, **13** (1958), 146.
11. Matijevic, E., Couch, J. B., and Kerber, M. J., *J. phys. Chem.*, **66** (1962), 111.
12. Aveston, J., Anakar, E. W., and Johnson, J. S., *Inorg. Chem.*, **3** (1964), 735.
13. Schugar, H. J., Walling, C., Jones, R. B., and Gray, H. B. *J. Am. chem. Soc.*, **89** (1967), 3712.
14. Biedermann, G., and Schindler, P., *Acta chem. scand.*, **11** (1957), 731.
15. van der Giessen, A. A., *J. inorg. nucl. Chem.*, **28** (1966), 2155.
16. Towe, K. M., and Bradley, W. F., *J. Colloid Interface Sci.*, **24** (1967), 384.
17. Francombe, M. H., and Rooksby, H. P., *Clay Miner. Bull.*, **4** (1959), 1.
18. Bernal, J. D., Dasgupta, D. R., and Mackay, A. L., *Clay Miner. Bull.*, **4** (1959), 15.
19. Richard, C. F., Gustafson, R. L., and Martell, A. E., *J. Am. chem. Soc.*, **81** (1959), 1033.
20. Anderegg, G., *Helv. chim. Acta,* **43** (1960), 1530.
21. Gustafson, R. L., and Martell, A. E., *J. phys. Chem.*, **67** (1963), 576.
22. Schugar, H. J., Rossman, G. R., and Gray, H. B., *J. Am. chem. Soc.* **91** (1969), 4564.
23. Lippard, S. J., Schugar, H. J., and Walling, C., *Inorg. Chem.*, **6** (1967), 1825.
24. Spiro, T. G., Allerton, S. E., Renner, J., Terzis, A., Bils, R., and Saltman, P., *J. Am. chem. Soc.*, **88** (1966), 2721.
25. Allerton, S. E., Renner, J., Colt, S., and Saltman, P., *J. Am. chem. Soc.*, **88** (1966), 3147.
26. Spiro, T. G., Pape, L., and Saltman, P., *J. Am. chem. Soc.*, **89** (1967), 5555.
27. Spiro, T. G., Bates, G., and Saltman, P., *J. Am. chem. Soc.*, **89** (1967), 5559.
28. Bates, G., Heggenauer, J. C., Renner, J., Saltman, P., and Spiro, T. G., submitted for publication.
29. Martin, R. L., and White, A. H. in *Transition Metal Chemistry*, vol. 4 (ed. R. L. Carlin), Dekker, New York (1968), p. 113.
30. Ewald, A. H., Martin, R. L., Sinn, E., and White, A. H., *Inorg. Chem.* **8** (1969), 1837.
31. Gray, H. B. in *Transition Metal Chemistry*, vol. 1 (ed. R. L. Carlin), Dekker, New York (1965), p. 256.
32. Earnshaw, A., *Introduction to Magnetochemistry*, Academic Press, New York (1968).
33. *Handbook of Chemistry and Physics* (editor-in-chief R. C. Weast), Chemical Rubber Company, Cleveland (48th edn 1967), p. E-109.
34. Evans, D. F., *J. chem. Soc.* (1959), 2003.
35. Live, D. H., and Chan, S. I., *Analyt. Chem.*, **42** (1970), 791.
36. Kambe, K., *J. phys. Soc. Japan*, **5** (1950), 48.
37. Sinn, E. *Coord. Chem. Rev.*, **5** (1970), 313.
38. Lewis, J., Mabbs, F. E., and Richards, A., *J. chem. Soc.* (1967), 1014.
39. Reiff, W. M., Baker, W. A., Jr, and Erickson, N. E., *J. Am. chem. Soc.*, **90** (1968), 4794.
40. Schugar, H. J., Rossman, G. R., Barraclough, C. G., and Gray, H. B., *J. Am. chem. Soc.*, **94** (1972), 2683.
41. Dunitz, J. D., and Orgel, L. E., *J. chem. Soc.* (1953), 2594.
42. Bauminger, R., Cohen, S. G., Marinov, A., and Ofer, S., *Phys. Rev.* **122** (1961), 743.

43. Ginsberg, A. P., Martin, R. L., and Sherwood, R. C., *Inorg. Chem.*, **7** (1968), 932.
44. Flood, M. T., Barraclough, C. G., and Gray, H. B., *Inorg. Chem.*, **8** (1969), 1855.
45. Dubicki, L., Kakos, G. A., and Winter, G., *Aust. J. Chem.*, **21** (1968), 1461.
46. Fisher, M. E., *Am. J. Phys.*, **32** (1964), 343.
47. Cattrall, R. W., Murray, K. S., and Peverill, K. I., *Inorg. Chem.*, **10** (1971), 1301.
48. *Magnetism*, vol. III (eds G. T. Rado and H. Suhl), Academic Press, New York (1963).
49. Néel, L., *J. phys. Soc. Japan*, **17** (1962), Supplement B-1, 676.
50. Jacobs, J. S., and Bean, C. P. in *Magnetism*, vol. III (eds G. T. Rado and H. Suhl), Academic Press, New York (1963), p. 271.
51. Boas, J. F., Ph.D. Thesis, Monash University, Victoria, Australia (1968).
52. Creer, K. M., *J. phys. Soc. Japan*, **17** (1962), Supplement B-1, 690.
53. Creer, K. M., *J. Geomagn. Geoelect.*, **13** (1962), 86.
54. Takada, T., and Kawai, N., *J. phys. Soc. Japan*, **17** (1962), Supplement B-1, 691.
55. Cohen, J., Creer, K. M., Pauthenet, R., and Srivastava, K., *J. phys. Soc. Japan*, **17** (1962), Supplement B-1, 691.
56. Ball, P. W., *Coord. Chem. Rev.*, **4** (1969), 361.
57. Kokoszka, G. F., and Duerst, R. W., *Coord. Chem. Rev.*, **5** (1970), 209.
58. Okamura, M. Y., and Hoffman, B. M., *J. chem. Phys.*, **51** (1968) 3128.
59. Harris, E. A., and Owen, J. *Proc. R. Soc.* **A289** (1965), 122.
60. Elliston, P. R., Ph.D. Thesis, Monash University, Victoria, Australia (1967).
61. Srivastava, K. G., and Srivastava, R., *Nuovo Cim.*, **39** (1965), 71.
62. Tanabe, Y., and Sugano, S., *J. phys. Soc. Japan*, **9** (1954), 753.
63. Rossman, G. R., Ph.D. Thesis, California Institute of Technology, Pasadena, California (1971).
64. Webb, J. M., Ph.D. Thesis, California Institute of Technology, Pasadena, California (1972).
65. Friedman, H. L., *J. Am. chem. Soc.*, **74** (1952), 5.
66. Ginsberg, A. P., and Robin, M. B., *Inorg. Chem.*, **2** (1963), 817.
67. Brown, D. H., *Spectrochim. Acta*, **19** (1963), 1683.
68. Birchall, T., Greenwood, N. N., and Reid, A. F., *J. chem. Soc.* (1969), 2382.
69. Faye, G. H., *Can. Mineralogist*, **10** (1969), 112.
70. Gray, H. B., *Adv. Chem. Ser.* **100** (1971), 365.
71. Pott, G. T., and McNicol, B. D., *J. chem. Phys.*, **56** (1972), 5246.
72. Thibeault, J., Powers, D., and Cowman, C., (1972), personal communication.
73. Johansson, G., *Acta. chem. scand.*, **16** (1962), 1234.
74. Ferguson, J., Guggenheim, H. J., and Tanabe, Y., *J. phys. Soc. Japan*, **21** (1966), 692.
75. Lohr, L. L., Jr, and McClure, D. S., *J. chem. Phys.*, **49** (1968), 3516.
76. Ferguson, J., *Aust. J. Chem.*, **21** (1968), 307.
77. Marzzacco, C. J., and McClure, D. S., *Faraday Soc. Symp.*, No. 3 (1969), 106.
78. Bearden, A. J., and Dunham, W. R., *Struct. Bond.*, **8** (1970), 1.
79. Walker, L. R., Wertheim, G. K., and Jaccarino, V., *Phys. Rev. Lett.*, **6** (1961), 98.
80. Bancroft, G. M., Maddock, A. G., Ong, W. K., and Prince, R. H., *J. chem. Soc.* (1966), 723.
81. Reiff, W. M., Long, G. J., and Baker, W. A., Jr, *J. Am. chem. Soc.*, **90** (1968), 6347.
82. Armstrong, R. J., Merrish, A. H., and Sawatzky, G. A., *Phys. Letters* **23** (1966), 414.
83. Rossiter, M. J., and Hodgson, A. E. M., *J. inorg. nucl. Chem.*, **27** (1965), 63.
84. Edwards, P. R., and Johnson, C. E., *J. chem. Phys.*, **49** (1968), 211.
85. Kundig, W., Bommel, H., Constabaris, G., and Lindquist, R. H., *Phys. Rev.*, **142** (1965), 327.
86. Brady, G. W., Kurkjian, C. R., Lyden, E. F. X., Robin, M. B., Saltman, P., Spiro, T., and Terzis, A., *Biochemistry*, **7** (1968), 2185.
87. Cohen, G. H., and Hoard, J. L., *J. Am. chem. Soc.*, **88** (1966), 3228.
88. Azaroff, L. V., and Buerger, M. J., *The Powder Method in X-Ray Crystallography*, McGraw-Hill, New York (1958).
89. Klug, H. P., and Alexander, L. E., *X-Ray Diffraction Procedures*, Wiley, New York, (1954), p. 504.
90. Guinier, A., and Fournet, G., *Small-angle Scattering of X-Rays*, Wiley, New York (1955).
91. Mikolaj, P. G., and Pings, C. J., *Phys. Chem. Liquids*, **1** (1968), 93.
92. Hewkin, D. J., and Griffith, W. P., *J. chem. Soc.* (1966), 474.
93. Khedekar, A. V., Lewis, J., Mabbs, F. E., and Weigold, H., *J. chem. Soc.* (1967), 1561.

94. Wing, R. M., and Callahan, K. P., *Inorg. Chem.*, **8** (1969), 871.

95. Nakamoto, K., *Infrared Spectra of Inorganic and Coordination Compounds*, Wiley, New York (1970).

96. Hair, M. L., *Infrared Spectroscopy in Surface Chemistry*, Dekker, New York (1967).

97. Webb J., Multani, J. S., Beach, N. A., Saltman, P., and Gray, H. B. *Biochemistry*, 12 (1973), 1797.

98. Multani, J. S., Cepurneek, C. P., Davis, P. S., and Saltman, P., *Biochemistry*, **9** (1970), 3970.

99. Greengard, O., Sentenac, A., and Mendelsohn, N., *Biochim. Biophys. Acta*, **90** (1964), 406.

100. Harrison, P. M., and Hoy, T. G. (1972) in *Inorganic Biochemistry* (ed. G. Eichhorn), in press.

101. Sturgeon, P., and Shoden, A., in *Iron Metabolism: An International Symposium* (ed. F. Gross), Springer, Berlin (1964), p. 121.

102. Marshall, P. R., and Rutherford, D., *J. Colloid Interface Sci.*, **37** (1971), 390.

103. Taborsky, G. *Biochemistry*, **2** (1963), 266.

104. Grant, C. T., and Taborsky, G. *Biochemistry*, **5** (1966), 544.

105. Rosenstein, R. W., and Taborsky, G., *Biochemistry*, **9** (1970), 649.

106. Mecham, D. K., and Olcott, H. S., *J. Am. chem. Soc.*, **71** (1949), 3670.

107. Joubert, F. J., and Cook, W. H., *Can. J. Biochem. Physiol.*, **36** (1958), 399.

108. Taborsky, G., and Mok, C.-C., *J. biol. Chem.*, **242** (1967), 1495.

109. Elwood, P. C., in *Iron in Flour: Reports on Public Health and Medical Subjects*, H.M.S.O., London, **117** (1968), 1.

110. Bearden, A. (1971), personal communication

111. Allerton, S. E., and Perlmann, G. E., *J. biol. Chem.*, **240** (1965), 3892.

112. Shainkin, R., and Perlmann, G. E., *J. biol. Chem.*, **246** (1971), 2278.

113. Taborsky, G. *J. biol. Chem.*, **243** (1968), 6014.

114. Perlmann, G. E. and Grizzuti, K., *Biochemistry*, **10** (1971), 258.

115. Davis, P. S., Luke, C., and Deller, D. J., *Nature, Lond.*, **214** (1967), 1126.

116. Davis, P. S., Multani, J. S., Cepurneek, C. P., and Saltman, P., *Biochem. biophys. Res. Commun.*, **37** (1969), 532.

117. Webb J. M., Multani, J. S., Saltman, P., and Gray, H. B., *Biochemistry*, **12** (1973), 265.

118. Aasa, R., Malmstrom, B. G., Saltman, P., and Vanngard, T., *Biochim. Biophys. Acta*, **80** (1963), 430.

119. Charley, P., Sarkar, B., Stitt, C., and Saltman, P., *Biochim. Biophys. Acta*, **69** (1963), 313.

120. Saltman, P., *J. chem. Educ.*, **42** (1965), 682.

121. Granick, S., *Chem. Rev.*, **38** (1946), 379.

122. Harrison, P. M., in *Iron Metabolism: An International Symposium* (ed. F. Gross), Springer, Berlin (1964), p. 40.

123. Crichton, R. R., *New Engl. J. Med.*, **284** (1971), 1413.

124. Bjork, I., and Fish, W. W., *Biochemistry*, **10** (1971), 2844.

125. Michaelis, L., Coryell, C. D., and Granick, S., *J. biol. Chem.*, **148** (1943), 463.

126. Granick, S., and Hahn, P. F., *J. biol. Chem.*, **155** (1944), 661.

127. Farrant, J. L., *Biochim. Biophys. Acta*, **13** (1954), 569.

128. Richter, G. W., *J. exp. Med.*, **109** (1959), 197.

129. Bessis, M., and Breton-Gorius, J., *C.r. hebd. Séanc. Acad. Sci., Paris, Ser. B.*, **250** (1960), 1360.

130. Muir, A. R., *Q. J. exp. Physiol.*, **45** (1960), 192.

131. Van Bruggen, E. F. J., Wiebenga, E. H., and Gruber, M., *J. molec. Biol.*, **2** (1960), 81.

132. Haggis, G. H., *J. molec. Biol.*, **14** (1965), 598.

133. Haydon, G. B., *J. Microsc.*, **89** (1969), 251.

134. Haydon, G. B., *J. Microsc.*, **91** (1970), 65.

135. Towe, K., *J. Microsc.*, **90** (1969), 279.

136. Crewe, A. V., and Wall, J., *J. molec. Biol.*, **48** (1970), 376.

137. Schoffa, G., *Z. Naturf.*, **20b** (1967), 167.

138. Blaise, A., Chappert, J., and Girardet, J.-L., *C.r. hebd. Séanc. Acad. Sci., Paris, Ser. B*, **261** (1965), 2310.

139. Blaise, A., Feron, J., Girardet, J.-L., and Lawrence, J.-J., *C.r. hebd. Séanc. Acad. Sci., Paris, Ser. B*, **265** (1967), 1077.

140. Boas, J. F., and Troup, G. J., *Biochim. Biophys. Acta*, **229** (1971), 68.

141. Webb, J., and Gray, H. B. (1972), unpublished observations.
142. Boas, J. F., and Window, B., *Aust. J. Phys.*, **19** (1966), 573.
143. Fischbach, F. A., and Anderegg, J. W., *J. molec. Biol.*, **14** (1965), 458.
144. Harrison, P. M., *J. molec. Biol.*, **6** (1963), 404.
145. Harrison, P. M., and Hoy, T. G., *J. Microsc.*, **91** (1970), 61.
146. Fischbach, G. A., Harrison, P. M., and Hoy, T. G., *J. molec. Biol.*, **39** (1969), 235.
147. Girardet, J.-L., and Lawrence, J.-J., *Bull. Soc. fr. Minér. Cristallogr.*, **91** (1968), 440.
148. Harrison, P. M., Fischbach, F. A., Hoy, T. G., and Haggis, G. H., *Nature, Lond.*, **216** (1967), 1188.
149. Arrhenius, G. (1971), personal communication.
150. van der Geissen, A. A., *J. inorg. nucl. Chem.*, **30** (1968), 1739.
151. McKay, R. W., and Fineberg, R. A., *Archs Biochem. Biophys.*, **104** (1964), 487, 496.
152. Ludewig, S., and Franz, S. W., *Archs Biochem. Biophys.*, **138** (1970), 397, and references therein.
153. Wohler, V. F., *Acta Haemat.*, **23** (1960), 342.
154. Richter, G. W., *Am. J. Path.*, **35** (1959), 690.
155. Fischbach, F. A., Gregory, D. W., Harrison, P. M., Hoy, T. G., and Williams, J. M., *J. Ultrastruct. Res.*, **37** (1971), 495.
156. Schwietzer, C. H., *Acta Haemat.*, **10** (1953), 174.
157. Fletcher, F., and London, E., *Br. med. J.*, **1** (1954), 984.
158. Cox, J. S. G., King, R. E., and Reynolds, G. F., *Nature, Lond.*, **207** (1965), 1202.
159. Ricketts, C. R., Cox, J. S. G., Fitzmaurice, C., and Moss, G. F., *Nature, Lond.*, **208** (1965), 237.
160. Nissim, J. A., *Lancet*, **253** (1947), 49.
161. Lindvall, S., and Anderson, N. S. E., *Br. J. Pharmac.*, **17** (1961), 358.
162. Nissim, J. A., and Robson, J. M., *Lancet*, **256** (1949), 686.

PART 5

Metal Ions as N.M.R. Probes in Biochemistry

S. J. FERGUSON

Department of Biochemistry, University of Oxford

23 METAL IONS AS N.M.R. PROBES IN BIOCHEMISTRY

23.1 INTRODUCTION

Metal ions occur widely in biological systems[1] and the increasing number of publications on the use of nuclear magnetic resonance (n.m.r.) in the study of their role is evidence for the power of n.m.r. methods. The aim of part 5 of the book is to indicate the type of information that can be obtained together with a brief outline of the different approaches available. For more complete surveys of specific areas the reader is referred to several reviews[2,3,4] and also to a recent book[5].

It is often the case that a naturally occurring metal ion does not possess useful n.m.r. properties but fortunately, as will be discussed, it often proves possible to substitute a metal ion with more attractive n.m.r. properties[6]. Additionally a metal ion can sometimes be introduced into a system that has no natural requirement for one so that the methods to be described can be entirely general.

23.2 WHAT INFORMATION IS SOUGHT FROM N.M.R. EXPERIMENTS?

One of the main attractions of n.m.r. is that it is the only general method that can give structural information in solution. Suppose, as an example, that an enzyme catalysing the following reaction is under study

$$A + B \rightleftharpoons C + D \qquad (23.1)$$

that catalytic activity is only exhibited in the presence of a metal ion and also that the enzyme is inhibited by a fifth compound E. Some of the questions about such a system that n.m.r. can help answer are:

(i) Do the ligands change their conformation on binding to the enzyme?
(ii) How are the ligands, including the inhibitor, bound with respect to each other?
(iii) Where is the metal bound with respect to A, B, C, D and E and therefore does it have a catalytic or structural role?
(iv) Does the enzyme conformation change either grossly or locally on binding the various ligands?
(v) Can the active-site amino-acid residues be identified?

Similar questions may be formulated in other aspects of biochemistry. For instance, the mechanism by which cations such as Na^+, K^+, Mg^{2+} and Ca^{2+} are transported across membranes might be further elucidated by n.m.r. investigation of the interaction of these ions with membranes. Such a problem might be investigated either by monitoring the n.m.r. properties of protons, the most commonly studied resonance, or by looking at the metal resonances themselves, although as will be explained metal-ion n.m.r. is often difficult.

23.3 SOME ESSENTIAL N.M.R. PRINCIPLES

It is assumed that the reader is familiar with the basic principles of n.m.r. (References 7 and 8 are good introductory texts). A significant feature of n.m.r. is that it involves transitions between states that differ only slightly in energy and thus by the Boltzmann law have almost equal populations. There are two important consequences of this. Firstly, a relatively large number of nuclei must be present for n.m.r. to be observed; that is, the method is inherently insensitive. For example, the lowest concentration of protons, the most sensitive nucleus that can be monitored, is of the order of millimolar while for other nuclei, as can be seen from table 28, higher concentrations are needed. A result of this is that n.m.r. experiments on biological material may have to be carried out at concentrations higher than is desirable.

TABLE 28 SOME NUCLEAR PROPERTIES

Nucleus	Natural abundance (%)	Sensitivity relative to hydrogen for equal numbers of nuclei at constant field	Spin (I)
^1H	99.9	1.00	$\frac{1}{2}$
^{13}C	1.1	1.6×10^{-2}	$\frac{1}{2}$
^{19}F	100	0.83	$\frac{1}{2}$
^{23}Na	100	9.27×10^{-2}	$\frac{3}{2}$
^{25}Mg	10.1	2.68×10^{-2}	$\frac{5}{2}$
^{31}P	100	6.64×10^{-2}	$\frac{1}{2}$
^{39}K	93.1	5.08×10^{-4}	$\frac{3}{2}$
^{43}Ca	0.1	6.39×10^{-2}	$\frac{7}{2}$
^{203}Tl	29.5	0.19	$\frac{1}{2}$
^{205}Tl	70.5	0.19	$\frac{1}{2}$

The second result of the small energy difference between states is that if a sample of nuclei absorbs energy in an n.m.r. experiment then it might be expected that the levels would rapidly equalise their populations and absorption would cease. However, there are relaxation processes, which oppose this effect and tend to return the sample to its original equilibrium Boltzmann distribution among the available energy levels.

It is usual to define the direction of the applied magnetic field as z. The equilibrium situation will then be a small magnetic moment in this z direction but zero magnetisation in the x and y directions, both of which are perpendicular to z; that is, the nuclei are regarded as aligning themselves with respect to the field. During the n.m.r. experiment, in which a small oscillating field is applied in the *xy* plane, the magnetisation in the *z* direction will decrease as the populations of the energy levels tend to equalise. Additionally, magnetisation will be induced in the *x* and *y* directions. It is usual to describe the rate of return to equilibrium by two relaxation times, which are essentially rate constants. For the *z* direction the constant is called the longitudinal or spin-lattice relaxation time T_1 and in the *x* and *y* directions the transverse or spin-spin relaxation time T_2.

The nature of the relaxation processes is important since studies of the relaxation times can provide structural information; a short and much simplified account follows. As a molecule rotates and diffuses it experiences fluctuating magnetic fields due to interactions with other nuclei. There is a wide spread of frequencies at which these fluctuations occur but if there is a component at the appropriate resonance frequency it can induce transitions between the energy levels. When a downward transition is induced in this way, energy is lost to the environment or lattice and as this occurs more often than the reverse process the net effect is to depopulate the higher energy levels. This mechanism then causes T_1 relaxation; that is, it tends to return the *z* direction magnetisation to its equilibrium value. A simple way of demonstrating the T_1 process is to consider what happens when a sample of magnetic nuclei is placed in a magnetic field. In the absence of the field the nuclei will have random orientation and on introducing the field there is an equal chance of any energy level being populated. Consequently there is initially no population difference and an n.m.r. signal cannot be observed. As, however, the system decays by the T_1 process to a Boltzmann distribution among the energy levels the n.m.r. signal can be seen to grow.

The T_2 relaxation process is more difficult to treat. The *xy* plane magnetisation of a whole sample is the vector sum of individual nuclear magnetic moments and any process that tends to reduce the coherence between these individual moments will therefore reduce the *xy* plane magnetisation of the sample. Transitions between the available energy levels (the T_1 process) result in a reduction in the *xy* plane magnetisation since the transitions tend to decrease this coherence between the individual magnetic moments in the *xy* plane. Additionally, different nuclei in a sample will experience small but different local fields in the *z* direction due to neighbouring nuclei. This again results in a loss of coherence of the individual nuclear moments so as to decrease the *xy* plane magnetisation.

Both of the above processes may also be viewed in terms of the Uncertainty Principle. Transitions between the energy levels result in a decrease in the lifetimes of the levels while the local fields have the effect of producing a spread in the energy of the levels available to the nuclei. T_2 can be related to the linewidth Δv of a magnetic resonance line, which is equivalent to saying that it is related to the uncertainty in the energy of the transition. The relationship is

$$\frac{1}{T_2} = \pi \Delta v \qquad (23.2)$$

where Δv is measured in hertz. Consequently broad n.m.r. lines arise when T_2 is short, that is relaxation is efficient.

23.4 TYPES OF METAL-ION PROBE

It is convenient to classify the use of metal ions as n.m.r. probes into those studies in which the n.m.r. of the metal itself is observed and those in which perturbations introduced by the metal on the resonances of other nuclei are measured.

23.4.1 Direct observation of the metal-ion n.m.r.

All the metal ions that occur biologically have either zero nuclear spin, in which case, of course, n.m.r. is not possible, or a nuclear spin of one or more. When the nuclear spin (I) is greater than one nuclei possess a quadrupole moment. This quadrupole moment can interact with local fields in a molecule so as to cause efficient nuclear relaxation. The consequence of this is to shorten T_2 or broaden resonance lines which, together with often inherently low sensitivity (table 28), can make working with the n.m.r. signals from metal ions difficult.

Although some ^{39}K, ^{43}Ca and ^{25}Mg studies have been carried out on few species of biological interest, ^{23}Na has been more widely studied; examples will be given later. For the potassium ion there is an alternative possibility, since it can be replaced by thallium(I), which has a nuclear spin of $\frac{1}{2}$ and a higher n.m.r. sensitivity. Some aspects of the inorganic chemistry of thallium(I) resemble those of potassium and it is therefore expected that thallium(I) will substitute for potassium in some biological systems.

23.4.2 The effect of metal ions on other nuclear resonances

Rather more use is made of this approach since the resonances that are commonly monitored (^{1}H, ^{31}P and increasingly ^{19}F and ^{13}C) arise from nuclei with nuclear spin = $\frac{1}{2}$, which can be studied at relatively low concentrations. The metal ions that perturb nuclear resonances in a way that can lead to structural information are paramagnetic, which can cause changes in the chemical shifts of nuclei and changes in nuclear-relaxation rates. The paramagnetic ions of most interest are those of the first transition and lanthanide series. Apart from the fact that divalent transition-metal ions occur frequently in biology they can often replace Mg^{2+} and Zn^{2+} without necessarily impairing biological activity.

The attraction of the lanthanides lies in the fact that they comprise a series of chemically similar but magnetically different cations. Additionally their similarity in size to Ca^{2+} offers the possibility of using them as probes for calcium binding sites.

The relaxation effects of paramagnetic metal ions are treated first if only because they have so far been more extensively employed than other metal-ion probes.

23.5 RELAXATION IN THE FIRST CO-ORDINATION SPHERE OF A METAL ION

In the brief account of relaxation processes it was pointed out that oscillating magnetic fields within a sample give rise to relaxation. Paramagnetic ions have large magnetic moments due to their unpaired electrons and thus produce large fields at neighbouring nuclei. The way in which these fields oscillate is important. Consider a nucleus bound to a paramagnetic metal ion where there is some process that causes a field from the metal to die away on average every 10^{-x} seconds. Thus the nucleus experiences a field that fluctuates at an average of 10^x times per second. This average time for a field to die away is known as a correlation time τ_c. However, for relaxation to occur the component of the oscillating field at the resonant frequency is required. A correlation function $f(\tau_c)$ can be formulated, which relates the average correlation time τ_c to the component at the resonant frequency ω_I.

The Solomon-Bloembergen equations indicate how relaxation times of nuclei in the first co-ordination sphere of a metal ion are related to the magnetic moment of the metal, the distance between the metal and the nucleus in question and correlation times

$$\frac{1}{T_{Im}} = \frac{2}{15} \times \frac{\gamma_I^2 g^2 S(S+1)\beta_e^2}{r^6} \left[\frac{3\tau_c}{1+\omega_I^2\tau_c^2} + \frac{7\tau_c}{1+\omega_s^2\tau_c^2} \right] + \frac{2}{3} S(S+1) \left(\frac{A}{\hbar}\right)^2 \left[\frac{\tau_e}{1+\omega_s^2\tau_e^2} \right]$$

(23.3)

and

$$\frac{1}{T_{2m}} = \frac{1}{15} \times \frac{\gamma_I^2 g^2 S(S+1)\beta_e^2}{r^6} \left[4\tau_c + \frac{3\tau_c}{1+\omega_I^2\tau_c^2} + \frac{13\tau_c}{1+\omega_s^2\tau_c^2} \right]$$

$$+ \frac{1}{3} S(S+1) \left(\frac{A}{\hbar}\right)^2 \left[\frac{\tau_e}{1+\omega_s^2\tau_e^2} + \tau_e \right] \quad (23.4)$$

where τ_c and τ_e are two different correlation times;
$\quad\quad$ g is the electronic g factor;
$\quad\quad$ γ_I is the magnetogyric ratio for the nucleus;
$\quad\quad$ β_e is the Bohr magneton for the electron;
$\quad\quad$ S is the spin quantum number of the metal ion;
$\quad\quad$ ω_I is the nuclear resonance frequency;
$\quad\quad$ ω_s is the electronic resonance frequency;
$\quad\quad$ $A/\hbar$ is the electron nuclear hyperfine coupling constant in hertz;
$\quad\quad$ r is the metal–nucleus distance.

The first term in each equation refers to a through-space dipole–dipole interaction while the second, scalar, term reflects the effect of electron delocalisation from the metal ion to the nucleus under study. The delocalisation usually occurs through chemical bonds. This scalar term can often be neglected and when this is the case these two relaxation equations show that if T_{jm} ($j = 1, 2$) and τ_c can be determined, distances can be estimated. Unfortunately the evaluation of these two parameters is not as straightforward as the following simplified account will show. For convenience the functions of τ_c in equations 23.3 and 23.4 will be abbreviated to $f_1(\tau_c)$ and $f_2(\tau_c)$.

23.6 CHEMICAL EXCHANGE AND THE DETERMINATION OF $1/T_{jm}(j = 1, 2)$

When ligands are exchanging between the first co-ordination sphere of a paramagnetic ion and an unbound site it is usually not possible to measure $1/T_{jm}$ directly since an averaged signal is seen. The observed paramagnetic contribution to the relaxation, $1/T_{jp}$, will in general depend on T_{jm}, τ_m (the lifetime of the ligand on the ion) and an outer-sphere contribution, which represents the effect of the paramagnetic centre beyond its first co-ordination sphere. Additionally $1/T_{2p}$ may depend on the chemical shift between the bound and unbound nuclei. However, for the metal ions that are most efficient at inducing relaxation (Mn^{2+}, Gd^{3+}) the outer sphere and shift terms are usually neglected and $1/T_{jm}$ is related to $1/T_{jp}$ by

$$\frac{1}{T_{jp}} = \frac{X_m}{T_{jm} + \tau_m} \tag{23.5}$$

where X_m refers to the mole fraction of the exchanging ligands that are bound at any one time. When exchange is very fast equation 23.5 reduces to

$$\frac{1}{T_{jp}} = \frac{X_m}{T_{jm}} \tag{23.6}$$

since $T_{jm} > \tau_m$. This case can be distinguished by studying the temperature dependence of $1/T_{jp}$. If the exchange is fast (T_{jm} dominant) then $1/T_{jp}$ will generally decrease with increasing temperature as $f_j(\tau_c)$ diminishes. In the case where τ_m is the dominant term $1/T_{jp}$ will increase with increasing temperature since the rate of relaxation is being governed by the rate of transport of ligands to the paramagnetic centre. Thus, if the fast-exchange condition holds, one of the unknowns, T_{jm}, in equation 23.3 and 23.4 can be determined by measuring the observed paramagnetic contribution to the relaxation $1/T_{jp}$ and the fraction bound (X_m) by any technique that will give binding constants.

Without wishing to delve too deeply into theory[9] there are factors that can complicate demonstration of fast exchange. Firstly, the $f_j(\tau_c)$ terms in equations 23.3 and 23.4 represent the way in which the correlation time affects relaxation times. τ_c usually diminishes on increasing the temperature, when $f_2(\tau_c)$ will always decrease but $f_1(\tau_c)$ in some circumstances will increase, thus giving an increase in $1/T_{1m}$ and $1/T_{1p}$ (see equation 23.3). Secondly, under some circumstances τ_c will increase with rise in temperature. To summarise, decrease in $1/T_{jp}$ with increasing temperature indicates fast exchange, while an increase in $1/T_{jp}$ may show slow exchange. Extensive studies of relaxation behaviour should distinguish between these possibilities. A further point is that individual atoms in the same molecule may satisfy different conditions. For example, a nucleus very close to the paramagnetic centre may have a very short T_{jm} such that τ_m dominates, while other, more distant, atoms will have the same τ_m but a longer T_{jm} so that $T_{jm} > \tau_m$. It follows that if it can be shown that the nearest atom satisfies $T_{jm} > \tau_m$ this will probably be the case for the whole molecule.

If the slow-exchange condition does hold—that is, $\tau_m > T_{jm}$—then an estimate of $1/T_{jm}$ can be made by assuming $1/T_{jp} = X_m/T_{jm}$, which will give an upper limit on T_{jm} and hence on any distance subsequently calculated.

23.7 CORRELATION TIMES AND THE DETERMINATION OF $f_j(\tau_c)$

The determination of the second unknown, τ_c, presents a more challenging problem and requires a fairly detailed knowledge of the nature of τ_c and $f_j(\tau_c)$. The following treatment is an attempt to briefly outline some of the basic ideas so that the n.m.r. experiments carried out involving this type of information may be understood, and perhaps criticised!

For relaxation induced by paramagnetic ions one or more of these terms may determine the correlation times

$$\frac{1}{\tau_c} = \frac{1}{\tau_s} + \frac{1}{\tau_m} + \frac{1}{\tau_r} \tag{23.7}$$

where τ_m is the lifetime of a nucleus in the bound state as before, τ_r is the rotational correlation time of the complexed paramagnetic ion and τ_s is the electron-spin-lattice (longitudinal) relaxation time of the metal ion. A comparison of $Mn(H_2O)_6{}^{2+}$ and $Co(H_2O)_6{}^{2+}$ shows how the relaxation behaviour is influenced by these terms. For Mn^{2+} τ_s is long ($\approx 10^{-8}$ s) and it is τ_r (3×10^{-11} s) which is the dominant term. For Co^{2+} τ_r is of the same order as for Mn^{2+} but, as τ_s is very short ($\approx 10^{-13}$ s), this is the important term. The nature of equations 23.3 and 23.4 immediately shows that the difference in τ_c alone will cause Mn^{2+} to be more effective than Co^{2+} at relaxing bound nuclei.

A number of methods are available for estimating τ_c but only one will be considered here since it has been the most widely used. Figure 64 shows a plot of

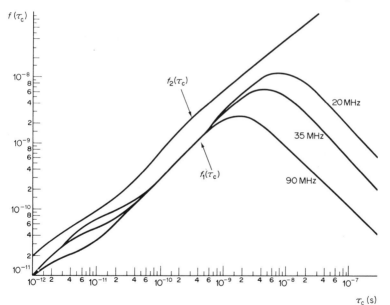

Figure 64 $f_1(\tau_c)$ at several frequencies and $f_2(\tau_c)$ at 60 MHz plotted against τ_c.

$f_1(\tau_c)$ of equation 23.3 against τ_c. For the $Mn(H_2O)_6{}^{2+}$ ion $\tau_c(\tau_r) = 3 \times 10^{-11}$ s. Hence, increasing τ_c will increase $f_1(\tau_c)$ and therefore shorten T_{1m} of the bound water molecules. In other words there would be a relaxation enhancement. This is exactly the effect observed when water-proton relaxation rates are studied in the presence of a Mn^{2+}-macromolecule (for example, protein) complex. Such a complex will clearly rotate more slowly than $Mn(H_2O)_6{}^{2+}$ so that τ_r will become longer. In fact, it may be so long that τ_s and/or τ_m become the dominant factors in τ_c. A term ϵ^*, the relaxation enhancement, is generally defined as

$$\epsilon^* = \frac{(1/T_1)_p^* - (1/T_1)^*}{(1/T_1)_p - (1/T_1)} \tag{23.8}$$

where $1/T_1$ is the observed relaxation rate of the water protons in the presence of p = paramagnetic ion, * = macromolecule. As the metal ions will exist in two environments, bound to macromolecule and free in solution, ϵ^* can be written as a sum of two terms

$$\epsilon^* = X_b \epsilon_b + X_f \epsilon_f \tag{23.9}$$

where X_b and X_f refer to the mole fractions of metal, bound and free, respectively, while ϵ_b and ϵ_f similarly refer to the enhancements of relaxation rates of water bound to Mn^{2+}-macromolecule and free manganese ion, respectively. ϵ_f is assumed to be one since the behaviour of free manganese ions is not considered to change when macromolecules are also present in solution. Hence if the dissociation constant of the manganese–macromolecule complex is known ϵ_b may be calculated.

ϵ_b is obtained from the measurement of $1/T_{1p}$, the observed paramagnetic contribution to the relaxation, and is the enhancement corresponding to a situation in which all the metal is bound to the macromolecule. From equation 23.5

$$\epsilon_b = \frac{q^*}{q} \times \frac{(T_{1m} + \tau_m)}{(T_{1m^*} + \tau_{m^*})} \tag{23.10}$$

where * indicates the Mn^{2+}-macromolecule complex, q^* is the number of water molecules in the first co-ordination sphere of Mn^{2+} when it is bound to a macromolecule and q is the number of water molecules in the first co-ordination sphere of the free Mn^{2+} ion. Clearly $q > q^*$. This equation reduces to

$$\epsilon_b = \frac{q^*}{q} \times \frac{(T_{1m})}{(T_{1m^*})} \tag{23.11}$$

when $T_{1m} > \tau_m$ and $T_{1m^*} > \tau_{m^*}$. This may now be written as

$$\epsilon_b = \frac{q^*}{q} \times \frac{f_1(\tau_c^*)}{f_1(\tau_c)}$$

This equation shows that, in principle at least, correlation times can be assessed if q^* and $f_1(\tau_c)$free are known and ϵ_b is calculated from ϵ^*. A correlation time obtained in this way relates to the metal–water interaction and in some circumstances can be applicable to other ligands. ϵ_b can respond to the metal environment since it will be altered by changes in the hydration number q^* or in τ_c and thus $f_1(\tau_c)$. Such effects can be caused by the binding of ligands to macromolecules so that monitoring water-proton relaxation rates in the presence of a Mn^{2+}-enzyme complex can give useful information about enzyme-ligand interactions.

Only spin-lattice (T_1) relaxation times are usually used in this way because for metal-water interaction the scalar term in the T_2 equation makes the shortening of T_2 less marked.

23.8 EXAMPLES OF THE USE OF PARAMAGNETICALLY INDUCED RELAXATION RATES

The earliest studies involved studying water-proton relaxation rates (p.r.r.) in the presence of a Mn^{2+}-enzyme complex. The approach has been fruitful in the study of enzyme mechanism. The enzyme creatine kinase, which catalyses the reaction

$$ADP + O=P-NH-C-N-CH_2-C \rightleftharpoons ATP + creatine \qquad (23.12)$$

phosphocreatine

provides an example of the type of information available from this type of experiment[4]. Creatine kinase is inactive unless a divalent metal ion is present. *In vivo* this is Mg^{2+} but in the isolated enzyme it can be replaced by Mn^{2+} so as to allow n.m.r. studies.

The questions that studies of longitudinal (T_1) relaxation rates of water may answer in this case include:

(i) Does the metal ion bind directly to the protein thus increasing ATP or ADP binding or is the binding of a metal-ADP complex to the protein involved?
(ii) Does binding of creatine to the metal-ADP-enzyme complex cause a detectable conformational change?

A first step is to observe water-proton relaxation rates in the presence of Mn^{2+} + enzyme, Mn^{2+} + enzyme + ADP and, as controls, Mn^{2+} alone, Mn^{2+} + ADP and enzyme alone. It turns out that there is only an appreciable proton relaxation-rate enhancement when Mn^{2+}, ADP and enzyme are all present. This then, is indicative of an enzyme-ADP-Mn^{2+} complex rather than an enzyme-metal-ADP species since it appears that Mn^{2+} only binds to the enzyme in the presence of ADP. It could, however, be argued that ADP binding increases the enzyme's affinity for Mn^{2+}. As the substrate is believed to be Mn-ADP, determining the number of water molecules bound to the Mn^{2+} in the enzyme-ADP-Mn^{2+} complex could indicate whether the metal is bound to the enzyme and ADP or just ADP.

When creatine is added to the enzyme-ADP-Mn^{2+} complex to give an abortive complex, further changes in the water-proton relaxation rates are observed. The nature of this effect is not easy to determine; it could arise from changes in $q*$ and τ_c* or the condition $T_{1m} > \tau_m$ might no longer be applicable. Extensive studies of the frequency and temperature dependence of water relaxation rates can distinguish between these possibilities.

Hence, simply by observing a property of solvent water molecules, information about the enzyme creatine kinase can be obtained. A very large number[10] of enzymatically catalysed phosphate-transfer reactions have now been studied by this method and it has been proposed that such reactions fall into two classes; Class I, enzyme–ATP–metal; Class II, enzyme–metal–ATP or enzyme $<\genfrac{}{}{0pt}{}{\text{metal}}{\text{ATP}}$. It must be pointed out that distinguishing between these three possibilities is by no means easy and this can often only be done rigorously when a study of water-proton relaxation rates is made at a range of temperatures and frequencies. E.P.R. studies on the bound Mn^{2+} can also be helpful. The complications that arise are dealt with elsewhere[5].

Sometimes a metal ion can be introduced as a probe into systems that normally have no metal-ion requirement. Such a case is glycogen phosphorylase b, which is responsible for catalysing the conversion of the polysaccharide glycogen into phosphorylated monosaccharide units—glucose-1-phosphate(G-1-P). It has been shown[11] that Mn^{2+} has a specific binding site on this enzyme by studying the p.r.r. of water in the presence of Mn^{2+} and phosphorylase. The significant feature of this enzyme is that it is inactive in the absence of another ligand, adenosine monophosphate (AMP), which increases the binding of substrates as well as increasing the maximal velocity at which the enzyme can function. Inosine monophosphate (IMP) activates the enzyme solely by increasing the maximal reaction velocity and this difference in activation is reflected by p.r.r. measurements. Adding AMP to Mn^{2+}-phosphorylase causes a change in the water relaxation rates while IMP essentially has no effect. Thus a conformational change of the protein caused by AMP seems to have been detected and this corresponds to a change in τ_c for the Mn^{2+}-H_2O interaction. Similar measurements involving studying G-1-P binding have allowed a pattern of conformational states of the enzyme to be postulated, again solely by studying water-proton relaxation rates[11].

The two examples described have essentially involved detection of ligand binding to enzymes and an interpretation of the observed changes of water p.r.r. in terms of conformational changes. It is clearly desirable to try to quantify the extent of any changes that are occurring. Examination of equations 23.3 and 23.4 shows that if the relaxation time of nuclei bound to a paramagnetic metal can be obtained, together with the correlation time for the metal–nucleus relaxation process, then metal–nuclear distances can be estimated. There is the possibility, therefore, of determining structures in solution and also the extent of conformational changes. Although several systems have now been investigated in this way the method is not easy and usually a number of approximations have had to be made, especially in trying to estimate the correlation time.

Work carried out on lysozyme, for which the determined X-ray structure allows evaluation of the validity of distances estimated from n.m.r. data, provides a good introduction to the n.m.r. method.

Lysozyme is inhibited by small molecules such as β-methyl-N-acetyl-glucosamine (βMeNAG) (see structure (XXIX)), which binds close to two amino acids that have been implicated in catalysis. Furthermore Mn^{2+} and Gd^{3+} (and many other metal

(XXIX)

ions) bind between these two amino acids so that the probe ion is strategically placed. βMeNAG gives two easily distinguishable proton n.m.r. signals from the glycosidic and the acetyl methyl groups, and the position of these two groups in the enzyme–inhibitor complex relative to the metal can be determined in the following way. The observed n.m.r. signal for both the groups will be averaged with contributions from lysozyme–βMeNAG, Gd^{3+}–lysozyme–βMeNAG and βMeNAG, the Gd^{3+}–βMeNAG concentration being negligible. Thus, by subtracting the contribution to relaxation by βMeNAG and βMeNAG–lysozyme the paramagnetic contribution, which only arises on the enzyme, can be estimated. Further by knowing the appropriate binding constants the fraction of βMeNAG that is bound as the Gd^{3+}–lysozyme–βMeNAG complex can be determined. It has already been pointed out that

$$\frac{1}{T_{1,2p}} = \frac{X_m}{T_{1,2m} + \tau_m}$$

and that if $T_{1,2m} > \tau_m$ measurement of $T_{1,2p}$ and X_m gives $T_{1,2m}$, which by equations 23.3 and 23.4 is related to a distance. Using n.m.r. the Gd^{3+}–N-acetyl methyl group distance on lysozyme was found to be 650 pm while the glycosidic methyl group was 540 pm from the Gd^{3+} [12]. These numbers should be compared with the X-ray values of 680 pm and 460 pm. The correlation time used to obtain the distances from n.m.r. data was not correct but the reasonable agreement with the X-ray data obtained is indicative of the value of n.m.r. for studying structural and biochemical problems in solution. Improved instrumental facilities make the evaluation of the correct correlation time possible and this difficulty should be largely overcome in the future [13].

Whereas in lysozyme the introduction of the metal ion inhibits the enzyme activity there are many other enzymes that require a metal ion for activity, and like creatine kinase, the naturally occurring metal can be replaced by others so as to facilitate n.m.r. experiments. Two illustrative cases are carboxypeptidase A and pyruvate kinase.

Carboxypeptidase A catalyses the hydrolysis of certain terminal peptide bonds in peptides and proteins and contains one atom of zinc per molecule. Substituting Mn^{2+} for Zn^{2+} introduces a paramagnetic probe with reduction but not total loss of activity. Observations of water p.r.r. showed that the binding of inhibitors such as bromoacetate displaced the water molecules from the enzyme-bound Mn^{2+}, thus indicating direct binding of the inhibitors to the Mn^{2+} [14]. By extending this work to the observation of the proton magnetic-resonance spectra of the inhibitors the hypothesis of direct binding to the Mn^{2+} was strengthened[15]. However, for most inhibitors it was not possible to calculate distances because $\tau_m \geqslant T_{1,2m}$ (see equation 23.5). Despite this difficulty upper limits on distances could be obtained by setting $T_{1,2m} = T_{1,2m} + \tau_m$. If τ_m and $T_{1,2m}$ are of the same order of magnitude it is possible, in principle at least, to measure τ_m independently, which would then allow $T_{1,2m}$ to be calculated. The condition that $\tau_m \geqslant T_{1,2m}$ is likely to occur fairly frequently in enzyme studies, since τ_m (that is, the lifetime of a metal–enzyme–ligand complex) is often of the order of 10^{-4} s, which is similar to typical values of $T_{1,2m}$.

Pyruvate kinase was studied[16] by replacing both the metal ions Mg^{2+} and K^+, which are needed for catalytic activity. The problem was to find out the distance between the two metals on the enzyme. This was achieved by replacing Mg^{2+} by Mn^{2+} and monitoring the effects of the latter on Tl^+–which replaced K^+–nuclear magnetic relaxation times. In so doing the Tl^+-Mn^{2+} distance was estimated to be 820 pm and, significantly, on adding a substrate, phosphoenolpyruvate, this changed to 490 pm, thus indicating a substrate-induced conformational change. The role of the two metal ions in pyruvate kinase catalysis is not understood, but clearly data of this kind are valuable considering the various possibilities.

In conclusion it has been demonstrated that structural information about enzymes can be obtained by use of nuclear magnetic-relaxation techniques, but there is still scope for refining the method. To summarise, the difficulties with the method are:

1. Evaluating the correlation time. Ideally this should be done for the metal–ligand interaction under study and not extrapolated from other measurements such as the metal–water relaxation rates.
2. Demonstrating the fast-exchange condition; that is, $T_{1,2m} > \tau_m$. This requires a study of the temperature variation of relaxation times and of all the binding constants in a system. This last point arises because a paramagnetic complex may simply dissociate on heating and affect relaxation-time measurements.
3. Determination of binding constants. This is a particular problem in multi-equilibria systems. Nevertheless, none of these points is insurmountable and the method promises to yield a good deal of structural information especially when combined with the use of paramagnetic metal ions as shift probes as described in the next section.

23.9 PARAMAGNETIC METAL IONS AS SHIFT PROBES

Certain paramagnetic metal ions, especially many lanthanides, give rise to changes in the chemical shifts of nuclear resonances of neighbouring nuclei, which are known as pseudocontact shifts. These arise from a through-space interaction, the extent of which depends on both the properties of the metal ion and the stereo-chemical relationship of the particular nucleus to the metal ion. Before considering this effect further the existence of a second shift mechanism must be recognised-the contact shift. This arises from a direct delocalisation of electrons via chemical bonds from the metal ion to the nucleus in question but is not distance dependent. It is not in general capable of yielding structural information, but it is important to eliminate it or prove its absence when attempting to use pseudocontact shift data.

The fractional pseudocontact shift of the nuclear-resonance frequency of a nucleus bound to a paramagnetic ion is, in cases of axial symmetry, given by

$$\frac{\Delta \nu}{\nu_0} = D \left\langle \frac{(3 \cos^2 \theta - 1)}{r^3} \right\rangle \tag{23.13}$$

where θ is the angle between the principal symmetry axis of the metal and the vector joining the metal to the nucleus in question and r is the length of the vector, that is the metal–nucleus distance. Clearly such measurements can yield valuable information about the stereochemistry of metal complexes. However, several difficulties arise. If the complex has a number of rapidly interconverting con-formations then the pseudocontact shifts may reflect only an average conformation, a problem which may also apply to relaxation measurements. Secondly the constant term D cannot be calculated although its form is known[17]. A further problem is that this equation is only valid in cases of axial symmetry about the metal ion. It is not altogether clear what this symmetry refers to in some compounds but it is possible that averaging through exchange of ligands may lead to effective axial symmetry. If axial symmetry cannot be demonstrated then a more general equation in θ, a second angle ϕ and r must be used[17].

Because the shift equation contains at least two unknowns, θ and r, plus any of the constant terms that cannot be calculated, stereochemical information cannot be derived directly from pseudocontact shifts. These problems have been neatly circumvented by use of a ratio method in order to define the conformation of a lanthanide–AMP complex in solution[18]. The ratios of the shifts for various protons are established relative to a given proton in the molecule using several different metals. Then if the ratios for the various protons are the same for each metal, shifts arising from contact terms can be eliminated and axial symmetry assumed to apply. The shift-ratio data can then be used as a filter in a computer search for the molecular conformation using basis data such as van der Waals contact distances to reduce the number of possible conformers. It has been shown that this method gives rise to a unique conformation of a lanthanide–AMP complex[18]. A general procedure for using lanthanide cations in this way has been proposed[19].

This method has enormous potential for the study of small molecules in solution, but in biochemistry the problem frequently concerns the structure of very large molecules. There are two immediate difficulties in extending the small-molecule method. Firstly, macromolecules usually give broad proton n.m.r. spectra with only a few resonances resolved, which are difficult to assign because of the large number of protons in a macromolecule. Secondly, if it were possible to measure the pseudocontact shifts of individual protons the computational difficulties in extending the method used for AMP are large.

The advent of n.m.r. spectrometers with on-line computers offers ways of surmounting some of the problems outlined above. For instance, by processing data stored in a computer it has proved possible to increase the resolution of macromolecule n.m.r. spectra considerably[20]. Once the amount of information available

Figure 65   Schematic diagram of a lecithin bilayer vesicle.

$$O \text{ represents the polar head } O-\overset{\overset{\displaystyle O}{\|}}{\underset{\underset{\displaystyle O^-}{|}}{P}}-O-CH_2-CH_2-\overset{+}{N}(CH_3)_3$$

and ≈ the hydrocarbon chains.

from n.m.r. spectra is increased it should be possible to use relaxation measurements to determine distances and pseudocontact shift data to obtain values of θ so as to obtain quite detailed structural information.

The quantitative use of pseudocontact-shift data relies on there being axial symmetry and on the absence or separation of any contact shifts. In cases where these conditions cannot be satisfied metal ions can be used empirically to shift nuclear resonances so as to aid assignments and increase resolution. For example, for a molecule in which a large number of resonances overlap it is possible to separate them by adding an appropriate metal ion, which causes shifts, and then to obtain structural information by studying the relaxation effects caused by adding a second metal ion. In effect an averaged shifted and broadened spectrum is seen as the two metal ions rapidly exchange on and off the molecule.

An example of how shifting of resonances via the introduction of the appropriate metal can be usefully used in biochemistry without trying to extract exact structural information is provided by work on lecithin vesicles (see figure 65). In these vesicles, which are formed by sonicating a dispersion of lecithin (XXX) in water, some phosphate groups are exposed while others are held inside the vesicle. The ^{31}P n.m.r. spectra from such vesicles comprise a single line but on adding Pr^{3+} a second line shifts away from the original line, which loses intensity[21]. The Pr^{3+} is binding

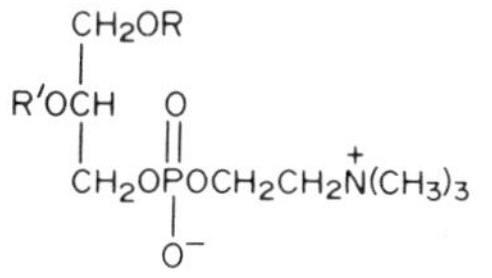

$$CH_2OR$$
$$|$$
$$R'OCH \quad O$$
$$| \quad\quad ||$$
$$CH_2OPOCH_2CH_2\overset{+}{N}(CH_3)_3$$
$$|$$
$$O^-$$

(XXX)

Lecithin (R and R' are two nonpolar hydrocarbon chains)

only to the exposed phosphate, and so only the resonances from these outer phosphorus atoms are shifted. Hence by measuring the integrated intensities of the two lines it is possible to determine the distribution of phosphate on the inside and outside.

23.10 GENERAL CONCLUSION ON THE USE OF PARAMAGNETIC METAL-ION PROBES

There is no doubt that valuable structural information can be obtained by the methods that have been described, but equally there is little doubt that these methods are not easy. A number of problems have been briefly discussed such as the necessity to demonstrate fast exchange and axial symmetry but several other difficulties must be pointed out. These methods rely on there being a single metal-binding site on a molecule, since a small fraction of a metal bound at secondary sites can give rise to meaningless results. For small molecules specific binding can usually be checked but in the case of macromolecules it can be rather harder to establish.

A limitation is that the effect of a paramagnetic centre can only be monitored up to distances of about 1.5 nm so that if the metal-binding site is not near the area of the enzyme or other macromolecule that is of interest these methods may be of restricted value. One way of overcoming this and other cases in which there is no metal-binding site is to introduce a metal-binding site by reacting a functional group, for example −SH on a protein, with a reagent such as

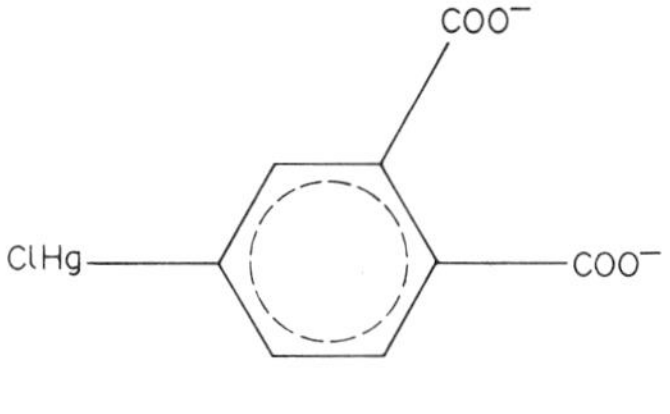

(XXXI)

which can chelate metals tightly. Finally it is worth noting that the replacement of Mg^{2+} by Mn^{2+} in biological systems, while apparently not greatly impairing biological function, may, in fact, cause larger changes than have been suspected if a recent study on pyruvate kinase is representative[22].

23.11 STUDIES ON METAL-ION RESONANCES

The observation of ^{205}Tl nuclear resonances as a probe for K^+ in pyruvate kinase has already been discussed. However, Tl^+ has $I = \frac{1}{2}$, and this section is devoted to considering briefly the type of information that can be gained from metal nuclear resonances where I is greater than $\frac{1}{2}$. As ^{23}Na n.m.r. has been used the most this will be taken as a typical example.

When nuclei have $I > \frac{1}{2}$, nuclear magnetic-relaxation processes are generally very efficient. This arises as a result of such nuclei possessing an electric quadrupole moment, which can interact with an electric field about the nucleus. Such an electric field will arise where there is less than cubic symmetry about the nucleus. The electric quadrupole moment will tend to align itself with respect to the electric-field gradient, which, because it is an intramolecular effect, will change direction with respect to an externally applied magnetic field as the molecule rotates. The nuclear magnetic moment tends to be drawn into the same direction as the quadrupole moment so that the effect will be to cause nuclear relaxation. This process is in general very efficient so that T_1 and T_2 for nuclei with $I > \frac{1}{2}$ are usually short. The immediate consequence of this is that resonance lines from such nuclei are broad, which, together with the low sensitivity of these nuclei, means that detection of their signals can be difficult. Consequently high concentrations have to be used, which reduces the value of their use in biological studies and in part explains why only a relatively small number of studies have been made. Two examples are considered here.

Some biological membranes that are not normally permeable to Na^+ will allow passage of this ion in the presence of cyclic antibiotics, which are generally called ionophores such as valinomycin. Two modes of action have been suggested, either that the ionophore interacts with the membrane so as to produce a 'channel' through which Na^+ can pass, or that a Na^+-ionophore complex permeates the membrane. ^{23}Na n.m.r. has been used to study this problem and the results favour the second interpretation, since the data is compatible with a 1:1 ratio of Na^+ and the complexing ionophore and a sufficiently rapid exchange rate of ^{23}Na on and off the ionophore for the observed rate of transport across a membrane[23].

The second application has involved a study of the state of sodium in biological materials, the results of which are open to several interpretations[5,24]. The ^{23}Na n.m.r. signal from a biological sample is less intense (about 40 per cent) than the signal from an aqueous solution containing the same amount of sodium. It was proposed that this intensity difference might be used to estimate the proportion of bound sodium in biological material, since only the unbound sodium would give an observable n.m.r. signal. However, it is possible that the intensity of the signal from the biological material is not just indicative of the amount of free sodium. There are three allowed transitions for ^{23}Na,

$$m_I, \frac{3}{2} \leftrightarrow \frac{1}{2}, \frac{1}{2} \leftrightarrow \frac{-1}{2}, \frac{-1}{2} \leftrightarrow \frac{-3}{2},$$

which are of equal energy in a symmetric environment. In an asymmetric environment, such as might be expected in biological material, these three transitions are no longer of equal energy, and the ^{23}Na n.m.r. spectrum will comprise a centre line with two satellites on either side. The separation between the outer and central resonances is dependent on the angle between the electric field gradient and the applied magnetic field. Hence in a non-orientated sample where all possible angles are considered, these outer lines are averaged over many angles to give very broad lines. Consequently only the central resonance of ^{23}Na in a biological sample would be observable. Therefore the single ^{23}Na n.m.r. signal seen from biological samples may be due entirely to bound sodium, or alternatively may be an average of resonances from bound and unbound sodium. Clearly this interpretation makes the use of ^{23}Na n.m.r. for detecting the amount of bound sodium difficult.

23.12 IRON PROTEINS

23.12.1 Haem proteins

The proton magnetic-resonance spectrum of a protein lies within a narrow frequency range, which means that there is a large number of overlapping resonances. This, coupled with the fact that the resonances tend to be broad, means that assignment of a resonance to a particular amino acid in the sequence of a protein is difficult, although increasingly methods are being devised which help overcome this[20]. Haem proteins have two properties that lessen these problems. Firstly, the haem group produces ring-current shifts in the same way as benzene[5,25], resulting in resonances of nuclei near to the haem ring shifting outside the main spectral envelope. Secondly, the haem iron can be paramagnetic, in which case it will cause shifts and broadenings of resonances corresponding to nuclei in the vicinity of the haem ring. In practice, it is found that the spectra of haem proteins can be spread over a frequency range approximately five times that of a normal protein.

Haem proteins have several functions; for instance haemoglobin transports oxygen, while many cytochromes are involved in electron-transport processes. An immediate problem therefore is to try to relate structural differences to the various functions of haem proteins. Associated with this are questions such as what changes in conformation of the three other subunits of the tetrameric protein haemoglobin are caused by the binding of oxygen to one haem group, or how an electron passes through a protein to haem iron and then travels from the iron to an electron acceptor. In order to attempt to answer such questions by the use of n.m.r. a first priority is to obtain structural information, which, of course, must involve assigning resonances to particular amino-acid residues. With this information it may be possible to attempt to elucidate mechanistic details of haem proteins.

The possibilities of studying haem proteins by n.m.r. are illustrated here mainly by reference to cytochrome c, while the attention of readers is drawn to the excellent review on haem-protein n.m.r. by Würtlich[25]. Cytochrome c, molecular weight 12 400 and containing a single haem group, is involved in the transfer of electrons along an electron chain from a hydrogen acceptor to oxygen. The haem iron is low spin in both the Fe(II) and Fe(III) states and so only the oxidised cytochrome contains a paramagnetic centre.

By 1969 the crystal structure of reduced cytochrome c had been essentially completed, but the X-ray data was unable to resolve the nature of one of the two axial ligands of the haem iron. The other ligand was identified as a histidine residue. A study of reduced cytochrome c from nine different sources revealed in each case a resonance shifted 3.3 p.p.m. upfield from the internal reference DSS[†][26]. The intensity of the peak, as judged by careful calibration with standards, indicated that it represented three protons which, together with its relatively narrow linewidth, suggested that the resonance originated in a freely rotating methyl group. The extent of the upfield shift meant that this methyl group must lie near the haem ring, and close to the haem symmetry axis along which the greatest shifts are expected. It was concluded that this resonance at +3.3 p.p.m., which appeared not to be coupled to any neighbouring protons, represented the methyl group of a methionine residue, whose sulphur atom was directly co-ordinated to the iron[26,27]. As methionine had already been identified as the sixth ligand in the crystal structure of oxidised cytochrome c, the important conclusion was made that there are no changes in haem-iron ligands between the oxidised and reduced states.

The paramagnetism of iron(III) results in the proton magnetic-resonance spectrum of oxidised cytochrome c being spread over a frequency range approximately five times that of the reduced form[25,28]. Redfield and Gupta[28] elegantly made use of the differences in these spectra to assign resonances and estimate the kinetics of electron transfer between the reduced and oxidised forms. Their method is best illustrated by considering the resonance at +3.3 p.p.m. in the spectrum of the reduced form and that at +23.4 p.p.m. in the spectrum of the oxidised state. Both are believed to be the resonances from the methyl group of the methionine ligand. The reasons for this assignment in the reduced state have already been explained, while for the oxidised form the reasoning is briefly as follows. The intensity of the resonance corresponds to three equivalent protons, while the linewidth is sufficiently large for the resonance to be relaxed by, and therefore correspond to, nuclei close to the iron atom. Additionally, the extent of the shift is also indicative of the resonance corresponding to nuclei near the iron. Redfield and Gupta took a 1:1 mixture of reduced and oxidised cytochrome c, and in their n.m.r. experiment irradiated the sample sufficiently strongly at +23.4 p.p.m. to saturate the signal at this position; that is, a double-resonance experiment was carried out so that the

† DSS = $(CH_3)_3Si.CH_2.CH_2.CH_2.SO_2^-Na^+$.

resonance at +23.4 p.p.m. disappeared. It was observed that this irradiation caused a diminution of the peak at +3.3 p.p.m. The implication of this was that electron exchange between the two forms of cytochrome c was faster than the rate at which the methyl protons of the reduced species relax to their equilibrium state in the magnetic field. In other words, the saturation of the methyl resonance of the oxidised protein is transferred to the resonance of the reduced form. This experiment confirms that these two resonances do indeed originate from the same methyl group. Two further points are worth noting. Firstly, if the reverse experiment is attempted by saturating the +3.3 p.p.m. resonance, then no effects on the +23.4 p.p.m. resonance are seen since this relaxes very rapidly. Secondly, two separate resonances will be seen in the mixture only if the rate of electron exchange between the two oxidation states is less than the frequency difference between the two resonances. The rate of electron transfer between reduced and oxidised cytochrome c was estimated by measuring the extent of the diminution of the 3.3 p.p.m. resonance together with the spin–lattice relaxation time (T_1) for this resonance, and applying simple rate theory[28, 29]. By studying the rate of exchange in the absence of small ions it was found that electron transfer occurs most rapidly at $\approx$ pH 10, the isoelectric point for cytochrome c, and that at this pH addition of small ions to the solution had no effect on the rate of electron exchange[30]. In contrast, at near pH this rate was dependent on ionic strength. From this it was concluded that the electron transfer did not require the mediation of small ions, but took place as a consequence of bimolecular collisions, which were most efficient when the protein was uncharged. An approximate calculation suggested that the rate of electron transfer at pH $\approx$ 10 was considerably less (by a factor of 10^5) than a diffusion-controlled collision rate would predict. The reasonable suggestion was made that only encounters with the correct orientation for electron transfer are effective[30].

Using the double-resonance technique outlined above, it proved possible to find resonances in the spectrum of reduced cytochrome c corresponding to eight resonances in the oxidised cytochrome c spectrum. It is therefore feasible with these resonances to measure the extent of these perturbations caused on oxidising the iron to a paramagnetic state. A knowledge of the extent of these perturbations, together with the position of the resonances in the reduced spectrum can be a considerable aid to assignment. A further help was the measurement of T_1 for the resolved resonances in the spectrum of oxidised cytochrome c, which in principle allows the calculation of distances between nuclei and the paramagnetic iron atom as described earlier in this chapter. Use of this method gave both corroborative, and in some cases primary evidence for assignments, while the distances obtained agreed well with those from the crystal structure[28].

In cytochrome c there are four methyl groups attached to the haem ring, and two of these in the oxidised-state spectrum are observed to be more highly shifted than the others. The orbital containing the unpaired electron, in which the transferable electron lies in the reduced state, is expected to show twofold symmetry about the haem axis[31]. Consequently the two most shifted methyl groups are diagonally opposite to each other and it is most likely, although not certain, that electron transfer in cytochrome c takes place in the direction defined by these two methyl groups. One of these methyl resonances can be assigned because it is unusually shifted in the reduced cytochrome-c spectrum by a ring-current shift from a nearby tryptophan, which is seen in the crystal structure.

The types of experiment described here should be applicable to a number of other haem-protein problems. It is worth pointing out, however, that the cyto-chrome-c work benefited from a knowledge of the crystal structure although clearly n.m.r. work contributed information unavailable from X-ray studies. N.M.R. methods should develop to such an extent that it will become possible to discern details of haem proteins in the absence of crystal structures. For instance, it has recently proved practicable to study the natural abundance (1 per cent) ^{13}C spectra of cytochrome c[32] despite the inherent low sensitivity of ^{13}C as an n.m.r. nucleus. Eventually the paths of the electron in cytochrome c might be traced by studying the contact shifts of the ^{13}C resonances.

Apart from cytochrome c, the haem protein most studied by n.m.r. is haemo-globin. The interest in this protein lies in trying to describe the nature of its co-operative oxygen binding in terms of conformational changes among the four subunits (two α chains and two β chains). Shulman and coworkers have studied this problem extensively by proton magnetic resonance. The essence of their approach is to try to detect changes in one type of chain by binding oxygen to the other. For instance, by producing oxidised (iron(III)) α chains and co-ordinating them with cyanide, the effects of binding oxygen to the β chains have been explored. The arguments are detailed, however, and the reader is referred to Shulman's papers for further information (see, for example, references 25, 33, 34, 35, 36).

Some experiments on the binding of carbon monoxide to haemoglobin will be described now, since they show promise as an approach for studying other haem proteins that are susceptible to inhibition by carbon monoxide. The ^{13}C resonance of ^{13}C-enriched carbon monoxide bound to haemoglobin is split into two peaks, which are separated by approximately 0.5 p.p.m.[37, 38, 39]. These two resonances are ascribed to carbon monoxide co-ordinated to the α and β chain haems. This has been verified in two ways. Pure α and β chains can be prepared and the two resonance positions of ^{13}CO bound to each of them measured. These positions correspond nearly, but not exactly, to the signals seen from haemoglobin[38]. There-fore it was suggested that there is some change at the haems as a result of tetramer formation. In a second approach a variant of haemoglobin was studied, which had its α chain in the oxidised iron (III) form so that only the β chain could bind carbon monoxide[39]. This allowed identification of the ^{13}C resonance from the β-chain carbon monoxide complex. Further experiments carried out with normal haemoglobin were designed to investigate whether α chains or β chains have a greater affinity for carbon monoxide. Preliminary indications were that on flushing with argon the α chains lost their carbon monoxide more rapidly. An important point about these ^{13}C experiments with haemoglobin is that they demonstrate the sensitivity of n.m.r. to small local changes in energy in a manner not always easily equalled by other spectroscopic techniques.

23.12.2 Iron–sulphur proteins

Iron–sulphur proteins occur commonly in biology as components of electron-transport chains, and have proved to be suitable for study by n.m.r. For example, two proteins, ferredoxin and rubredoxin, have been isolated from *Clostridium pasteurianum* and their properties compared using n.m.r. experiments. The ferre-doxin from this source contains six to eight iron atoms per molecule and undergoes a two-electron oxidation–reduction reaction. The sulphurs of eight cysteine residues have been implicated in the binding of iron together with acid-labile sulphur. The first n.m.r. experiments determined the magnetic susceptibilities of the oxidised and reduced ferredoxin by monitoring the positions of the DSS and tetramethylammonium ion resonances in the presence and absence of the protein[40]. By following the temperature dependence of the resonances it is possible to distinguish two types of magnetic-susceptibility behaviour. The Curie law is followed for the reduced protein while the oxidised ferredoxin has a susceptibility that increases on raising the temperature. This has been interpreted as indicating extensive antiferromagnetic coupling between the iron atoms. As both the oxidised and reduced forms of ferredoxin are paramagnetic, the proton magnetic-resonance spectrum of the protein is expected to show perturbations caused by the iron centres. In the spectrum of ferredoxin there are eight resonances well shifted to low field whose intensity can, by reference to standards, be equated with eight protons per ferredoxin molecule. The temperature dependence of the positions of these resonances parallels that of the magnetic susceptibility, so the shifts probably originate in a contact mechanism. Normally nuclear-resonance frequencies are temperature independent. The eight shifted resonances were assigned to the β methylene groups of four cysteines whose sulphur atoms, it is postulated, form a bridge between two iron centres[40]. The two protons of each of these β methylene groups are apparently nonequivalent because of constrictions on rotations, which result in slightly different hyperfine shifts for each. Another eight, but less shifted, resonances were detected and assigned to the remaining four cysteine β methylene groups. In this case the cysteine residues are thought to have their sulphur atoms co-ordinated to only one iron atom and so their proton resonances are less shifted. From these observations a structure of the iron–sulphur centre was suggested[40].

Rubredoxin from *Clostridium pasteurianum* showed different behaviour compared to ferredoxin on examination by proton magnetic resonance[41]. Rubredoxin contains one iron atom per molecule together with four cysteine residues. The iron can undergo a single one-electron oxidation-reduction reaction. Measurements of the magnetic susceptibility by the n.m.r. method indicated that both the oxidised (iron (III)) and reduced (iron (II)) forms of the protein are paramagnetic and high spin. Furthermore this paramagnetism showed a Curie-law behaviour in its tem-perature dependence, although the magnetic moment was greater than that for ferredoxin. The proton magnetic-resonance spectrum of rubredoxin showed, as expected, resonances shifted out of the main spectral envelope. However, compared with ferredoxin, the shifts observed are less and the linewidths greater for rubredoxin. It appears likely that the β methylene protons belonging to the cysteine ligands of the iron are too broad to be seen, since the larger magnetic moment of the iron will be more efficient at relaxing the nuclei. The shifts that are observed are probably pseudocontact shifts of nuclei further away from the iron.

As a final example, ferredoxin from parsley is considered. This protein contains two atoms of iron per molecule, and the oxidised form can undergo a one-electron reduction to produce a paramagnetic species. In the proton magnetic-resonance spectrum of the reduced ferredoxin there are eight highly shifted resonances, which again are assigned to four methylene groups of cysteine residues acting as ligands to the iron. The shifts of these resonances, however, show different temperature dependencies, which, it is suggested, indicates that the unpaired electron is distributed unequally between the two iron atoms[42]. Any exchange of the electron between the two iron centres must take place at a slow enough rate to avoid exchange broadening or averaging of the two signals.

A last n.m.r. method that is applicable to the study of iron–sulphur (and haem) proteins is the measurement of solvent-water proton-relaxation rates as discussed elsewhere in this chapter. Such experiments have been carried out for a number of iron–sulphur proteins with the result that only very small water relaxation-rate enhancements are seen[10]. This is consistent with a model for iron–sulphur proteins in which the iron is not accessible to solvent water, but is closely surrounded by a number of sulphur ligands.

Nuclear magnetic-resonance methods have already made a significant contribution to the study of iron-containing proteins, and with the rapid development of new techniques (see reference 20, for example) it seems certain that n.m.r. will play an increasing role in studies of these proteins.

Note added in proof

Since completion of this chapter a study of cytochrome c$_3$ from *Desulfovibrio vulgaris* has been published[43] which further illustrates the value and potential of n.m.r. studies on haem proteins. From the amino acid sequence, and n.m.r. data obtained using the resolution enhancement technique[20], an outline structure for this cytochrome could be proposed and information on the rate of electron transfer obtained.

REFERENCES

1. Williams, R. J. P., *Endeavour*, **26** (1967), 96.
2. Mildvan, A. S., and Cohn, M., *Adv. Enzymol.*, **33** (1970), 1.
3. Dwek, R. A., Williams, R. J. P., and Xavier, A. V., *Metal Ions in Biological Systems*, vol. 4 (ed. H. Sigel), Dekker. In press.
4. Cohn, M., *Q. Rev. Biophys.*, **3** (1970), 61.
5. Dwek, R. A., *N.M.R. in Biochemistry*, O.U.P., London (1973).
6. Williams, R. J. P., *Q. Rev. chem. Soc.*, **24** (1970), 331.
7. McLauchlan, K. A., *Magnetic Resonance*, O.U.P., London (1972).
8. Lynden-Bell, R. M., and Harris, R. K., *Nuclear Magnetic Resonance Spectroscopy*, Nelson, London (1969).
9. Dwek, R. A., *Adv. molec. Relaxation Processes*, **4** (1972), 1.
10. Mildvan, A. S., *The Enzymes*, vol. 2 (ed. P. D. Boyer), Academic Press, New York (3rd edn, 1972), p. 445.
11. Birkett, D. J., Dwek, R. A., Radda, G. K., Richards, R. E., and Salmon, A. G., *Eur. J. Biochem*, **20** (1971), 494.
12. Morallee, K. G., Nieboer, E., Rossotti, E. J. C., Williams, R. J. P., Xavier, A. V., and Dwek, R. A. *Chem. Commun.* (1970), 1132.
13. Dwek, R. A., Ferguson, S. J., Radda, G. K., Williams, R. J. P., and Xavier, A. V., *Proceedings of 10th Rare Earth Conference Arizona 1973* (ed. C. J. Kevane and T. Moeller), United States Atomic Energy Commission, Oak Ridge, Tennessee, p. 111.
14. Shulman, R. G., Navon, G., Wylunda, B. J., Douglass, D. C., and Yamane, T., *Proc. natn. Acad. Sci. U.S.A.*, **56** (1966), 39.
15. Navon, G., Shulman, R. G., Wylunda, B. J., and Yamane, T., *Proc. natn. Acad. Sci. U.S.A.*, **60** (1968), 86.
16. Reuben, J., and Kayne, F. J., *J. biol. Chem.*, **246** (1971), 6227.
17. Bleaney, B., *J. magn. Resonance*, **8** (1972), 91.
18. Barry, C. D., North, A. C. T., Glasel, J. A., Williams, R. J. P., and Xavier, A. V., *Nature, Lond.*, **232** (1971), 236.
19. Bleaney, B., Dobson, C. M., Levine, B. A., Martin, R. B., Williams, R. J. P., and Xavier, A. V., *Chem. Commun.* (1972), 791.
20. Campbell, I. D., Dobson, C. M., Williams, R. J. P., and Xavier, A. V., *J. magn. Resonance*, **11** (1973), 172.
21. Bystrov, V. F., Shapiro, Yu. E., Viktorov, A. V., Barsukov, L. I., and Bergelson, L. D., *Fed. Eur. biochem. Socs Lett.*, **25** (1972), 337.
22. Kayne, F. J. and Price, N. C., *Biochemistry*, **11** (1972), 4415.
23. Hughes, D. H., Pressman, B. C., and Kowalsky, A., *Biochemistry*, **10** (1971), 852.
24. Shporer, M., and Civan, M. M., *Biophys. J.*, **12** (1972), 114.
25. Würtlich, K., *Struct. Bond.*, **8** (1970), 53.
26. McDonald, C. C., Phillips, W. D., Vinogradov, S. N., *Biochem. biophys. Res. Commun.*, **36** (1969), 442.
27. Würtlich, K., *Proc. natn. Acad. Sci. U.S.A.*, **63** (1969), 1071.
28. Redfield, A. G., and Gupta, R. K., *Cold Spring Harb. Symp. quant. Biol.*, **36** (1971), 405.
29. Gupta, R. K., and Redfield, A. G., *Biochem. biophys. Res. Commun.*, **41** (1970), 273.
30. Gupta, R. K., Koenig, S. H., and Redfield, A. G., *J. magn. Resonance* **7** (1972), 66.
31. Shulman, R. G., Glarum, S. H., and Karplus, M., *J. molec. Biol.*, **57** (1971), 93.
32. Oldfield, E., and Allerhand, A., *Proc. nat. Acad. Sci. U.S.A.*, **70** (1973), 3531.
33. Shulman, R. G., Ogawa, S., and Hopfield, J. J., *Cold Spring Harb. Symp. quant. Biol.*, **36** (1971), 337.
34. Ogawa, S., Shulman, R. G., and Yamane, T., *J. molec. Biol.*, **70** (1972), 291.
35. Ogawa, S., Shulman, R. G., Fujiwara, M., and Yamane, T., *J. molec. Biol.*, **70** (1972), 301.
36. Ogawa, S., and Shulman, R. G., *J. molec. Biol.*, **70** (1972), 315.
37. Moon, R. B., and Richards, J. H., *J. Am. chem. Soc.*, **94** (1972), 5093.

38. Antonini, E., Brunori, M., Conti, F., and Geraci, G., *Fed. Eur. biochem. Soc. Lett.*, **34** (1973), 69.
39. Vergamini, P. J., Matwiyoff, N. A., Wohl, R. C., and Bradley, T., *Biochem. biophys. Res. Commun.*, **55** (1973), 453.
40. Poe, M., Phillips, W. D., McDonald, C. C., and Lovenberg, W., *Proc. natn. Acad. Sci. U.S.A.*, **65** (1970), 797.
41. Phillips, W. D., Poe, M., Weiher, J. F., McDonald, C. C., and Lovenberg, W., *Nature, Lond.*, **227** (1970), 574.
42. Poe, M., Phillips, W. D., Glickson, J. D., McDonald, C. C., and San Pietro, A., *Proc. natn. Acad. Sci. U.S.A.* **68** (1971), 68.
43. Dobson, C. M., Hoyle, N. J., Geraldes, C. F., Wright, P. E., Williams, R. J. P., Bruschi, M., and LeGall, J., *Nature, Lond.*, **249** (1974), 425.

AUTHOR INDEX

Numbers in square brackets, [ ], refer to the reference number; numbers in parentheses refer to pages in the text containing the particular reference. The final number refers to the page containing the full reference citation.

Tohjo, H. [2.154] (167) 202; [2.219] (170, 171) 203.
Tomkinson, J. C. [3.25] (214) 264.
Totter, J. R. [3.55] (225) 265.
Towe, K. M. [4.16] (272, 295, 297) 301; [4.135] (292) 303.
Traylor, T. G. [2.146] (183) 202.
Treitel, I. M. [2.220] (181) 203.
Troup, G. J. [4.140] (293) 303.
Trudinger, P. A. [3.102] (237) 266.
Truter, M. R. [1.291] (90) 105.
Tsuchida, R. [1.216] (68) 104.

Ugo, R. [2.74] (181) 200.
Ullrich, V. [2.222] (189) 203.
Underwood, E. J. [3.12] (208) 264.
Urry, D. W. [1.293] (91) 105.

Vallarino, L. M. [1.77b] (17) 101.
Vallee, B. L. [1.4] (5) 99; [1.84] (18, 64, 65, 73, 76) 101; [1.85] (18, 64, 65, 69, 70, 71, 73, 84) 101; [1.190] (60, 64, 65) 103; [1.191] (60, 64, 65) 103; [1.200] (64) 103; [1.203] (65, 66, 67, 68, 69, 87) 104; [1.204] (65) 104; [1.233] (73) 104; [1.234] (65, 76) 104; [1.235] (65) 104; [2.221] (139, 190) 203; [1.269] (87) 105.
Valdermo, C. [3.64] (228) 265.
Van Bruggen, E. F. J. [2.78] (118) 200; [4.131] (292) 303.
Van der Giessen, A. A. [4.15] (272, 295, 297) 301; [4.150] (297) 304.
Van der Helm, D. [1.292] (90) 105.
Vanngard, T. [3.45] (219) 265; [3.52] (224) 265; [3.53] (224) 265; [4.3] (271) 301; [4.118] (292) 303.
Vaska, L. [1.183] (59) 103.
Vergamini, P. J. [5.39] (326) 330.
Vevers, G. [2.82] (116, 117, 118) 200.
Viktorov, A. V. [5.21] (321) 329.
Vinogradov, S. N. [5.26] (324) 329.
Vishniac, W. [3.102] (237) 266.
Vivaldi, G. [1.178] (58) 103; [2.153] (130, 146, 149) 202.
Voelter, W. [3.26] (215) 264.
Volpin, M. E. [2.223] (182) 204.
Vonderschmitt, D. [2.224] (119) 204.
Von Zelewsky, A. [2.250] (119) 204.

Waara, I. [1.37] (7, 15, 78, 79, 81, 83, 87, 88) 100; [1.38] (7) 100; [1.252] (78, 79, 80, 81, 83, 87, 88) 105; [1.278] (88) 105.
Wacker, W. E. L. [1.4] (5) 99.
Waldschmidt-Leitz, E. [1.192] (60) 103.
Wall, J. [4.136] (292) 303.
Walling, C. [4.13] (272) 301; [4.23] (272, 283) 301.

Walker, F. A. [2.225] (119) 204.
Walker, G. W. [3.76] (232) 265.
Walker, L. R. [4.79] (282, 283) 302.
Walsh, K. A. [1.31] (7) 99; [1.197] (61) 103.
Wang, J. H. [1.184] (59) 103; [1.255] (80, 81, 87) 105; [1.272] (87) 105; [2.114] (161) 201; [2.119] (124, 146) 201; [2.122] (117, 131, 134) 201; [2.155] (124, 131) 202; [2.226] (179) 204; [2.227] (125, 148) 204; [2.228] (136) 204; [2.229] (117, 125, 146, 148) 204; [2.230] (148) 204.
Wang, R. H. [4.7] (271) 301.
Ward, K. B. [2.144] (117, 134, 143) 202.
Ward, M. A. [3.93] (236) 266.
Ward, R. C. [1.254] (79, 83, 88) 105.
Warren, S. G. [1.36] (7) 100.
Watenpaugh, K. D. [1.28] (6, 15, 16) 99; [1.75] (15, 16) 101.
Watson, H. C. [1.11] (6, 21, 23, 35, 37, 46, 47, 48) 99; [1.66] (13, 21, 70) 100; [1.70] (14, 21, 35, 54) 101; [1.91] (21, 35, 36) 101; [1.92] (21, 35, 36) 101; [1.142] (44, 57, 58) 102; [1.276] (88) 105; [2.12] (136) 199; [2.162] (143) 202; [2.198] (136) 203; [2.211] (130, 143) 203; [2.231] (129, 143) 204; [2.232] (129, 138, 143) 204.
Watts, D. C. [3.13] (208) 264.
Weakliem, H. A. [1.207] (66, 68, 88) 104.
Weast, R. C. [1.78] (17) 101.
Weaver, L. H. [1.311] (96, 97) 106.
Webb, E. C. [1.1] (3) 99.
Webb, J. M. [4.64] (280, 282, 293, 295, 296, 297, 298) 302; [4.97] (286, 287, 290, 300) 303; [4.117] (291, 292, 300) 303; [4.141] (293, 295, 296, 297) 304.
Weber, B. H. [1.42] (7) 100.
Weber, J. H. [2.233] (119) 204.
Wedler, F. [1.42] (7) 100.
Weigold, H. [4.93] (284) 302.
Weiker, J. F. [5.41] (327) 330.
Weisberger, F. [2.99] (181) 201.
Weiss, J. J. [1.179] (58) 103.
Werner, P. E. [1.35] (7) 100.
Wertheim, G. K. [4.79] (282, 283) 302.
Whelan, R. [2.1] (119) 199; [2.2] (119) 199; [2.3] (120) 199.
White, A. H. [1.165] (33) 103; [4.29] (273, 293) 301; [4.30] (273) 301.
Whitney, P. L. [1.282] (89) 105.
Wickoff, H. W. [1.34] (7) 100.
Wiebenga, E. H. [4.131] (292) 303.
Wiekliem, H. A. [1.129] (38) 102.
Wilchek, M. [1.306] (95) 106.
Wiley, D. C. [1.36] (7) 100.
Wilkins, R. G. [2.234] (119, 149) 204.
Wilkinson, P. J. [2.183] (149) 203.

SUBJECT INDEX